VISUALIZATION IN SCIENCE EDUCATION

Brian Fleming Research & Learning Library
Ministry of Education
Ministry of Training, Colleges & Universities
900 Bay St. 13th Floor, Mowat Block
Toronto, ON M7A 1L2

Models and Modeling in Science Education

VOLUME 1

Series Editor: Professor J.K. Gilbert
Institute of Education
The University of Reading
UK

Editorial Board:
Professor D.F. Treagust
Science and Mathematics Education Centre
Curtin University of Technology
Australia

Assoc. Professor J.H. van Driel
ICLON
University of Leiden
The Netherlands

Dr. Rosária Justi
Department of Chemistry
University of Minas Gerais
Brazil

Dr. Janice Gobert
The Concord Consortium
USA

Visualization in Science Education

Edited by

JOHN K. GILBERT
The University of Reading, UK

A C.I.P. Catalogue record for this book is available from the Library of Congress.

ISBN-10 1-4020-5882-9 (PB)
ISBN-13 978-1-4020-5882-0 (PB)
ISBN-10 1-4020-3612-4 (HB)
ISBN-13 978-1-4020-3612-5 (HB)
ISBN-10 1-4020-3613-2 (e-book)
ISBN-13 978-1-4020-3613-2 (e-book)

Published by Springer,
P.O. Box 17, 3300 AA Dordrecht, The Netherlands.

www.springer.com

All Rights Reserved
© 2007 Springer
No part of this work may be reproduced, stored in a retrieval system, or transmitted
in any form or by any means, electronic, mechanical, photocopying, microfilming, recording
or otherwise, without written permission from the Publisher, with the exception
of any material supplied specifically for the purpose of being entered
and executed on a computer system, for exclusive use by the purchaser of the work.

CONTENTS

Acknowledgements	vii
Introduction	1
Section A: The significance of visualization in science education	7

1. **Visualization: A metacognitive skill in science and science education** — 9
   John K. Gilbert

2. **Prolegomenon to scientific visualization** — 29
   Barbara Tversky, Stanford University, USA

3. **Mental Models: Theoretical issues for visualizations in science education** — 43
   David Rapp, University of Minnesota, USA

4. **A model of molecular visualization** — 61
   Michael Briggs, George Bodner

5. **Leveraging technology and cognitive theory on visualization to promote students' science learning and literacy** — 73
   Janice D Gobert

Section B: Developing the skills of visualization — 91

6. **Teaching and learning with three-dimensional representations** — 93
   Mike Stieff, Robert Bateman, David Uttal

7. **Modelling students becoming chemists: Developing representational competence** — 121
   Robert Kozma, Joel Russell

8. **Imagery in physics: From physicists' practice to naïve students' learning** — 147
   Galit Botzer, Miriam Reiner

9. **Imagery in science learning in students and experts** 169
   John Clement, Aletta Zietsman, James Monaghan

Section C: Integrating visualization into curricula in the sciences 185

10. **Learning electromagnetism with visualization and active learning** 187
    Yehudit Judy Dori, John Belcher

11. **Teaching visualizing the science of genomics** 217
    Kathy Takayama

12. **Models visualization in undergraduate geology courses** 253
    Stephen J Reynolds, Julia K Johnson, Michael D Piburn, Debra E Leedy, Joshua A Coyan, Melanie M Busch

Section D: Assessing the development of visualization skills 267

13. **Evaluating the educational value of molecular structure representations** 269
    Vesna Ferk Savec, Margareta Vrtacnik, John K Gilbert

14. **Assessing the learning from multi-media packages in chemical education** 299
    Joel Russell, Robert Kozma

Endpiece: Future research and development on visualization in science education 333
John K. Gilbert

Appendices to Chapter 13 337

About the Authors 345

Index 351

ACKNOWLEDGEMENTS

The Editor and the Authors would like to thank Mrs. Helen Apted for her excellent work in meeting the diverse challenges of the template and pagination to bring this book together as a coherent whole.

Some authors have thanked specific sponsors. These statements are made at the end of the relevant chapters.

JOHN K GILBERT

INTRODUCTION

Institute of Education, The University of Reading, UK

The roles of models and of visualization in science and science education have gained theoretical and practical saliency separately over the last decade or so. Two social trends seem to have underlain their emergence as focuses for research and development work. First, the much greater emphasis being placed on introducing students, at all levels of the education system, to the nature and processes of science. Being 'able to think like a scientist' has become much more important. Second, the ready availability of powerful computers with which models, especially dynamic models and simulations, can be displayed and manipulated in a virtual format. The rapid development of these closely allied trends has taken place in the hands of four very different groups of people: computer software specialists, who have a command of what might technically be done by way of representation; scientists, who are concerned that science education best prepares students to be scientists; educationalists, who have a strong interest in what ought to be done in that preparation; cognitive scientists, who are interested in finding out the consequences for learning of the actions taken. Hitherto, these groups have pursued their somewhat separate agendas. This book is an attempt to bring together some practitioners and researchers from these groups. The main purpose has been to disseminate their achievements so far. However, as will be discussed in the 'Endpiece' to the book, some desirably trends for future research and development have also been identified.

In order to generate an overview of any emergent field of endeavour, the simplest way to do so is to build the work around self-evident themes. In this case, the book has been structured into four sections, following the classical educational design structure of aims, teaching methods, curriculum, and assessment. The first is concerned with 'The significance of visualization in science education': justifications for why the volume was worth assembling. The second is concerned with 'Developing the skills of visualization': accounts of what has been done with students and to what effect. The third is concerned with 'Integrating visualization into curricula in the sciences': accounts of how attempts to develop the skills of visualization have been placed within course structures to good effect. Lastly, a concern with 'Assessing the development of visualization': case studies of how changes in students' capacity to visualize have been evaluated. Work on visualization in science education has so far, perhaps inevitably, been development-lead, with most effort being placed on producing systems that can be readily accessed by students. The gradual accumulation of experience is leading to an awareness of the need for research to underpin further development. The field of research into 'models and visualization' is, in Thomas Kuhn's terms, gradually emerging from a pre-paradigmatic state: there has so far not been a clear agreement on what to study, how to do so, what constitutes good practice, or how to evaluate the outcomes of that practice. Inevitably, there is therefore a degree of overlap between chapters – and indeed sections – as authors present their attempts to render the field coherent. But the four sections do address some key ideas, summarized below, and provide something of the basis for a Kuhnian 'normal science' paradigm for the field.

Section A, *'The significance of visualization in science education'*, consists of five chapters. The centrality of models in science, science education, and in theories of cognition, is established. The idea that most important ontological status for 'model' is 'mental model': that which each individual constructs for themselves, is supported. It is suggested that existing mental models play three key roles in learning. First, they facilitate cognitive engagement by the learner with what is being taught. Second, in doing so, they enable each student to interact with the lesson as it progresses. . These are the central tenets of all the various forms of constructivism. Third, by being able to respond to a variety of mutually supportive media formats, they enable diverse types of information to be assimilated. These functions are performed with the aid of visualization.

The two main meanings of 'visualization' are set out: 'external visualization', in which models are represented for visual perception, and 'internal visualization', in which the results of that perception are represented in the mind. Of these, the most readily addressed in research and development is external visualization. It is argued that, in terms of learning, text offers the least support for the development of visualizations, being linear in structure. Two-dimensional visualizations, for example diagrams, and static 3D models, provide more support for the development of visualizations where spatial aspects of paramount. There is a focus on the issues involved in of understanding 2D representations. Where causal relations are the aspects of a model to be understood – as is the case in the most important models in science and science education – simulations and dynamic representations offer the greatest support for visualization. Computers provide the most effect way to run simulations, including those based on data downloaded from the worldwide web, and enable communication between learners to take place as they are becoming ubiquitous in many societies. Two case studies are included of how access to external visualizations that are valuable in learning can be readily provided via computers. At the end of the day, however, it is the generation of internal visualizations that is of the greatest significance in science education. It is argued that they are so important that direct effort must be made to cultivate what is termed the 'metavisual capacity' in learners. A case study of internal visualization being engaged provides an exciting window through which a 'model of the operation of a mental model' can be glimpsed and an avenue to the development of metavisual capacity perhaps identified.

Section B, 'Developing the skills of visualization' which consists of four chapters, follows on from Section A. It is pointed out that, historically, advances in chemistry took place as new representational tools became available which supported more insightful visualizations. It is therefore suggested that making a broader range of representational tools available to students should facilitate their learning both in chemistry and in the other sciences. Ethnographic case studies of expert chemists and novice chemists solving problems in the laboratory reveal the gap between them in their manner of use of representations. It is suggested that deliberate measures be taken to improve novice's 'representational competence' and a progressive scale for the attainment of that competence (echoing the attainment of 'metavisual capacity' mentioned in Section A) is proposed. A structure for student enquiry work is proposed in which this attainment is embedded.

The notion that students' spatial cognition can and must be systematically developed is central to these chapters. It is suggested that the use of the broad suite of pedagogies called 'constructivism' is a pre-requisite for success. Of particular interest is the suggestion that constructionism – learning as a consequence of an overt requirement to develop personal models – is likely to be highly effective. The visualization tools that can be made available to support this development are classified as 'content specific' e.g. eChem™ and as providing a 'general learning environment' e.g. NetLogo™, WorldWatcher™, and Geo3D™, the use of each of which is exemplified and reviewed.

The existing literature on the effectiveness of efforts to improve metavisual competence and their impact on learning is reviewed. However, the need to develop cognitive models that depict this development and its consequences runs throughout the section. It does seem likely that such models will emerge from an analysis of the protocols of 'think aloud' studies as individuals solve problems that do or might involve visualization. A case study in one chapter identifies the progressive employment of three types of representation that merge from the engagement of visualization: sensory-based, pure imaginistic representation, and formalism-based. Other case studies elsewhere identify four observable indicators of visualization: personal action projection, the use of depictive hand gestures, the production of overt reports of imagery, the making of reference to perceptions. The merging of these two sets of cases studies may be able to provide the basis for a testable cognitive model of visualization in 'real world contexts'.

Tools to encourage the use of visualization and the deployment of a range of forms of representation are becoming widely available, mainly via the use of computers. Some understanding of their educational significance and mode of operation is gradually emerging. Section C, 'Integrating visualization into curricula in the sciences', consists of three case studies where this potential and understanding were exploited to improve the quality of science education. Inevitably, these new technologies are being used to address existing problems. The case study from physics that is presented is based on the pressing need to decrease failure and improve understanding in large-scale university preliminary courses. Evidence is presented that this was achieved in the case of a course on electromagnetism by the strategic inclusion of visualizations for abstract ideas within a pedagogy that emphasised more active learning and mutual student support. Preliminary courses in geology place great emphasis on students becoming able to recognise the shape of the landscape from 2D topographical maps, being able to translate between 2D and 3D representations, and being able to unravel the layers existing in 3D structures, all within the wide diversity of timeframes that the subject covers. Visualization was supported through animations provided via the internet within a constructivist approach to course design. A comparative study showed the general value of the approach and its capacity to remove any perceptual inequalities between males and females.

In the case of genomics, the problem is new and pressing, for the rapid increase in the number of genomes that are being sequenced means that students have to develop a general capacity to understand any example that they may encounter. This is done by introducing them to structured contextual exemplars that are, from the start of the course, inevitably complex. Metavisual literacy in genomics means the attainment of visual cognition in respect of suitable representations for gene structure, gene orientation and organisation, gene relationships, and genome relations, and their integration into an overview. In the case study presented, a cognitive apprenticeship approach to enquiry-based learning was taken by student teams in different countries collaborating via the internet. This entailed the design and use of a wide variety of visualization tools. The evaluation of the course showed the pattern of interactions within the groups, in which the visualizations played central roles, and the changed function for the instructors who supported their learning.

Section D, concerned with 'Assessing the development of visualization skills', consists of two chapters, both concerned with chemistry students. In one of them, students were given access to various combinations of 2D representations, 3D models, and computer-based 'virtual 3D' models, to support their solving of visualization tasks that entailed perception, rotation, and reflection. The impact of these aids was evaluated by standardised tests administered before and after the tasks were attempted. It was found that visualization was most effectively supported by 3D models, through virtual 3D models, to computer-displayed static 2D models. It was also found that the provision of too many forms of representation hindered, rather than supported, visualization. In the other, the design of multi-media test items to evaluate the attainment of conceptual understanding and of visualization skills is discussed and the outcome of studies involving the educational packages SMV:Chem™ and ChemSense™ presented. As standardised visualization tests are normally administered by a psychologist, it does seem desirable that a broad range of such tests be developed that can be administered by chemistry teachers.

In summary, the chapters of this book provide support for some assertions about the field of 'visualization in science education' at the moment:

- models and modelling play key roles in science and hence should do so in science education;
- the formation and use of mental models are central to the learning of science;
- visualizations can be both external and internal to the learner of science;
- external visualizations provide support for all perception, including that in science;
- internal visualization plays a key role in all cognition, including that in science;

- there is a need to develop systematically the skills of visualization (the 'metavisual capacity') of all students ('novice scientists') up to the standard of that of practitioners ('expert scientists);
- the significance of visualization in science education is being realised by a diverse group of specialists;
- collaborative work between these groups of specialists has begun;
- the pedagogies of constructivism and constructionism both provide frameworks within which the development of metavisual capacity can be facilitated;
- keys to success in acquiring metavisual capacity are involvement in classroom / laboratory arrangements that promote an active, social, engagement in learning;
- the various forms of 2D and 3D representation support an understanding of different, yet overlapping, aspects of models;
- the capacity to mutually translate between 2D and 3D representations is a vital part of a developed metavisual capacity;
- there are many distinct systems of representation each of which has its strengths and weaknesses;
- computer-based 'virtual model' systems of representation are of rapidly growing importance in the development of metavisual capacity and the promotion of understanding;
- 'learning experiments' and 'think-aloud' case studies are providing evidence on how visualizations are used in thinking and hence on the educational value of the different systems of representation;
- a systematic focus on visualization is raising existing attainment standards in science education;
- education in those sciences where the pace of research in fast e.g. genomics, based on a 'cognitive apprenticeship model', is placing great emphasis on the provision and generation of visualizations;
- standardised tests for aspects of visualization (perception, rotation, reflection) have been used in the assessment of students under the supervision of psychologists;
- content - specific tests that also address visualization are being developed by science educators for use in their teaching.

SECTION A

THE SIGNIFICANCE OF VISUALIZATION IN SCIENCE EDUCATION

JOHN K GILBERT

CHAPTER 1

VISUALIZATION: A METACOGNITIVE SKILL IN SCIENCE AND SCIENCE EDUCATION

Institute of Education, The University of Reading, UK

Abstract. The range of terminology used in the field of 'visualization' is reviewed and, in the light of evidence that it plays a central role in the conduct of science, it is argued that it should play a correspondingly important role in science education. As all visualization is of, and produces, models, an epistemology and ontology for models as a class of entities is presented. Models can be placed in the public arena by means of a series of 'modes and sub-modes of representation'. Visualization is central to learning, especially in the sciences, for students have to learn to navigate within and between the modes of representation. It is therefore argued that students –science students' especially - must become metacognitive in respect of visualization, that they must show what I term 'metavisual capability'. Without a metavisual capability, students find great difficulty in being able to undertake these demanding tasks. The development of metavisual capability is discussed in both theory and practice. Finally, some approaches to identifying students' metavisual status are outlined and evaluated. It is concluded that much more research and development is needed in respect of visualization in science education if its importance is to be recognised and its potential realised.

THE NATURE OF VISUALIZATION

The Concise Oxford Dictionary gives the following two definitions for the verb 'visualize:

'1. form a mental image of; imagine. 2. Make visible to the eye.' (Pearsall, 1999).

The distinction between these definitions in maintained in discussions about the nature of visualization and its role in accounts of the development of understanding. Tufte uses the word 'visualization' to mean the systematic and focused visual display of information in the form of tables, diagrams, and graphs (Tufte, 2001). Other writers are concerned with the reception and processing of that information by the brain. Reisberg, for example, distinguishes between 'visual perception', as meaning that image of an object achieved when and as it is seen, 'visual imagery' as meaning the mental production of an image of an object in its absence, and 'spatial imagery', as meaning the production of a mental representation of an object by tactile means (Reisberg, 1997). The link with brain activity is emphasised by Kosslyn's use of the phrase 'mental imagery' instead of 'visual imagery' (M. S. Kosslyn, 1994). Just to 'muddy the water' still further, 'visualization' is also often used just to cover 'visual imagery' e.g by (NSF, 2001).

John K. Gilbert (ed.), Visualization in Science Education, 9 27.
© 2007 *Springer*.

The use of 'visualization' to mean just an array of information (Tufte, 2001) seems to imply a naïve realist view of the world: what is 'out there' must have the same impact on all brains. However, the possibility of a personal construction of knowledge is supported by what is known of how the brain deals with optical phenomena. The close association, if not conflation, of terms associated with brain activity is hardly surprising as there is evidence that visual perception and visual imagery involve similar mental processes and that they are mutually supportive (Reisberg, 1997). Thus they both preserve the spatial layout of an object / image. This is because the speed with which a person is able to scan it (change the focus of attention in the object / image), zoom relative to it (appear closer to or further from it), and rotate it (move it through 360 degrees along any axis), are constant and identical in both cases. Moreover, both provide greater discrimination of detail (i.e. they show greater 'visual acuity') at the centre of the object / image than elsewhere in it. This similarity of processes stops short of them being identical operations, for:

> '(visual) images---have some pictorial properties, but they are of limited capacity and are actively composed' (S. M. Kosslyn, Pinker, Smith, & Shwartz, 1982) (p.133)

It does seem that visual perception is selective, this selectivity being responsible, in part, for the qualitative differences in any subsequently produced visual image. Additionally, differences in the purposes for which and the contexts within which visual perceptions and visual images are produced leads to the latter being active creations that are partial and selective even in respect of the former. In short, 'reality', the products of 'visual perception' and of 'visual imagery', may differ quite a lot.

Whilst the distinctions between 'visual perception' and 'visual imagery' are of great importance to psychologists, they are probably of a lesser importance to practising scientists and science educators. The word 'visualization' may, for convenience, be taken in this book to cover them both.

VISUALIZATION IN SCIENCE AND SCIENCE EDUCATION

Science seeks to provide explanations for natural phenomena: to describe the causes that lead to those particular effects in which scientists are interested. However, 'phenomena' are not ready – made: we impose our ideas of what might be important on the complexity of the natural world. Scientists then investigate these idealisations, what may be called 'exemplar phenomena', at least at the outset of their enquiries in any given field. Early chemists preferred to work with solutions of pure substances, not with the mixtures found in nature. Early physicists opted for the study of the movement of objects where there was little friction. Early biologists chose systems where tidy crosses of physical characteristics occurred in the initial study of what would become genetics. These exemplar phenomena have one thing in common: they are simplifications chosen to aid the formation of visualizations (visual perceptions) of what was happening at the macro level. Such a descriptions and/or simplification of a complex phenomena is usually called a 'model', this corresponding to the everyday meaning of that word (Rouse & Morris, 1986). As scientific enquiry proceeds in any given field, the complexity of the models of exemplar phenomena that are addressed increases progressively, and the aims of the enquiry become ever more ambitious.

This process of simplification and representation within the scope of human senses with the aid of models becomes of greater importance as, later in a sequence of enquiries, explanations for exemplar phenomenon are sought at the sub-micro level. Models then become vital if the visualization (visual imagery) of entities, relationships, causes, and effects, within exemplar phenomena is to take place. The development of models and representations of them are in crucial in the production of knowledge. A classic example is Kekule's dream about the structure of the benzene molecule being like a snake biting its tail (Rothenberg, 1995). Models also play central roles in the dissemination and acceptance of that knowledge: for example, that of the double helix of DNA has now reached icon status, such that an abbreviated version of it is instantly recognized (Giere, 1988; S. W. Gilbert, 1991; Tomasi, 1988). Models can function as a bridge between scientific theory and the world-as-experienced ('reality') in two ways. They can act, as outlined above, as simplified depictions of a reality-as-observed (exemplar phenomena), produced for specific purposes, to which the abstractions of theory are then applied. They can also be idealisations of a reality-as-imagined, based on the abstractions of theory, produced so that comparisons with reality-as-observed can then be made. In this latter way they are used both to make abstractions visible (Francoeur, 1997), and, crucially, to provide the basis for predictions about, and hence scientific explanations of, phenomena (J. K. Gilbert, Boulter, & Rutherford, 2000).

This wide range of function is made possible because models can depict many different classes of entities, covering both the macro and sub-micro levels of representation. Many models are of objects which are viewed as having either an independent existence (e.g. a drawing of a reaction flask, of an atom) or as being part of a system (e.g. a drawing of a reaction flask in an equipment train, of an atom in a molecule). A model can be smaller than the object that it represents (e.g. of a whale) or larger than it (e.g. of a virus). Some models are representations of abstractions, entities created so that they can be treated as objects (e.g. flows of energy as lines, forces as vectors). Inevitably, a model can include representations both of abstractions and of the material objects on which they act e.g. of the forces thought to act within a structure. A model can be of a system itself, a series of entities in a fixed relation to each other (e.g. of carbon atoms in a crystal of diamond). It can be of an event, a time-limited segment of behaviour of a system (e.g. of the migration of an ion across a semi-permeable membrane). Lastly, it can be of a process, where one or more elements of a system are permanently changed (e.g. of a catalytic converter of hydrocarbons in operation).

Many of the examples given in the paragraph above were drawn from chemistry. This is not surprising – and is evident in the balance of contributions to this book - for the key role of models in the development of chemical knowledge was recognised by the mid-twentieth century (Bailer-Jones, 1999; Francoeur, 1997). Indeed, they have become 'the dominant way of thinking' (Luisi & Thomas, 1990) in chemistry, something that chemists do 'without having to analyse or even be aware of the mechanism of the process' (Suckling, Suckling, & Suckling, 1980). The development and widespread use of computer-based systems for generating and displaying models had its initial impact on chemistry, where visualization is so vital.

However, as later chapters show, this capability is now being fully exploited in physics, biology, and the earth science. Indeed, Martz and Francoeur have produced and regularly update a web-site on the history of the representation of biological macromolecules (Martz & Francoeur, 2004), whilst Martz and Kramer provide a similar service in respect of the teaching resources available (Martz & Kramer, 2004).

If models play important roles in science, it therefore follows that they should play equally important roles in science education. Those students who may become scientists must understand the nature and significance of the models that played key roles in the development of their chosen subject. They must also develop the capacity to produce, test, and evaluate, both exemplar phenomena and explanatory models. Models are equally important in the education of the majority who will need some level of 'scientific literacy' for later life (Laugksch, 2000).

These roles for models in science education are not easy to discharge, for models can attain a wide diversity of epistemological status. A *mental model* is a private and personal representation formed by an individual either alone or in a group. All students of chemistry must have a mental model, of some kind, of an 'atom', all those of physics a mental model of a 'force', all those of biology a mental model of a 'gene', all those of earth science a mental model of a 'tectonic plate'. By its very nature, a mental model is inaccessible to others. However, in order to facilitate communication, a version of that model must be placed in the public domain and can therefore be called an *expressed model*. Any social group, for example a science education class, can agree on an (apparently!) common expressed model that therefore becomes a *consensus model*. Where that social group is of scientists, working in a given subject area, and the consensus model is one in use at the cutting edge of science, it can be termed a *scientific model* e.g. the Schrödinger model of the atom, the Watson - Crick model of DNA. A superseded scientific model can be called an *historical model* e.g. the Bohr model of the atom, the Pauling model of DNA (J. K. Gilbert, Boulter, & Elmer, 2000). Historical models remain in use where they can provide the basis of explanations that are adequate for a given purpose. Historical models also find their final resting place in the science curriculum!

On major aspect of 'learning science' (Hodson, 1992) is the formation of mental models and the production of expressed models by individual students that are as close to scientific or historical models as is possible. To this end, simplified versions of scientific or historical models may be produced as *curricular models* (for example, the widely used dot-and-cross version of the Lewis-Kossel model of the atom) that are then taught. Specially developed *teaching models* are created to support the learning of particular curricular models (for example, the analogy 'the atom as the solar planetary system' used in the lower secondary / junior high school) (J. K. Gilbert, Boulter, & Rutherford, 2000). Sometimes teachers employ curricular models which can be called *hybrid models* because they merge the characteristics of several historical models, this having first been recorded in respect of chemical kinetics (Justi & Gilbert, 1999b). In respect of 'the atom', the dominant model on which school chemistry is based is the Bohr model (an historical model) whilst the dominant model in higher education is based on the Schrödinger 'probability envelope' model (the scientific model).

A further complication for science education is that any version of a model of a phenomenon in the public domain (i.e. an expressed, scientific, historical, curricular, or hybrid, model) is placed there by use of one or more of five *modes of representation*.

- The *concrete (or material) mode* is three-dimensional and made of resistant materials e.g. a plastic ball-and-stick model of an ion lattice, a plaster representation of a section through geological strata.
- The *verbal mode* can consist of a description of the entities and the relationships between them in a representation e.g. of the natures of the balls and sticks in a ball-and-stick representation. It can also consist of an exploration of the metaphors and analogies on which the model is based, e.g. 'covalent bonding involves the *sharing* of electrons' as differently represented by a stick in a ball-and-stick representation and in a space-filling representation. Both versions can be either spoken or written.
- The *symbolic mode* consists of chemical symbols and formula, chemical equations, and mathematical expressions, particularly mathematical equations e.g. the universal gas law, the reaction rate laws.
- The *visual mode* makes use of graphs, diagrams, and animations. Two-dimensional representations of chemical structures ('diagrams') are universal examples. Those pseudo three - dimensional representations produced by computers, that figure so prominently in this book, which may be termed 'virtual models', also fall into this category.
- Lastly, the *gestural mode* makes use movement by the body or its parts e.g. of ions during electrolysis by school pupils moving in counter - flows.

These canonical modes are often combined (Buckley, Boulter, & Gilbert, 1997) e.g. in a verbal presentation of the visual representation of the Krebs' cycle.

In the case of chemistry, and perhaps all the major sciences, the concrete, visual, and symbolic, modes predominate. There are many sub-modes in use within each mode. Taken overall, these modes and sub-modes can be referred to as constituting a 'spatial language' (Balaban, 1999). They occupy the region between the extremes marked by the arbitrary relationship that exists between words and ideas, on the one hand, and the isomorphism that exists between pictures and their referents, on the other (Winn, 1991).

Each of these sub-modes of representation has, to a first approximation, a 'code of interpretation'. This is a series of conventions by means of which those entities and relationships in the model that are capable of effective representation in the sub-mode are depicted. Alas, the problem becomes even more complex. For example, chemical equations, even the two parts to a chemical equation, can be represented in a wide range of ways (sub-sub-modes?) e.g. zinc + hydrochloric acid, Zn + HCl, $Zn_{(s)}$ + $H^+_{(aq)}$ + $Cl^-_{(aq)}$. Learning these 'codes of representation' is a major task for students: moving between modes is intellectually demanding. Worse still, where the codes are intermingled, as is sometimes the case in textbooks, e.g. $Zn + H^+_{(aq)}$, confusion can reign.

The intellectual demand of moving between modes and sub-modes of representation is high, particularly for chemistry, where a full understanding of a chemical phenomenon involves the ability to move fluently between the, confusingly termed, three *levels* of representation of it (Johnstone, 1993; Gabel, 1999; Treagust & Chittleborough, 2001). These are:

- the macroscopic level. This is met directly during observational experience in the laboratory and everyday life, for example colour change or precipitate formation in a chemical reaction, or in pictures of such situations;
- the sub-microscopic level. This is met during the representation of the inferred nature of chemical entities (as atoms, ions, or molecules) and the relationships between them, for example those involved in a chemical reaction. These representations are expressed in the concrete, visual, or verbal, modes;
- the symbolic level. This is the representation of the identities of entities (atoms, ions, or molecules), for example those involved in a chemical reaction (producing a 'chemical equation') or of the quantitative relationships between them (producing a 'mathematical equation', for example in calculating equilibrium constants).

Transitions between these levels of representation are found difficult by students to make, as is born out by research. Undergraduate chemists have been found able to identify both the macroscopic manifestations of chemical phenomena and to produce symbolic level representations for what they interpreted as happening in those phenomena, whilst having a poor understanding of them at the sub-microscopic level (Hinton & Nakhleh, 1999). In short, they were not able to move into and between the modes of representation with the fluency that is expected of them. Expert chemists, by definition, do achieve this fluency (Kosma, 2003; Kosma, Chin, Russell, & Marx, 2000; Kosma & Russell, 1997).

All expert scientists - chemists, physicists, biologists, earth scientists - must be readily able to visualize a model when it is met in any one of the modes, or sub-modes, of representation and at any level of representation. As one might expect, there is a correlation between the level of what might be termed the 'visuospatial skill' that a person displays and the capacity to solve problems requiring an overt component of visualization. What is, however, very unexpected is that there is also a correlation with success in respect of problems - at least in chemistry - that *do not* require visualization (Bodner & McMillen, 1986). In an excellent review of the overall field of visualization in the learning of chemistry, Wu and Shah put forward a range of explanations for the latter. The most likely explanation is that 'non-spatial requirement' problems are more effectively addressed by the insertion of the skill of visualization, especially where diagrams are physically drawn to help the student in the process of finding a solution (Wu & Shah, 2004). There is a steadily growing body of research that suggests that student achievement in science is generally supported by direct access to multi-media modes of representation e.g.(Ardac & Akaygun, 2004).

METACOGNITION IN VISUALIZATION

The processes of visualization are, as we have seen, widely used throughout science and science education. Their attainment and fluent use must, I suggest, entail 'metacognition': the ability to 'think about ones thinking' (P. S. Adey & Shayer, 1994). In formal terms

> 'Metacognition is probably best conceptualised as a set of interrelated constructs pertaining to cognition about cognition' (Hertzog & Dixon, 1994)

whilst, in more accessible terms:

> 'A metacognitive learner is one who understands the tasks of monitoring, integrating, and extending, their own learning' (Gunstone, 1994)

Why should metacognition in respect of visualization exist? First, because the existence of modern technology has provided so many important images that they cannot easily be learnt separately by an individual. Yet most people are able to navigate through these shoals. Second, because, from that range of images, there is no way for a person to know which one(s) will be of importance in the future (after (Kluwe, 1987)). We cannot safely learn to interpret just a few such types of image. In view of the opinion of Favell that many aspects of cognition may attain 'meta-' status (Flavell, 1987), I suggest that 'metacognition in respect of visualization' be referred to as 'metavisualization'.

What evidence is there that metavisualization can exist? There are three sources of evidence. First, a general 'spatial intelligence' does seem to exist i.e. one that applies across all fields of knowledge. If it is of universal applicability, then a fluency of competence – of metacognition - must be capable of acquisition. Second, a general model of memory exists that is capable of application to visualization and which represents the development of metacognitive competence. Third, there is evidence that visualization is central in the processes of thinking, in which memory is inevitably employed, and which must therefore be acquired by all. Taking these in turn:

General spatial intelligence

Gardner (1983) suggests that the mind consists of a series of distinctive 'intelligences'. The indicators for the existence of a given intelligence are that:
- it resides in a mental faculty, located in a specific area of the brain, that can be damaged and even destroyed;
- it has traceable evolutionary antecedents. It should be possible to infer how the intelligence has come about, to deduce the consequences of that process, and to gain evidence of its consequences over time;
- a particular set of operations are employed to process input and to encode that which is learnt. A given class of stimulae are treated in an identifiable and distinctive way;
- there is an identifiable developmental trajectory for individuals in respect of the intelligence. It should therefore be possible to say

where a person is in the development of the intelligence by their identifying current performance;
- specific tests can be developed to identify how and to what extent it operates;

Applying these criteria, Gardner concludes that there is a specific 'spatial intelligence' such that:

> 'Central to spatial intelligence are the capabilities to perceive the visual world accurately, to perform transformations and modifications upon one's visual experience, even in the absence of physical stimulation' (Gardner, 1983) (p.173)

The majority opinion amongst psychologists is that the capacity to visualize is derived from the right-hemisphere of the brain e.g. (McGee, 1979) although some believe that the left-hemisphere is involved in some operations (S. M. Kosslyn, 1987). There is evidence of the evolution of sight and of the capacity for visual imagery (Morris, 1998) (p. 8). The issues of a developmental trajectory and of the use of specific tests to identify it are dealt with later in this chapter. How it operates is the second piece of evidence.

General model of memory

A model for the operation of memory and hence for the performance of metacognition (Nelson & Narens, 1994) can be represented as:

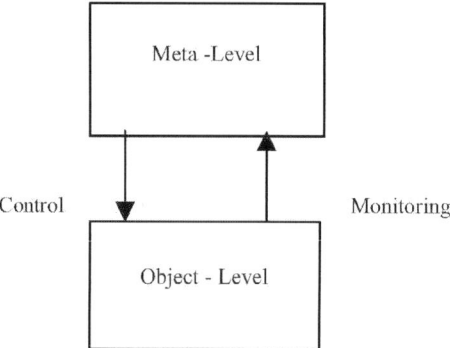

Visual perception monitors events taking place at the object – level, providing information that causes a model of the perceived object to be initially attained, and then either retained or amended, at the meta-level in the brain. Control, exerted by the meta-level, causes either the unchanged retention of, or changes in, what is perceived at the object-level. Monitoring and control are assumed to act simultaneously.

The Nelson & Narens model of the operation of memory – here we are concerned with the retention of an image – has three stages: acquisition, retention, and retrieval. In the development of metavisualization – what might helpfully be called the development of 'metavisual capability' – the learner becomes increasingly aware of *monitoring* what image is being learnt, of how to retain that image, and how to retrieve it. I suggest that a fourth stage might be added to this model:

amendment, the production of a version of the stored image that is retrieved for a specific purpose: a visual image. In acquiring metavisual capability, the learner:
- in respect of the acquisition stage, becomes increasingly able to make 'ease -of -learning judgements i.e. to be able to state what has been learnt and to predict the difficulty of and likely success in future learning. It becomes progressively easier to say, with certainty, what images are known and how difficult it will be to successfully acquire other specified images;
- in respect of the retention stage, becomes increasingly consciously able to mentally rehearse the acquired memory. It becomes progressively easier to retain specific images in memory;
- in respect of the retrieval stage, becomes increasingly convinced both that what has been learnt will be remembered in the future and that knowledge is held accurately. It becomes increasingly easy to retrieve accurate images;
- in respect of the proposed amendment stage, becomes increasingly able to consciously amend retrieved information for particular purposes. It becomes progressively easier to make changes to a retrieved image in order to meet any specific demands made of it and so produce a visual image.

Visualization and thinking

The third piece of evidence is based on the four categories of relationship that visualization has to thinking (Peterson, 1994). Thus:
- Reasoning. One form of reasoning involves the generation of new images by recombining elements of existing images. This is the basis of visual analogy. For example, the perception of waves on water led historically first to the development of the wave model of light and later to the wave model of sound.
- Learning a physical skill. In learning a skill, a person first produces a visual perception that defines the nature of the physical movement entailed in the exercise of the skill. This is done by observing an expert demonstrating the skill. This model is used by the learner to guide the personal development of the physical movement until, when perfected (an ideal situation!), the original visual perception is matched by the visual image that has evolved. For example, this is done in learning to use a pipette, to dissect a carcass, to tune a radio circuit.
- Comprehending verbal descriptions. Visual memory is distinct from linguistic memory (Haber, 1970). However, visualizations can be generated from a series of propositional statements, a process that, for many, makes an understanding of the relationships between the latter easier to acquire. For example, the structure of a crystalline substance can be understood by producing a mental image after reading a description of it. As will be argued implicitly throughout

this book, the availability of virtual representations may be making this process obsolescent, for 'visual understanding' will be acquired directly.
- Creativity. This can take place either by the reinterpretation of the meaning of an existing image or by a change in the frame of reference within which an image is set (Reisberg, 1997). The literature of the history of science is replete with examples of how major scientific advances have been made in these ways e.g. by Faraday, Maxwell, Tesla, Feynman, and, as has already been said, Kekule (Shepard, 1988).

THE CONSEQUENCES FOR LEARNING OF NOT HAVING A METAVISUAL CAPABILITY

If visualization is an important aspect of learning – especially in the sciences, where the world-as-perceived is the main focus of interest – then not possessing, having failed to develop, metavisual competence will have serious consequences.

Although many of the studies into the consequences of poor metavisual skills have taken place with secondary (high) school students, it does seem likely that similar problems will be faced by some university students. Wu identify several classes of problems in the field of chemistry (Wu & Shah, 2004), of which the most significant are:
- that whilst chemical phenomena can be represented at the macroscopic level, students find it difficult to do so for the same phenomena at the sub-micro and symbolic levels (Ben-Zvi, Eylon, & Silberstein, 1988);
- that students find difficulty in understanding the concepts represented in a given sub-mode at the sub-micro and symbolic levels (Kosma & Russell, 1997). In particular, they find difficulty with the interpretation at the sub-micro level of a reaction represented at the symbolic level (Krajcik, 1991);
- moving between the modes and sub-modes of representation a given molecule, what Siegel delightfully refers to as 'transmediation' (Siegel, 1995), is found problematic (Keig & Rubba, 1993).

Thus developing the skills of visualization is important if progress is to be made in learning science.

DEVELOPING METAVISUALIZATION CAPABILITY

A person with metavisual capability in the area of science will have a range of knowledge and skills in respect of the specific conventions associated with the modes and sub-modes of representation used there, together with more general skills of visualization *per se*.

These 'codes of representation' can best be discussed with use of the idea of semiotics – the study of signs and their meaning (Buchler, 1940). A consensus model has an identified relationship to that which it represents (the referent) such that there is societal agreement on the meaning that it conveys (the 'mental model 1' is evoked)

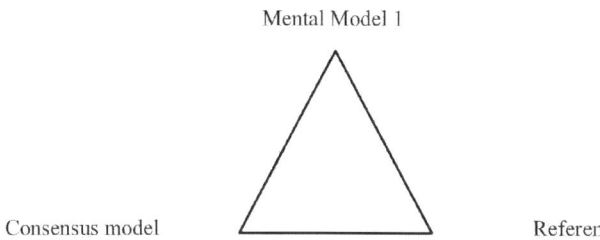

Figure 1: The semiotic triangle for a model

As we have seen, the relationship between the referent and the model is governed by the nature of the simplifications made to the former and the purposes and assumptions embodied in the creation of the latter.

A given consensus model is represented (i.e. produced into the public arena as an expressed model) through the use of a particular mode of representation. A mode of representation is produced by the operation of analogy on a source that is a commonly experienced phenomenon. A given mode of representation is useful in that it has a clear relationship to the model that it represents (now 'mental model 1') and to the source from which it is drawn such that there is agreement on the meaning that it conveys (the 'mental model 2').

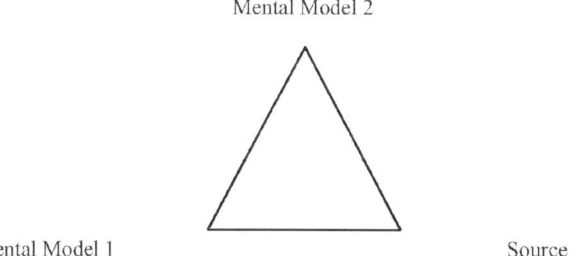

The ideal representation, given above, would include all aspects of Mental Model 1 in Mental Model 2. In reality, this is not achievable because each mode of representation has a specific scope and limitation, so that several modes of representation are usually used, each conveying specific aspects of Mental Model 1.

Representation is made more complicated by the fact that the drawing of any analogy with a source (e.g. of the referent in respect of Mental Model 1, and of Mental Model 1 in respect of Mental Model 2) is itself a complex business. Hesse (1966) argued that all analogies consist of three components. The 'positive analog' is those aspects of the source that are thought to be similar to aspects of that-which-is-to-be-represented. The 'negative analog' is those aspects of the source that are known to have no similarity to that - which - is - to - be -represented. The 'neutral analog' is those aspects of the source whose similarity status to that - which - is - to - be - represented is unknown.

Take, as a very simple example, the simplest material sub-mode of representation used to express atoms / ions at school level. It uses polystyrene balls to represent them in the commonly used 'ball-and-stick' mode (J.K.Gilbert, 1993) (p.16) (see Fig 1).

Feature of Polystyrene balls	Positive analog	Negative analog	Neutral analog
Variable colour	*		
Finite size	*		
Variable size	*		
Spherical shape	*		
Solid surface			*
Rough surface			*
Low density		*	
Homogeneous		*	
Aerated texture		*	
Compressible		*	
Can be pierced		*	
Soluble		*	
Flammable		*	

Figure 2: The analog status of polystyrene balls

If a student incorrectly assumed that a negative analog (e.g 'homogeneous nature') was in fact a positive analog (i.e. that an atom/ion is of a uniform composition), then a misunderstanding, a misconception, would be generated.

The 'diagram', the most commonly used form of two-dimensional visualization, is equally demanding of students. A diagram can, in the most general terms, be described as a series of nodes connected by lines. The nodes can be of a wide range of types, from pictures, to sketches, to icons, to symbols. The connections between them can be lines or surfaces, indicating spatial, temporal, or propositional, relationships. There seem to be no generic forms of diagram, with textbooks often conflating different types (Unsworth, 2001). In the case of the 'virtual mode' of visual representation, heavily used within this book e.g. IsisDraw, RasWin,ChemDraw,Chime, each of the several trademarked systems all seem to have its own convention.

A necessary condition for students to understand a diagram i.e. be able to interpret specific aspects of a model from it, is that the convention of representation should be both stated and adhered to by the producer of the diagram, e.g. a textbook writer. This is not always done. A sufficient condition for students to understand a diagram is that they have explicitly learned these conventions. Again, this is apparently not currently done in any systematic way. Drawing an analogy to work on the learning of the 'nature of science' (Abd-El-Khalick & Lederman, 2000), I suggest that this could be done by a mixture of direct instruction and ample opportunity to use the conventions in practice.

In summary, in order to become metacognitively capable in respect of visualization, students should:
- know the conventions of representation, both for the modes and major sub-modes of representation that they are likely to encounter;
- know the scope and limitations of each mode and sub-mode i.e. what aspects of a given model each can and cannot represent;

According to Barnea (2000), there are three complementary skills associated with what I now call a metavisual capability i.e.

> '1. Spatial visualization: the ability to understand three-dimensional objects from two-dimensional representations of them (and vice versa). ;
>
> 2. Spatial orientation: the ability to imagine what a three-dimensional representation will look like from a different perspective (rotation);
>
> 3. Spatial relations: the ability to visualize the effects of the operations of reflection and inversion'

Perhaps, in the light of what has been said earlier, 'spatial visualization' might now be called 'spatial interpretation' to avoid confusion.

These three skills respectively entail:
- being able to 'translate' (transmediate) between modes or sub-modes e.g to be able to move fluently between two-dimensional and three-dimensional representations of a given model, that is, to be able to produce a material mode presentation from a virtual mode representation, and vice versa ;
- being able to mentally change the perspective from which a given three-dimensional representation is viewed;
- being able to operate on the representation itself, particularly in terms of taking mirror images of it.

The development of a full range of metacognitive skills is considered so important that there has been a strong advocacy for the reformulation of the school curriculum in general into the 'thinking curriculum' (Fisher, 1998). In the area of metavisualization, the systematic cultivation of these specific skills, within a broad envelope of 'visual literacy skills', has been suggested by Christopherson (1997). But how can this be done? One approach is by general good practice in the use of representations by teachers and in textbooks, whilst the other is by the specific cultivation of the skills involved. The 'general good practice' approach involves (Hearnshaw, 1994):
- starting any sequence of representations with the most regular, geometrically simple forms available. This will enable students to 'get their eye in';
- using as full a range of modes /sub-modes of representation as is possible, introducing them deliberately, systematically, and steadily. This will encourage students to engage their knowledge of the codes of representation;
- maximizing the salience of shapes, edges, shadings, and patterns, within any representation. This will enable students to distinguish the structure of the representation. . This might even be preceded by

teaching students the 'master images' used in representations (Christopherson, 1997);
- using a range of degrees of illumination for different sections of the representation. This should enable students to more readily perceive contrasts;
- making the full use of colour effects, in terms of saturation, hue, and lightness, of a full range of blues, reds, and greens. Again, this will maximize contrasts.

The specific development of the skills of visualization in the subject of chemistry has been reviewed by Tuckey & Selvaratnam (1993). They identified three approaches, respectively using:
- stereodiagrams. These consist of pairs of drawings or photographs, one giving the view of a model as it would appear in the left eye and the other as it would appear in the right eye. The illusion of a three-dimensional image is produced viewing these two images with a device such that the right eye only sees the right-eye view and the left eye only sees the left-eye view;
- teaching cues. All diagrams, including the virtual mode, that purport to show three-dimensions, do so by the use of specific cues e.g. the overlap of constituent entities, the foreshortening /extension of lines of show below-surface / above -surface inclination, the distortion of bond angles, the emphasis of the relative size of constituent entities (atoms, ions, molecules);
- systematically teaching rotation and reflection through the use of a series of diagrams.

Evidence exists from of specific studies to show that each of these approaches can be successful (Tuckey & Selvaratnam, 1993). It should be noted that all the studies were completed before the widespread advent of the personal computer. This must surely have made the task easier, if only because of the 24/7 availability of any teaching material.

It does seem that skills of visualization improve with age during childhood and adolescence, with relevant experience playing a major role in that development. With the use of the 'Cognitive Acceleration in Science Education' programme, Adey et al (Adey, Robertson, & Venville, 2002) were able to develop metacognitive skills in general in UK school pupils aged 5-6 years. Studies of the relationship between gender and metavisual capability seem inconclusive: any possible initial advantages for boys can readily be nullified by providing suitable experience for all from which girls seem to benefit most (Tuckey & Selvaratnam, 1993).

Attention has recently been paid to 'intentional conceptual change': the bringing about of learning by an individual internally initiating thought, by then acting in a goal-directed manner, whilst exerting conscious control on both throughout. Hennessey has argued that developing a capacity to undertake intentional conceptual change is intertwined with the development of metacognitive capabilities generally (Hennessey, 2003). By extension, this work suggests that, where the learning involves 3D structures, e.g. of molecules, it will be mutually supportive of the development of metavisual capability.

EVALUATING THE ATTAINMENT OF METAVISUALIZATION

Assessing an individual's performance on visualization tasks, the necessary first step to the evaluation of their status in respect of metavisualization, is difficult for two reasons. First, for any particular skill of visualization that is becoming metacognitive, the person concerned is undertaking a process of which there may already be an inner awareness but which is not yet evident in overt behaviour. Second, that individual may not be aware that these processes are taking place. Three general ways of obtaining insight into a person's status in respect of any metacognitive skill have been suggested (Garner & Alexander, 1989): asking them about it; having them think out loud whilst doing a task thought to involve the skill; asking them to teach another person a way of successfully tackling such a task. These approaches do assume that the person has the verbal skills necessary to explain what they are doing and that they are not too bound up in the immediacy of the task.

The task of the assessment of competence in visualization may be made easier by the suggestion that there are two 'levels' of metavisualization. At the lower level, an individual is:

> 'capable of reflecting about many features of the world in the sense of considering an comparing them in her (sic) mind, and of reflecting upon her means of coping with familiar contexts. However--she is unlikely to be capable of reflecting about herself as the intentional subject of her own actions' (Von Wright, 1992) (p.60-61) (quoted in Georghiades, 2004)

whilst at the upper level:

> 'Reflecting about one's own knowledge or intentions involves an element which is absent from reflections about the surrounding world. Self-reflection presupposes, in the language of mental models, a 'metamodel': in order to reason about how I reason, I need to access to a model of my reasoning performance' (Von Wright, 1992) (p.61)(quoted in Georghiades, 2004).

There are a series of general tests available for assessing competence in some of the key aspects of visualization at the lower level of metavisualization. These are in respect of three spatial ability factors (Carroll, 1993):

- 'spatial visualization'. Defined by Carroll as tests that 'reflect processes of apprehending, coding, and mentally manipulating spatial forms' (Carroll, 1993), one well-known example is the 'Purdue Visualization of Rotation' test (Bodner & McMillen, 1986);
- 'closure flexibility'. This is concerned with the speed with which a person identifies and retains a visual pattern in the presence of distractions. One such scheme is the 'Find-a-Shape-Puzzle' (Pribyl & Bodner, 1987);
- 'spatial relations'. This is concerned with a person's ability to judge which figure is the same as a target figure. On example of such a test is the 'card rotation' task (Barnea & Dori, 1999).

These tests have been used in the field of chemistry, for which additional specialised paper-and-pencil tests are also available. These are for an understanding of and capability to use:

* the conventions for representing 3D structures in 2D, the use of 'depth cues'(Tuckey & Selvaratnam, 1993)(p.101-102);
* the relationship between diagrams (2D) and material models(Tuckey & Selvaratnam, 1993)(p.104);
* the operations of rotation, reflection, and inversion (Tuckey & Selvaratnam, 1993)(p.104-108). Ferk has produced computer-based versions of tests of these skills that relate to the 'virtual' mode of representation (Ferk, 2003).

Assessment of performance at the upper level of metavisual competence could be made by interview as these tasks are being completed.

CONCLUSION

A case has been made out for the existence of 'metavisualization' or 'metavisual capability'. For this to be substantiated, there is a need for a systematic programme of research into the role that visualization plays in learning, into the scope and limitations of the various sub-modes of representation, into the ways that the learner navigates between the three levels of representation. It is only then that we can embark on an informed programme of curriculum development and teacher education to maximize the attainment of metavisual capability by all students of science.

REFERENCES

Abd-El-Khalick, F., & Lederman, N. (2000). Improving science teachers' conceptions of the nature of science: a critical review of the literature. *Interntional Journal of Science Education, 22*, 665-702.

Adey, P., Robertson, A., & Venville, G. (2002). Effects of a cognitive acceleration programme on Year 1 pupils. *British Journal of Educational Psychology, 72*(1), 1-25.

Adey, P. S., & Shayer, M. (1994). *Really Raising Standards: Cognitive Acceleration and Academic Achievement*. London: Routledge.

Ardac, D., & Akaygun, S. (2004). Effectiveness of multimedia instruction that emphasises molecular representations on students' understanding of chemical change. *Journal of Research in Science Teaching, 41*(4), 317-337.

Bailer-Jones, D. M. (1999). Tracing the development of models in the philosophy of science. In L. Magnani, N. J. Nersessian & P. Thagard (Eds.), *Model-based reasoning in scientific discovery* (pp. 23-40). New York: Kluwer Academic Publishers.

Balaban, A. T. (1999). Visual chemistry: Three-dimensional perception of chemical structures. *Journal of Science Education and Technology, 8*(4), 251-255.

Barnea, N. (2000). Teaching and learning about chemistry and modelling with a computer-managed modelling system. In J. K. Gilbert & C. Boulter (Eds.), *Developing Models in Science Education* (pp. 307-324). Dordrecht: Kluwer.

Barnea, N., & Dori, Y. (1999). High-school chemistry students' performance and gender differerences in a computerized molecular learning environment. *Journal of Science Education and Technology, 8*(4), 257-271.

Ben-Zvi, R., Eylon, B., & Silberstein, J. (1988). Theories,principles and laws. *Education in Chemistry*(May), 89-92.

Bodner, G. M., & McMillen, T. L. B. (1986). Cognitive restructuring as an early stage in problem solving. *Journal of Research in Science Teaching, 23*(8), 727-737.

Buchler, J. (Ed.). (1940). *The philosophy of Peirce:Selected writings.* London: Routledge and Kegan Paul.
Buckley, B., Boulter, C. J., & Gilbert, J. K. (1997). Towards a typology of models for science education. In J. Gilbert (Ed.), *Exploring models and modelling in science and technology education: contributions from the MISTRE group* (pp. 90-105). Reading: School of Education, The University of Reading.
Carroll, J. B. (1993). *Human cognitive abilities:A survey of factor-analytic studies.* New York: Cambridge University Press.
Christopherson, J. T. (1997). *The growing need for visual literacy at the university.* Paper presented at the Visionquest:Journeys Towards Visual Literacy. 28th. Annual Conference of the International Visual Literacy Association, Cheyenne,Wyoming.
Ferk, V. (2003). *The significance of different kinds of molecular models in teaching and learning chemistry.* Unpublished Ph.D., University of Ljubljana, Ljubljana.
Fisher, R. (1998). *Teaching thinking.* London: Cassel.
Flavell, J. H. (1987). Speculations about the nature and development of metacognition. In F. E. Weinert & R. H. Kluwe (Eds.), *Metacognition,motivation and understanding* (pp. 21-29). Hillsdale,NJ: Erlbaum.
Francoeur, E. (1997). The forgotten tool: the design and use of molecular models. *Social Studies of Science, 27,* 7-40.
Gabel, D. (1999). Improving teaching and learning through chemical education research: a look to the future. *Journal of Chemical Education, 76,* 548-554.
Gardner, H. (1983). *Frames of Mind.* New York: Basic Books.
Garner, R., & Alexander, P. A. (1989). Metacognition: Answered and unanswered questions. *Educational Psychologist, 24*(2), 143-158.
Georghiades, P. (2004). From the general to the situated: three decades of metaocognition. *International Journal of Science Education, 26*(3), 365-383.
Giere, R. (1988). *Explaining Science.* Chicao: University of Chicago Press.
Gilbert, J. K. (1993). *Models and Modelling in Science Education.* Hatfield: The Association for Science Education.
Gilbert, J. K., Boulter, C. J., & Elmer, R. (2000). Positioning models in science education and in design and technology education. In J. K. Gilbert & C. J. Boulter (Eds.), *Developing Models in Science Education* (pp. 3-18). Dordrecht: Kluwer.
Gilbert, J. K., Boulter, C. J., & Rutherford, M. (2000). Explanations with Models in Science Education. In J. K. Gilbert & C. J. Boulter (Eds.), *Developing Models in Science Education* (pp. 193-208). Dordrecht: Kluwer.
Gilbert, S. W. (1991). Model building and a definition of science. *Journal of Research in Science Teaching, 28*(1), 73-79.
Gunstone, R. F. (1994). The importance of specific science content in the enhancement of metacognition. In P. J. Fensham, R. F. Gunstone & R. T. White (Eds.), *The content of science: A constructivist approach to its teaching and learning.* (pp. 131-146). London: Falmer.
Haber, R. N. (1970). How do we remember what we see? *Scientific American, 222*(5), 104-112.
Hearnshaw, H. (1994). Psychology and displays in GIS. In H. Hearnshaw & D. J. Unwin (Eds.), *Visualization in Geographic Information Systems* (pp. 193-211). Chichester: Wiley.
Hennessey, M. G. (2003). Metacognitive aspects of students' reflective discourse: implications for intentional conceptual change teaching and learning. In G. M. Sinatra & P. R. Pintrich (Eds.), *Intentional conceptual change* (pp. 103-132). Mahwah,NJ: Erlbaum.
Hertzog, C., & Dixon, R. A. (1994). Metacognitive development in adulthood and old age. In J. Metcalfe & A. P. Shinamura (Eds.), *Metacognition* (pp. 227-251). Cambridge,MA: MIT Press.
Hesse, M. (1966). *Models and Analogies in Science.* London: Sheen and Ward.
Hinton, M. E., & Nakhleh, M. B. (1999). Students' microscopic, macroscopic, and symbolic representations of chemical reactions. *The Chemical Educator, 4*(4), 1-29.
Hodson, D. (1992). In search of a meaningful relationship: an exploration of some issues relating to integration in science and science education. *International Journal of Science Education, 14*(5), 541-562.
Johnstone, A. H. (1993). The development of chemistry teaching: a changing response to a changing demand. *Journal of Chemical Education, 70*(9), 701-705.
Justi, R., & Gilbert, J. K. (1999b). A cause of ahistorical science teaching: the use of hybrid models. *Science Education, 83*(2), 163-178.

Keig, P. F., & Rubba, P. A. (1993). Translations of representations of the structure of matter and its relationship to reasoning, gender,spatial reasoning, and specific prior knowledge. *Journal of Research in Science Teaching, 30*(8), 883-903.

Kluwe, R. H. (1987). Executive decisions and regulation of problem solving behavior. In F. E. Weinert & R. H. Kluwe (Eds.), *Metacognition, motivation, and understanding* (pp. 31-64). Cambridge,MA: MIT Press.

Kosma, R. (2003). The material features if multiple representations and their cogntive and social affordances for science learning. *Learning and Instruction, 13*, 205-226.

Kosma, R., Chin, E., Russell, J., & Marx, N. (2000). The role of representations and tools in the chemistry laboratory and their implications for chemistry, learning. *The Journal of The Learning Sciences, 9*(2), 105-143.

Kosma, R., & Russell, J. (1997). Multimedia and understanding: expert and novice responses to different representations of chemical phenomena. *Journal of Research in Science Teaching, 34*(9), 949-968.

Kosslyn, M. S. (1994). *Image and brain:The resolution of the imagery debate*. Cambridge,MA: MIT Press.

Kosslyn, S. M. (1987). Seeing and imagining in the cerebral hemispheres: a computational approach. *Psychological Review, 94*, 148-175.

Kosslyn, S. M., Pinker, S., Smith, G. E., & Shwartz, S. P. (1982). On the demystification of mental imagery. In N. Block (Ed.), *Imagery* (pp. 131-150). Cambridge,MA: MIT Press.

Krajcik, J. S. (1991). Developing students' understanding of chemical concepts. In Y. S. M. Glynn, R. H. Yanny & B. K. Britton (Eds.), *The psychology of learning science: International perspectives on the psychological foundations of technology-based learning environments* (pp. 117-145). Hillsdale,NJ: Erlbaum.

Laugksch, R. C. (2000). Scientific literacy: a conceptual overview. *Science Education, 84*(1), 71-94.

Luisi, P. L., & Thomas, R. M. (1990). The pictographic molecular paradigm: pictorial communication in the chemical and biological sciences. *Naturwissenschaften, 77*(67-74).

Martz, E., & Francoeur, E. (2004). *History of biological macromolecules*, from http://www.umass.edu/microbio/rasmol/history.htm

Martz, E., & Kramer, T. D. (2004). *World index of molecular visualization resources*, from http://molvis.sdsc.edu/visres/index.html

McGee, M. (1979). Human spatial abilities, psychometric tests and environmental, genetic, hormonal, and neurological influences. *Psychological Bulletin, 86*, 889-918.

Morris, S. C. (1998). *The crucible of creation*. Oxford: Oxford University Press.

Nelson, T. O., & Narens, L. (1994). Why investigate metacognition? In J. Metcalfe & A. P. Shinamura (Eds.), *Metacognition* (pp. 1-25). Cambridge,MA: MIT Press.

NSF. (2001). *Molecular Visualization in Science Education*. Washinton,D.C.: National Science Foundation.

Pearsall, J. (Ed.). (1999). *The Concise Oxford Dictionary*. Oxford: Oxford University Press.

Peterson, M. P. (1994). Cognitive issues in cartographic visualization. In A. M. Maceachren & D. F. Taylor (Eds.), *Visualization in Modern Cartography* (pp. 27-43). Oxford: Pergamon.

Pribyl, J. R., & Bodner, G. M. (.1987). Spatial ability and its role in organic chemistry; A study of four organic courses. *Journal of Research in Science Teaching, 24*(3), 229-240.

Reisberg, D. (1997). *Cognition*. New York: Norton.

Rothenberg, A. (1995). Creative cognitive processes in Kekule's discovery of the structure of the benzene molecule. *American Journal of Psychology, 108*(3), 419-438.

Rouse, W. B., & Morris, N. M. (1986). On looking into the black box: prospects and limits in the search for mental models. *Psychological Bulletin, 100*(3), 349-363.

Shepard, R. (1988). The imagination of the scientist. In K. Egan & D. Nadaner (Eds.), *Imagination and Education* (pp. 153-185). Milton Keynes: Open University Press.

Siegel, M. (1995). More than words: the generative power of transmediation for learning. *Canadian Journal of Education, 20*(4), 455-475.

Suckling, C. J., Suckling, K. E., & Suckling, C. W. (1980). *Chemistry through Models*. Cambridge: Cambridge University Press.

Tomasi, J. (1988). Models and modelling in theoretical chemistry. *Journal of Molecular Structure, 179*, 273-292.

Treagust, D. F., & Chittleborough, G. (2001). Chemistry:a matter of understanding representations. In *Subject-Specific Instructional Methods and Activities* (Vol. 8, pp. 239-267). New York: Elsevier.

Tuckey, H., & Selvaratnam. (1993). Studies involving three-dimensional visualisation skills in chemistry. *Studies in Science Education, 21*, 99-121.

Tufte, E. R. (2001). *The visual display of quantitative information* (2 nd ed.). Cheshire,CT: Graphics Press.
Unsworth, L. (2001). *Teaching Multiliteracies Across the Curriculum*. Buckingham: Open University Press.
Von Wright, J. (1992). Reflection on reflections. *Learning and Instruction, 2*(1), 59-68.
Winn, W. (1991). Learning from maps and diagrams. *Educational Psychology Review, 3*(211-247).
Wu, H. K., & Shah, P. (2004). Exploring visuospatial thinking in chemistry learning. *Science Education, 88*(3), 465-492.

BARBARA TVERSKY

CHAPTER 2

PROLEGOMENON TO SCIENTIFIC VISUALIZATIONS

Stanford University

Abstract. Visualizations are central to many tasks, including instruction, comprehension, and discovery in science. They serve to externalise thought, facilitating memory, information processing, collaboration and other human activities. They use external elements and spatial relations to convey spatial and metaphorically spatial elements and relations. The design of effective visualizations can be improved by insuring that the content and structure of the visualization corresponds to the content and structure of the desired mental representation (Principle of Congruity) and the content and structure of the visualization are readily and correctly perceived and understood (Principle of Apprehension). Visualizations easily convey structure; conveying process or function is more difficult. For conveying process, visualizations are enriched with diagrammatic elements such as lines, bars, and arrows, whose mathematical or abstract properties suggests meanings that are often understood in context. Although animated graphics are widely used to convey process, they are rarely if ever superior to informationally equivalent static graphics. Although animations use change in time to convey change in time, they frequently are too complex to be apprehended. Moreover, because people think of events over time as sequences of discrete steps, animations are not congruent with mental representations. Visualizations, animated or still, should explain, not merely show. Effective visualizations schematize scientific concepts to fit human perception and cognition.

INTRODUCTION

People invent tools to enhance their physical comfort--clothing, shelter, implements for obtaining and preparing food. People are not unique in creating tools for food or shelter. People, however, seem to be unique in creating tools that enhance their mental well-being; they keep track of things by counting on their fingers or on calculators, they remember their ways by notching trees or sketching a route, they convey ways to others by drawing them in the sand or on paper. Altering the external world to facilitate memory, information processing, and communication is ancient, preceding written language. The modern visualizations critical to scientific understanding, explanation, and discovery are an extension of these ancient devices. How do they do their job?

Maps serve as a paradigm, an instructive example. They are ancient and modern, they appear in cultures all over the world, they are created by children and adults, both schooled and unschooled. Effective maps schematize, they are not "realistic." They select the information that is needed for the task at hand, simplifying, even distorting, it to make it more accessible. Roads, for example, are not large enough to appear in many road maps if they were drawn to scale. The zigs and zags of crooked roads are simplified. Effective maps omit the information that is not needed, so churches appear in tourist maps but not in maps for drivers, topography appears in maps for hikers but not for drivers. Tourist maps aid sightseers by presenting impossible perspectives, overviews of roads, frontal view of destinations. Maps typically add information that is not visual, place names, boundaries, distance scales, heights, and depths.

Until the late 18th century, most visualizations conveyed information that was naturally visual, maps, architectural plans, flora and fauna, mechanical devices. Only recently have visualizations been designed to convey concepts that are not inherently visual, such as balance of trade and population growth (Beniger and Robyn, 1978; Tufte, 1983). Two centuries later, most graphs depict what they did when invented, change over time (Cleveland and McGill, 1985). Until recently, most diagrams conveyed only the structure of things, often exquisitely. Depicting how structure changes, that is, how things function, is a more contemporary phenomenon. Witness the paucity of arrows in earlier diagrams and their proliferation now (e. g., Gombrich, 1990; Horn, 1998). Perhaps not coincidentally, arrows entered diagrams to convey motion at about the same time as graphs portraying abstract information. Arrows, as we shall see, readily convey function.

COMMUNICATION: SPATIAL RELATIONS

Maps and other visualizations, like spoken language, are structured; they consist of elements and the spatial relations among them. In maps, the elements may be dots or lines or other shapes meant to be cities or streets or building or countries; the spatial relations on paper reflect the distances and directions among the elements in actual space. Contrast maps with tree diagrams, such as corporate charts or evolutionary trees or linguistic trees. For these, the elements are the nodes, the corporate roles or species of plants or animals or languages. The spatial relations among the elements in a tree are typically not metric distance; rather, they convey order or subset relations among the entities. Thus for visualizations of things not inherently visualizable, the spatial relations stand for abstract relations that are metaphorically spatial.

The spatial relations in visualizations preserve different levels of information from abstract relations. Many bar graphs and tables map only categorical information, for example, the number of cases in each category, as in the numbers of students in each discipline or the numbers of plants of each variety. Trees and some graphs may map abstract relations ordinally, for example, kinds of kinds of kinds or rankings of hues by wave length or risks by fatalities. Finally, visualizations may preserve information at the interval level, where not only the order of elements but also the distances between elements are meaningful, or at the ratio level, where zero, as well as order and interval, are meaningful. Graphs of all sorts are typically used to convey interval and ratio relations.

People seem to spontaneously think about abstract relations in spatial terms. Languages are packed with spatial metaphors, we say we feel *close* to friends or to solving a problem, that a new *field* is *wide open*, that a student is at the *top of the heap*. Not only is spatial distance used to convey abstract distance, but also certain directions, namely the vertical, are loaded. Upwards is used to convey better, more, stronger. Gestures reflect spatial thinking as well, good things get a thumbs up or a high five, bad things get a thumbs down. The space of visualizations conveys meaning in exactly the same way, distance on paper reflects distance on abstract dimensions, and upwards reflects positive dimensions. A survey of common visualizations in science textbooks confirms this (Tversky, 1995). All but one or two of the diagrams of the evolutionary tree had man (yes, *man*) at the top, and those of geologic eras had the present at the top.

The prevalence of spatial metaphors in language and gesture suggests that mapping abstract relations onto spatial ones is natural and spontaneous. Querying children is one way to address this. Children from pre-school through university from three language cultures, English, Hebrew, and Arabic, were asked to place stickers on paper to indicate the meals of the day or various sized containers of candy or books or liked or disliked food and TV shows (Tversky, Kugelmass, and Winter, 1991). These are concepts that can be readily ordered by time or quantity or preference. Would the children order them such on paper? Would their placement of stickers reflect distance on these dimensions? The mappings of stickers to concepts of even the youngest children reflected order on each of these dimensions. However, the mappings reflected distance or interval in only older children. What about direction of the orderings? For quantitative and preference, children of all languages mapped increased left to right, right to left, or bottom to top; they avoided mapping increases downwards. For temporal concepts, direction of increasing value followed direction of writing.

Spontaneous mappings of abstract relations onto space are neither random nor arbitrary. Rather they reflect *meanings* that are consistent across cultures and across age. As shall be seen, meanings of elements are often readily interpretable as well.

COMMUNICATION: ELEMENTS

Icons. Visualizations use elements as well as spatial relations to convey their messages. One time-tested kind of element is an icon, a depiction that resembles the thing that it represents. Written languages all over the world began this way. Not every concept can be depicted, of course. Common in ideographic languages are figures of depiction, such as synecdoche, where a part represents a whole as in the head of a cow to stand for a cow, or metonymy, where a symbol represents a whole as in the staff of office to represent a king or scales to stand for justice. These figures of depiction are as modern as those in computer menus, scissors for delete, a trashcan for eliminating files, a floppy disk (remember those?) for saving files.

Morphograms. Visualizations use another kind of element for conveying meanings, simple schematic, geometric figures, something we termed *morphograms*. Examples include lines, crosses, arrows, boxes, and blobs. Their geometric forms and Gestalt properties suggest general meanings, which contexts can clarify. Lines are one dimensional, they connect or form paths from one point to another. As such, they suggest a relationship between the points. Arrows are asymmetric lines, suggesting an asymmetric relationship. Blobs are amorphous and two-dimensional, suggesting areas where exact shape is irrelevant. Let us now turn to research illustrating how these are understood in context.

Graphs: Bars and Lines. Bar graphs and line graphs are popular both in scientific and lay publications. They are often used interchangeably, though purists recommend reserving lines for interval data. People's interpretations of the forms of representation are not interchangeable; rather, they depend on geometric properties of the forms (Zacks and Tversky, 1999). Bars are containers; they separate. Lines are links; they connect. Bars for X's and Y's suggest that all the X's share a property and all the Y's share a different property. A line connecting X and Y, however, suggests that X and Y share a dimension but have different values on that dimension.

If people respond to those geometric properties, then their interpretations of data presented as bars should be as discrete comparisons and their interpretations of data presented as lines should be as trends. In fact, when asked to interpret an unlabeled bar graph, people said that there are more Y's than X's or that the Y's are higher than the X's. For unlabeled line graphs, people said that there's an increase from X to Y or a rising trend from X to Y. When the graphs were labelled with continuous variables, such as the height of 10 and 12 year olds, or with discrete variables, such as the height of women and men, the graphic form played a larger role in interpretations than the underlying nature of the data. Some students interpreted a line graph connecting the height of women and men as, "if you get more male, you get taller." Form also overrode content when students were asked to produce graphs from descriptions of data. Students produced bar graphs for data described as discrete comparisons and line graphs for data described as trends. Geometric form and conceptual interpretations of bar and line graphs are tightly linked.

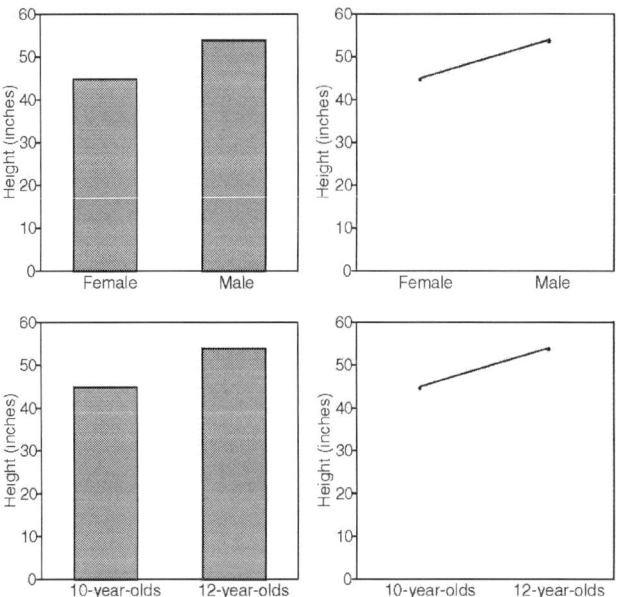

Figure 1. *Examples of bar and line graphs used by Zacks and Tversky (1999).*

Route Maps: Lines, Curves, Crosses, and Blobs. The visual devices of route maps are also tightly linked to linguistic devices. To compare route maps and route descriptions, we asked students outside a dormitory if they knew how to get to a nearby fast-food restaurant. If they did, we asked them to either sketch a map or write directions to get there (Tversky and Lee, 1998). We got a broad range of responses, some long, some short, some overflowing in detail, some crisp and elegant. Underneath the variability, however, was a structure common both to sketch maps and to written directions.

The structure underlying maps and directions extended a scheme developed by Denis (1997) for a large corpus of route directions. He found that directions consisted of strings of segments with four components: a start point, a reorientation, a progression on a path, and an end point. Like Denis' corpus, our corpus of directions consisted of segments with the same four components, though in many cases, some were implicit rather than explicit. For example, if the previous segment ended in an end point, the next segment often began with a reorientation, under the assumption that the end point of one segment served as the start point of the subsequent segment. Sketch maps also consisted of strings of segments with the same components, but the pragmatics of sketch maps, unlike the pragmatics of words, do not allow ellipsis.

Table 1. Examples of Route Directions - (From Tversky & Lee, 1998)

DW 9
From Roble parking lot
R onto Santa Theresa
L onto Lagunita (the first stop sign)
L onto Mayfield
L onto Campus drive East
R onto Bowdoin
L onto Stanford Ave.
R onto El Camino
Go down few miles. It's on the right.
BD 10
Go down street toward main campus (where most of the buildings are as
opposed to where the fields are) make a right on the first real street
(not an entrance to a dorm or anything else). Then make a left on the
2nd street you come to. There should be some buildings on your right
(Flo Mo) and parking lot on your left. The street will make a sharp
right. Stay on it. That puts you on Mayfield road. The first
intersection after the turn will be at Campus drive. Turn left and stay
on campus drive until you come to Galvez Street. Turn Right. Go down
until you get to El Camino. Turn right (south) and Taco Bell is a
few miles down on the right.
BD 3
Go out St. Theresa
Turn Rt.
Follow Campus Dr. way around to Galvez
Turn left on Galvez.
Turn right on El camino.
Go till you see Taco Bell on your Right

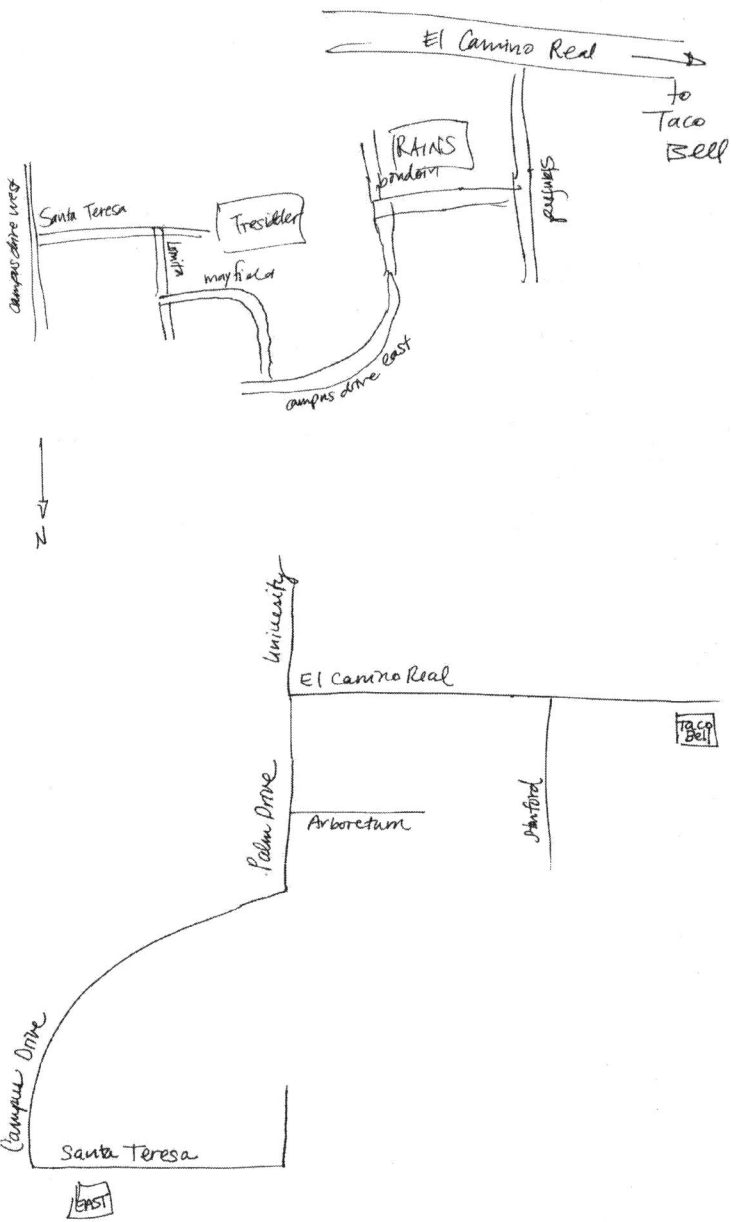

Figure 2. *Sketch maps from Tversky & Lee (1998)*

Although differing in pragmatics, the semantics and syntax of the route descriptions and the route depictions had noticeable correspondences. Start points and end points were landmarks in both, sometimes a street name, sometimes a building, named in directions, depicted by a blob in depictions. Reorientations disregarded amount of turn in both cases. In maps, they were +'s or T's or L's or Y's depending on the actual shape of the intersections. In directions, they were indicated by "take a," "make a," or "turn," followed by "left" or "right." Road shape was either straight or curved in depictions; straight corresponded to "go down" in directions, and curved corresponded to "follow around." It is important to note here that although the route maps could be analog, they were not. In fact, they made the same distinctions that language did for the most part. Similarly, exact distance was not represented in either. Distance in both seemed to reflect complexity. Descriptions got longer for complicated reorientations just as depictions got larger. Long, straight stretches on the highway didn't take space in either depictions or descriptions.

The correspondence between elements of directions and elements of depictions suggest that they both derive from the same underlying cognitive structure. The structure of routes is a sequence of actions at intersections or links and nodes, where exact reorientation and exact distance are not important. Why can this information, which seems critical, be omitted? Most likely because the information is sufficient for the situations in which the directions are used. If the angle of the turn is unspecified or different from the angle in the world, the traveller will follow the road. Similarly, the traveller will reorient when the landmark signifying reorientation appears, irrespective of the distance. In fact, when the distance is long, people indicate that on both maps and directions by adding landmarks along the route that are not associated with reorientations. Significantly, the schematisation apparent in route maps and directions parallels the schematisation of memory (Tversky, 1981). People remember turns as closer to right angles, roads as closer to parallel, roads as straighter than they actually are.

The near sufficiency of these semantic elements was demonstrated in a task in which students were asked to use verbal or pictorial toolkits consisting of these elements to construct a large number of routes, short and long, simple and complex (Tversky and Lee, 1999). They were told that they would probably have to supplement the tool kits with elements of their own design. In fact, most students succeeded in generating verbal and visual directions with only the tool kit provided.

The common underlying structure was instantiated as cognitive design principles to guide development of an algorithm to automatically generate route maps on demand (Agrawala and Stolte, 2001). Users reported vastly preferring these maps to the more typical output from websites, highway maps with routes overlaid. The common underlying structure also raises the possibility of automatic translation between route directions and route maps.

Mechanical Diagrams*: Arrows*. Arrows are lines, connectors, but asymmetric ones, so they suggest asymmetric relationships. To assess what arrows communicate, we asked students to interpret diagrams with or without arrows (Heiser and Tversky, 2004). The diagrams were of mechanical systems that would be familiar to students, a bicycle pump (see Figure 3), a car brake, and a pulley system. Each student interpreted a single diagram. The arrows led to striking differences in interpretation for all three systems. When the diagrams had no arrows, students wrote structural descriptions, that is a description of the parts of the system and how the parts were connected. When the diagrams had arrows, they wrote causal, functional descriptions, that is, a description of the sequence of actions of parts and the effects of those actions. As for the previous examples, we asked new groups of students to produce diagrams given either structural or functional descriptions. For structural descriptions, students did not use arrows, but for functional descriptions, they did.

In a comprehensive survey of scientific diagrams (MacKenzie and Tversky, 2004), we have found (as have others, e. g., Gombrich, 1990; Horn, 1998; Westendorp and van der Waarde, 2000/2001; Winn, 1987), many different uses of arrows. A common use is to label or point at something, a function served early on by hands in diagrams. Other uses are to indicate direction of movement, manner of movement, sequence, causality, dependency, and more (it is reported that there are close to a dozen uses in chemistry diagrams alone, Peter Mahaffy, personal communication).

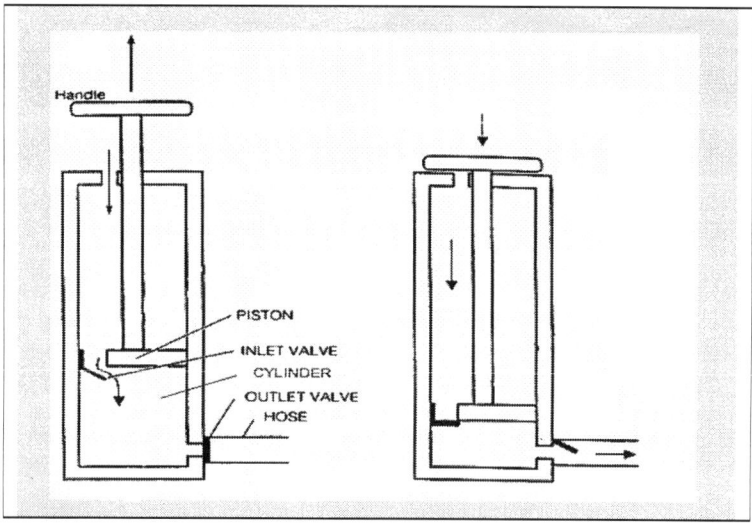

Figure 3. *Bicycle pump with arrows (from Heiser & Tversky, submitted, adapted from Morrison, 2001, adapted from Mayer & Gallini, 1990).*

Morphograms such as lines, blobs, crosses, and arrows are among many simple geometric forms that appear in visualizations of all kinds. Their meanings are often clear in context from their geometric or Gestalt properties. They can be combined not randomly but systematically to create complex graphical messages. As such, they share similarities with words such a *line* or *relationship* or *direction*, which also carry meanings that require context to disambiguate and which can be combined systematically to convey complex meanings. Morphograms, along with icons, figures of depiction, and metaphoric uses of spatial relations explain why many visualizations are easily produced and readily interpretable.

COGNITIVE DESIGN PRINCIPLES

The previous review and analysis suggests two cognitive principles for designing effective visualizations (Tversky, Morrison, and Betrancourt, 2002). According to the *Congruence Principle*, the structure and content of the visualization should correspond to the desired mental structure and content. According to the *Apprehension Principle*. The structure and content of the visualization should be readily and accurately perceived and comprehended. Using diagrammatic space to reflect conceptual space, as in mapping increases upwards, illustrates the Congruence Principle, as do successful uses of icons and figures of depictions and morphograms. Route maps are a subtler, deeper example of the Congruence Principle. Spontaneous route sketches do not convey distance and direction accurately. Not incidentally, mental representations of maps schematise the information in the same way. In memory, turns are remembered as closer to right angles than they actually are, and roads as more parallel than they actually are (Tversky, 1981). The much-lauded and much-imitated London subway map makes the same simplifications, and more. Schematising information to reflect schematic cognitive structures facilitates apprehension as well. They simplify the information, but the simplification is systematic, that is, schematic. Schematic visualizations preprocess the actual information, extracting what is needed even distorting it for emphasis, and eliminating what is non-informative. Schematic visualizations remove the irrelevant information that interferes with finding the relevant information.

DIAGRAM NARRATIVES: STRUCTURE AND PROCESS

What kinds of stories do scientific visualizations tell? To answer this, MacKenzie and Tversky (2004) conducted a survey of visualizations in textbooks for a range of disciplines in science. Two types of visualizations dominated: structure and process. Structural diagrams show the parts of a system and their spatial or conceptual relations. Process diagrams show change over time; they often show structure incidentally. Many visualizations combine or expand these types, for example, visualizations that show structure to function or that show structural variants of a category or that show structural hierarchies, parts and subparts. Static diagrams are ideal for conveying structure; they map the elements and spatial relations of a system onto the elements and spatial relations in diagrammatic space.

CONVEYING PROCESSES

Conveying process in static diagrams is not as straightforward. Process or function normally entails change in structure, as in the operation of a pump or cell meiosis or molecular changes, or in the part of the structure that is active as in a circuit diagram or nerve conduction or the nitrogen cycle. Although students high in mechanical ability or expertise are able to infer action or change from static diagrams, students low in mechanical ability/expertise (but high in other abilities) are unable to infer action or change from static diagrams. These students have no trouble understand action from verbal explanations (Heiser and Tversky, 2004). The finding that expertise or ability is needed to infer function from structure is a general finding. Experienced architects can infer change or function, such as traffic patterns and changes in light throughout the day and seasons, from architectural sketches, but novice architects cannot (Suwa and Tversky, 1997).

Fortunately, there are a number of different techniques for conveying process. A frequent one is use of arrows (Heiser and Tversky, 2004). But a close examination of arrows across a range of diagrams reveals many different senses, often in the same diagram, and often not disambiguated. Another is a sequence of static diagrams. A third is animation.

The Principle of Congruence suggests that animations are a natural way of conveying processes, change over time. This is undoubtedly one of the reasons for the enthusiasm for animations. The "gee whiz" factor is another; many animations are esthetic. But are animations effective in instruction? A broad survey of dozens of studies comparing animated graphics to informationally-equivalent static ones did not turn up a single study where animations were superior (Tversky, et al., 2002). This result has been resisted, and requires reflection. On reflection, animations violate both design principles. They are all too frequently too complex to be adequately perceived. They often have many moving parts; what is key is often the exact timing of the changes of the parts, and the eye and the mind cannot grasp them. Beginners don't even know where to look. People do not know how to parse or perceive the animations that life naturally provides. The art museums of the world are filled with paintings of horses galloping with their legs incorrect configured. It was Muybridge's stop-gap photography that revealed the correct configuration (Solnit, 2003). However, even animations portraying a single moving dot are not superior to a static graphic of the path (Morrison and Tversky, in preparation).

On closer inspection, animations may fail for a deep cognitive reason. People discretize continuous events that take place over time. They think about animated events as sequences of discrete steps (e. g., Hegarty, 1992; Zacks, Tversky, and Iyer, 2001). What's more, the segmentation into steps is systematic and predictable; for example, in the case of mundane human activities, such as making a bed, segmentation is by objects and object parts at a coarse level, and by articulated actions on objects and parts at a fine level (Zacks, et al., 2001). Recall routes; they are segmented by turns at landmarks. If people think about continuous actions as sequences of discrete steps, then visualizations of processes may better serve users by breaking them into the significant steps. Frequently, those steps are marked by changes in object and/or action. In fact, there is evidence that infants, children, and adult novices use large changes in physical actions to infer changes in goals and causes (e. g., Baldwin, Baird, Saylor, and Clark, 2001; Martin and Tversky, 2004; Woodward, Sommerville, and Guarjardo, 2001).

Route maps suggest yet another technique for producing better visualizations of processes that occur in time. Route maps distort space in order to enhance communication. They shorten long straight distances and enlarge short ones with tricky turns; they present turns of all angles as right angles (or diagonals), all in the interest of facilitating navigation. Animations could do the same for time; use time in ways that reflect expert understanding of processes, start, stop, slow down, speed up. Time and space could be altered together to allowing zooming, enlargement, change in perspective—spatial variations—cued by abrupt or continuous temporal changes. But this is not all.

Throughout evolution, humanity has witnessed change, process. The world does not sit still, it is always in flux. Watching things change does not tell us *how* or *why* things change. If it did, there would be little need for science and little scientific progress. How many generations watched water rise in the bathtub or apples fall from trees or the paths of the stars without any *eureka*s? All too many animations just show change. They need to explain it. Concomitant verbal explanations do help students learn from them (Mayer, 2001), but that is not enough. Good explanations do more than annotate the step-by-step action of a mechanical device or biochemical cycle.

VISUAL NARRATIVES

Insights into designing scientific visualizations can come from thinking more broadly about visual narratives. As for other external representations, they are ancient, like the remnants of the frescoes and friezes in Crete and Babylonia, and more recent, like the stained-glass windows and tapestries, and modern, like comic books and children's stories. Each medium tells stories in pictures or in pictures artfully combined with words.

What do good explanations do? Good explanations of the new are based in the old. That is, good explanations capitalize on what their audience already knows. They put things in context. Good explanations interweave the formal information with examples and analogies that elucidate aspects of the formal information. Contrast this to the typical animation, simply showing a process. Showing a process can be thought of as a series of stills snapshots, perhaps at a rate that is perceived as continuous, with temporal links between the stills. Thinking broadly, an explanation can be thought of as a series of stills with many different kinds of links, some temporal, some spatial, some, examples, some analogs, and so on. Verbal explanations can be thought of in the same way, as a string of concepts and relations. In fact, as was shown for routes, analyzing depictions and descriptions of the same content is an effective way of revealing the underlying cognitive structures that need to be communicated. Analyzing experts' depictions and descriptions of scientific concepts should be an effective means of discovering the content and structure that needs to be communicated.

The *Cartoon Guides* that Gonick and his collaborators have written for a variety of scientific and other disciplines are instructive. *The cartoon guide to physics* (Gonick and Huffman, 1990) for example, explains concepts like mechanics and electricity by sequences that zig-zag from general principles articulated in words, to equations, to visualizations of equations that are concrete or in graphs, to depictions of physical examples, and of course, to jokes. The conceptual links are varied and rich; only a minority are temporal. That these guides have been adopted as textbooks in serious courses in first-rate universities is some testimony to the success of this kind of visual explanation. These techniques are waiting to be exploited in scientific animations.

VISUAL COMMUNICATION

Visualizations are an essential element of teaching, understanding, and creating scientific ideas. Visualizations are not unique to the sciences; they belong to a large class of cognitive tools that have been crafted by people from all cultures and all eras for remembering, for reasoning, for discovering, and for communicating a wide range of ideas. Their effectiveness derives from cognitively compelling mappings from real and conceptual elements and spatial relations to elements and spatial relations on paper (or sand). They capitalize on people's extensive experience and facility in making spatial comparisons and inferences.

Visualizations, like language and other cognitive and communicative tools, vary in effectiveness. Effective visualizations take into account human perceptual and cognitive capacities. That means selecting the essential information, removing the irrelevant information, and structuring the essential information so that it can be readily and easily and accurately grasped and understood. Easier said than do, of course. Clarity is paramount for communication. Not so for visualizations for discovery and insight. For these, it cannot be known ahead of time what information is essential nor how to structure it; rather, these are what needs to be discovered. Clutter rather than brevity, ambiguity rather than clarity, excess rather than essence may encourage insight and discovery.

FOOTNOTE

The author is grateful to her collaborators on the projects described, including Sonny Kugelmass, Atalia Winter, Paul Lee, Jeff Zacks, Masaki Suwa, Julie Morrison, Mireille Betrancourt, and Julie Heiser. Portions of the research reported were supported by Office of Naval Research, grants NOOO14-PP-1-O649 and N000140110717 to Stanford University.

REFERENCES

Agrawala, M. and Stolte, C. (2001). Rendering effective route maps: Improving usability through generalization. *Proceedings of SIGGRAPH '01,* 241-250.
Baldwin DA, Baird JA, Saylor MM, Clark MA. (2001). Infants parse dynamic action. *Child Development*, 72: 708-717.
Beniger, J. R., & Robyn, D. L. (1978). Quantitative graphics in statistics. *The American Statistician,* 32, 1-11.
Cleveland, W. S., & McGill, R. (1985). Graphical perception and graphical methods for analyzing scientific data. *Science*, 229, 828-833.
Denis, M. (1997). The description of routes: A cognitive approach to the production of spatial discourse. *Cahiers de Psychologie Cognitive*, 16, 409-458.
Gonick, L. and Huffman, A. (1990). *The cartoon guide to physics.* New York: HarperCollins.
Gombrich, E.(1990). Pictorial instructions. In C. Blakemore, H. Barlow, and M. Weston-Smith (Editors), *Images and understanding.* Cambridge: Cambridge University Press.
Hegarty, M. (1992). Mental animation: Inferring motion from static displays of mechanical systems. *Journal of Experimental Psychology: Learning, Memory, and Cognition* 18, 1084-1102.
Heiser, J. and Tversky, B. (2004). Descriptions and depictions of complex systems: Structural and functional perspectives. Manuscript submitted for publication.
Horn, R. E. (1998). *Visual language.* Bainbridge Island, WA: MacroVu, Inc.
MacKenzie, R. and Tversky, B. (2004). Diagrammatic narratives: Telling stories effectively with scientific diagrams. Manuscript submitted for publication.
Martin, B. and Tversky, B. (2004). Making sense of abstract events: Building event schemas. Manuscript submitted for publication.
Mayer, R. E. (2001). *Multimedia learning.* Cambridge: Cambridge University Press.
Mayer, R. E. and Gallini, J. K. (1990). When is an illustration worth ten thousand words? *Journal of Educational Psychology,* 82, 715-726.
Morrison, J. B. and Tversky, B. (2004). Does animation facilitate learning? An evaluation of the congruence hypothesis. Manuscript submitted for publication.
Solnit, R. (2003). *River of shadows.* N. Y.: Viking Press.
Suwa, M., & Tversky, B. (1997). What architects and students perceive in their sketches: A protocol analysis. *Design Studies, 18,* 385-403.
Tufte, E. R. (1983). *The Visual Display of Quantitative Information.* Cheshire, CT: Graphics Press.
Tversky, B. (1981). Distortions in memory for maps. *Cognitive Psychology, 13,* 407-433.
Tversky, B. (1995). Cognitive origins of graphic conventions. In F. T. Marchese (Ed.)., *Understanding Images* (pp. 29-53). New York: Springer-Verlag.
Tversky, B. (2000) Some ways that maps and graphs communicate. In Freksa, C. Brauer, W., Habel, C and Wender, K. F.. (Eds.), *Spatial cognition II: Integrating abstract theories, empirical studies, formal methods, and practical applications.* Pp. 72-79. N. Y.: Springer.
Tversky, B. (2001). Spatial schemas in depictions. In M. Gattis (Ed.), *Spatial schemas and abstract thought..* Pp. 79-111. Cambridge: MIT Press.
Tversky, B., Kugelmass, S., & Winter, A. (1991) Cross-cultural and developmental trends in graphic productions. *Cognitive Psychology,* 23, 515-557.
Tversky, B., & Lee, P. U. (1998). How space structures language. In C. Freksa, C. Habel, & K. F. Wender (Eds.), *Spatial cognition: An interdisciplinary approach to representation and processing of spatial knowledge* (pp. 157-175). Berlin: Springer-Verlag.
Tversky, B., & Lee, P. U. (1999). Pictorial and verbal tools for conveying routes. In C., Freksa, & D. M., Mark, (Eds.), *Spatial information theory: Cognitive and computational foundations of geographic information science* (pp. 51-64). Berlin: Springer.
Tversky, B., Morrison, J. B., and Betrancourt, M. (2002). Animation: Can it facilitate? *International Journal of Human Computer Systems,* 57, 247-262.
Westendorp, P. and van der Waarde, K. (2000/2001). Icons: Support or substitute? *Information Design Journal,* 10, 91-94.
Winn, W. D. (1987). Charts, graphs and diagrams in educational materials. In D. M. Willows and H. A. Haughton (Eds.). *The psychology of illustration.* Pp. 152-198. N. Y.: Springer-Verlag.

Woodward, A. L., Sommerville, J. A., and Guajardo, J. J. (2001). How infants make sense of intentional action. In B. Malle and L. Moses (Editors). *Intentions and intentionality: Foundations of social cognition.* Pp. 149-169. Cambridge, MA: MIT Press.

Zacks, J., & Tversky, B. (1999). Bars and lines: A study of graphic communication. *Memory and Cognition*, 27, 1073-1079.

Zacks, J., Tversky, B., & Iyer, G. (2001). Perceiving, remembering and communicating structure in events. *Journal of Experimental Psychology: General,* 136, 29-58.

DAVID N RAPP

CHAPTER 3

MENTAL MODELS: THEORETICAL ISSUES FOR VISUALIZATIONS IN SCIENCE EDUCATION

University of Minnesota

Abstract. Mental models have been outlined as internal representations of concepts and ideas. They are memory structures that can be used to extrapolate beyond a surface understanding of presented information, to build deeper comprehension of a conceptual domain. Thus, these constructs align with the explicit objectives of science education; instructors want students to understand the underlying principles of scientific theories, to reason logically about those principles, and to be able to apply them in novel settings with new problem sets. In this chapter, I review cognitive and educational psychological research on mental models. Specific attention is given to factors that may facilitate students' construction of mental models for scientific information. In addition, these factors are related directly to the use (and potential) of visualizations as educational methodologies. The chapter concludes with several challenges for future work on visualizations in science education.

INTRODUCTION

There are many ways to demonstrate comprehension of a scientific principle. In school settings, an exam score is usually considered an acceptable measure of scientific understanding. However, students often succeed at tests by simply recalling what they have read or heard (for example, on an essay test), or recognizing what they have previously seen (for example, with a multiple choice) without necessarily understanding the material. To safeguard against this possibility, some tests require students to think beyond the course material to develop new hypotheses or apply their knowledge to new domains. Students with a deep understanding of a scientific principle can apply this knowledge to generate inferences about alternative possibilities *they have not yet considered*. Comprehension, according to this view, is the capacity to think critically about material above and beyond what has been presented; it is the *potential* knowledge students can generate and apply when they have adequate theoretical grounding in a topic, or to apply existing knowledge in new ways (Kintsch, 1998).

To assess the ways in which we learn and apply knowledge, psychologists have focused on cognitive processes including how information gets into memory (encoding), how information is accessed from memory (retrieval), and so on. A main tool for evaluating these processes is empirical investigation, in the tradition that chemists, physicists, and biologists empirically test their scientific hypotheses through rigorous experimentation. The large body of existing research on human memory seeks to outline the underlying mechanisms of thought by detailing the types of mental representations that are encoded and retrieved during everyday experiences. Similarly, research on learning has assessed how individuals build and apply these representations to solve problems (such as test questions). One construct that has received considerable attention with respect to memory is the *mental model*. Mental models are internal representations of information and experiences from the outside world. Indeed, mental models have been discussed beyond psychology proper; they are often invoked by science educators to describe the types of representations that equate with adequate comprehension of educational material.

The goal of this chapter is to provide a theoretical description of mental models based on work from the domains of cognitive and educational psychology. The chapter begins by outlining some of the inherent challenges in defining mental models. Despite these challenges, a tentative set of defining characteristics is provided. Next, focus is placed on three specific areas in cognitive psychology that have relied on mental models to describe successful comprehension. Attention to these research areas is intended to provide historical and theoretical context for the empirical study of mental models.

A working familiarity with mental models will, hopefully, afford the development of hypotheses about their impact on learning. As a step in that direction, the next section of the chapter describes three instructional features that impact the construction of mental models (i.e., engagement, interactivity, and multimedia). Of course, any discussion of these features should lead directly into consideration of whether current trends in education are consistent with theory. In line with the theme of this book, I will discuss visualizations in science education, and relate visualizations to our focused set of instructional features. By considering features of effective learning situations we can begin to judge, from a theoretical viewpoint, the appropriateness of visualizations for science education. The chapter then concludes with a discussion of future challenges for assessing the educational implications of visualizations. These challenges are designed to provide avenues of inquiry that more firmly establish linkages (both theoretical and practical) between cognitive, education, and visualization sciences.

DIFFICULTY IN DEFINING MENTAL MODELS

Mental models are challenging to define for a variety of reasons. First, they are not actual physical entities, and cannot be revealed through neuroimaging, surgery, or introspective interview. They are abstract concepts that we cannot directly observe; we can only make claims about them through logical associations with observable behavior. Second, mental models are purely *abstract* descriptions of memory. They are dynamic representations that can change over time. They are not single, immutable entities that remain invariant across (or even within) students. This makes them less amenable to a simple, concrete definition. Third, mental models have been defined in a variety of ways across different research traditions. In many cases there has also been little interaction between those traditions. It is not uncommon to read differing accounts of mental models in disparate research areas including, but not limited to, human factors design, computer programming, educational assessment, and developmental psychology.

A fourth issue, deserving of extended attention, is that although mental models are invoked to describe comprehension processes, mental models can both facilitate and inhibit successful performance. Individuals often develop and rely on mental models that are faulty or inaccurate (Guzzetti, Snyder, Glass, & Gamas, 1993; Kendeou, Rapp, & van den Broek, in press; McCloskey, 1983; Nussbaum & Novak, 1976; Vosniadou, 2003). These faulty models can lead to flawed reasoning. For example, children often possess faulty mental models for the structure of the earth; one documented model involves the belief that the earth is a hollow sphere with a flat interior (Vosniadou & Brewer, 1992; 1994). This model, when applied by the children, leads to inaccurate expectations and/or predictions about the Earth. Note that these children are using their inaccurate representation of the Earth in an appropriate way; they are constructing inferences by relying on a model (however inappropriate) stored in memory. Thus while underlying comprehension *processes* are the same regardless of the accuracy of the belief, the mental model (and the *products* resulting from its use) are not. Faulty models can occur across ages, experience levels, and topic domains (Carey, 1985; Diakidoy & Kendeou, 2001; Lewis & Linn, 1994; Osborne & Freyberg, 1985). Thus, mental models are representations that rely on a person's individual understanding, but are not always valid or reliable.

Given these limitations, how can we characterize mental models?

DEFINITIONAL CHARACTERISTICS OF MENTAL MODELS

In the most general sense, mental models are conceptual organizations of information in memory. Early descriptions outlined mental models as internal representations of the external world that allow for the generation of inferences (Craik, 1943). The notion that mental models could be used to move beyond some specifically experienced stimuli (such as the material in a lecture or textbook) led to the hypothesis that mental models are *mental simulations* (Kahneman & Tversky, 1982). In much the same way a film or software program can be run to review material or focus on specific components, a mental model can be 'run,' to generate hypotheses, solve problems, and transfer knowledge to new domains. Mental models combine our stored knowledge with our immediate experiences. Thus it is clear why good students might be described as possessing mental models of class concepts. These students can reason and problem solve beyond presented course material. But despite mental models seeming to evoke notions of 'perfect memory,' evidence suggests that for the most part, they are actually tacit and incomplete (e.g., Franco & Colinvaux, 2000; Norman, 1983). Indeed to speak of memory in general, individuals' mental representations are often fragmented, malleable, and subject to reconstructive processing (e.g., Loftus & Palmer, 1974; Loftus, Miller, & Burns, 1978). While mental models are useful for reasoning beyond some learned information, individuals' mental models are certainly not direct replicas or simulacra of experienced stimuli. Instead they are abstract representations that store the spatial, physical, and conceptual features of those experiences, useful for retrieval in the service of problem solving, inference generation, and decision making.

Most educational conceptualizations of mental models stop here as this definition addresses the important components of successful learning. However, research on human cognition has gone beyond these definitional characteristics to evaluate the underlying nature of mental models. Again, a mental model represents the perceptual and conceptual features of the external world, but is not an exact replica of that world (Barsalou, 1999). As an example, my mental model for a car's brake system contains the components, structure, and relationships of that system as a function of my particular experiences with brakes. This might lead one to believe that my mental model for brakes is visual in nature; indeed phenomenologically, cognitive activity often feels this way. Several researchers have proposed instead that mental models are *imagistic* rather than image-based (see Kosslyn, 1994, and Pylyshyn, 2002, for a discussion of these issues). Imagistic representations maintain the visual characteristics and physical feature-based relationships of objects and concepts, but are not inherently visual. For example, a mental model for celestial bodies in the Milky Way would not simply be a mental picture or video of the information, but an abstraction of the universe that conveys the organized relationships for objects based on size, distance, and so forth (perhaps through hierarchical organization or some other association-based system). The underlying imagistic nature of mental models suggests that although they are certainly not mental pictures, they are both useful and necessary for considering the visuospatial characteristics of a concept or system.

In summary, mental models are internalized, organized knowledge structures that are used to solve problems. They are encoded with respect to the spatial, temporal, and causal relationships of a concept. They can be run to simulate that concept, in order to assess alternative viewpoints and consider possibilities not readily available or apparent. Mental models are not exact internal replicas of external information, but are rather piecemeal, incomplete chunks of information that are retrieved as a function of user tasks or goals (e.g., Tversky, 1993). They are accessed and implemented in a wide variety of circumstances, as a function of integration between presented information and existing knowledge. Both successful and unsuccessful comprehension involves processes relevant to mental model construction and application.

HISTORY AND EVIDENCE FOR MENTAL MODELS

Mental models have been critical for theories of cognition. Work on discourse processing (van Dijk & Kintsch, 1978), logical reasoning (Johnson-Laird, 1983), and the comprehension of physical systems (Gentner & Stevens, 1983) have relied on mental models to describe comprehension. We will examine these three views to demonstrate how they have established an implicit consilience with respect to what a mental model is – a representation not only of what is described by a stimulus (be it a narrative, a syllogism, or the workings of an electrical circuit), but also what can be inferred from that stimulus based on prior knowledge (also see Tversky, in press, for a discussion of similarities and differences in conceptualizations of mental models). Work in each of these areas is directly relevant for science education; the processes involved in constructing meaning from language, evaluating logical premises, and understanding conceptual/physical systems are critical in a variety of scientific learning experiences. In the following section, evidence for the construction and application of information from mental models is provided by focusing on these three cognitive domains.

Beginning with the work of van Dijk and Kintsch (1978), researchers have proposed that readers can build several levels of representations following a text experience. One type of representation, the mental or situation model, is a memory trace for the information described by a text, but not necessarily contained within a text. It involves the construction of inferences and the application of expectations or preferences in the development of comprehension for text. Put simply, comprehension necessitates the encoding of text information, and the linking of current information to existing knowledge to build upon the experience. The result is a mental model (Zwaan & Radvansky, 1998). Evidence suggests that readers track the relationships between characters, objects, and environments using mental models (e.g., Franklin & Tversky, 1990; Rapp, Gerrig, & Prentice, 2001; Zwaan, Magliano, & Graesser, 1995). As one example of this type of research, Bower & Morrow (1990) showed that readers develop inferences about object and character locations based on information only implied by the text, applying that information to make decisions and predictions about story events (Morrow, 1994; Morrow, Bower, & Greenspan, 1989).

In an example of this work (Morrow, Greenspan, & Bower, 1987), participants were asked to memorize a spatial map. The map described various rooms in a building, with each room containing several objects. Following this memorization task, participants read a story presented sentence by sentence on a computer. The story described a character moving around the memorized environment of the map. At several points in the story, the narrative was interrupted and a pair of objects was presented on the computer screen. Participants were asked to identify whether the objects in these pairs were located in either the same room or in different rooms. At these interruption points, the objects had not yet been explicitly mentioned, and thus recognition of object locations can suggest how easily retrievable those objects were from memory (e.g., from a mental model of the learned map). Participants were faster to make correct decisions about objects pairs when those pairs were from locations currently occupied by the protagonist at the moment the story was interrupted. Participants took longer to correctly identify objects from rooms adjacent to the current location of the protagonist; objects from distances of two rooms away were even less accessible from memory. Readers applied what they knew (the memorized map layout) to generate spatial inferences about the relations between protagonists and objects in stories. Reader knowledge for the map was a function of current narrative focus (with respect to the location of the protagonist) and prior knowledge (memorized object locations from the studied map). This evidence demonstrates that readers can construct inferences as they read text, in ways that go beyond the explicit information in the narrative. To do this, readers simulate the layout of the building and consider where objects are to be found in relation to the protagonist. Readers similarly track information about time (e.g., Rapp & Gerrig, 2002; Zwaan, 1996), causality (e.g., van den Broek, 1988; van den Broek & Trabasso, 1986), and characters (e.g., Gernsbacher, Hallada, & Robertson, 1998; Rapp et al., 2001).

While this evidence on the surface might seem far afield from the notion of mental models in science education, these studies provide an indication as to how models are applied during processes of comprehension. Certainly the more general notion that readers actually encode information (i.e., the ongoing narrative), connect text statements to information from background knowledge (i.e., the studied map stored in memory), and integrate these sources to conduct mental simulation, is directly applicable to science learning. For science education, students must also encode information (e.g., the ongoing lecture or text lesson), access existing knowledge in long-term memory (e.g., information they have already learned), and integrate those two sources. In both cases, readers and learners actively attempt to develop these associations to achieve their comprehension goals (van den Broek, Risden, & Husebye-Hartmann, 1995).

In a similar way, work by Johnson-Laird (1983) has demonstrated that mental models are necessary for resolving logical syllogisms and spatial relationships. For example, consider the following logical premises: *Imagine a table setting. The plate is to the left of the fork. The knife is to the left of the plate.* Now evaluate the validity of a final premise: *The fork is to the left of the knife.* To evaluate such statements, readers must develop mental a model that tests the boundaries of the existing premises. That is, they internally represent the relations, and then simulate those mental organizations to evaluate the problem. For the above example, the final premise is false. The first three premise statements did not explicitly detail the relationship between the knife and fork (that relationship was implicit with respect to the position of the plate to *both* the knife and fork separately). This necessitates integration of the premises in a mental model. Evidence demonstrates that the more models an individual must construct to evaluate plausible premise relationships (that is, the more simulations that must be run), the longer it takes to resolve those problems (Johnson-Laird, Legrenzi, & Girotto, 2004). Simply put, the more premises and premise-based inconsistencies, the more alternative models that must be constructed, making the task more difficult. Mental simulations of logic require the development of multiple perspectives for successful comprehension. Developing a logical inference from a set of premises necessitates similar processes and memory structures as developing a narrative inference from a text description.

Individuals learning about physical systems, such as the operation of a pulley connection, or conceptual systems such as Newton's first law, must also rely on similar processes to move beyond a basic, surface-level understanding of material. Yet again, learning about the parts and their relations, and connecting that information with existing knowledge in memory, facilitates the construction of a mental model and comprehension of the material. But what happens when students possess faulty prior knowledge about conceptual (e.g., scientific topics) and physical systems (e.g., mechanical mechanisms)? In classroom settings, instructors try to help students develop new knowledge. At times existing knowledge is already in place, but is incorrect and in need of revision. Physics is a domain in which students often possess preexisting, faulty knowledge (diSessa, 1982). Students' flawed models of motion and movement are due to the expectations they generate as a function of their interactions in the environment and their beliefs about the world. Misconceptions are often systematic, and provide insight into the components of mental models that may be incorrect. (It is important to note that misconceptions can be due to both a faulty mental model, or to the misapplication of that model in the service of problem solving.) Norman (1983, 1988) has argued that even beyond classroom settings, designers must be aware of the types of mental models that users may possess when they think about and use physical and conceptual systems (for example computer interfaces, everyday appliances, and scientific concepts like physical laws). Educational designers, as a specific example, must ensure that students have little difficulty developing a correct mental model that affords an understanding of the possible uses and inferences engendered by a concept or topic.

FACTORS INFLUENCING LEARNING

Given this background, it is clear why science educators would be interested in the notion of mental models. The goal of an educational experience is for students to develop an understanding of some principle or concept, and to be able to apply that information to resolve an extended range of problems in a variety of situations. Mental models capture a type of memory that instructors want students to build. However, defining the circumstances that may or may not lead to the construction of an accurate mental model is not simple. Yet, given what we know about how memory functions, and the success of particular instructional methodologies, it is entirely reasonable to hypothesize mechanisms that may foster mental model construction. The next section of this chapter will describe three qualities of educational situations that influence learning in line with principles associated with mental models.

Cognitive Engagement

Engagement implies extended focus and thought on a topic. One way to operationalize engagement is to describe it as actively focusing and attending to stimuli, while also generating connections between that stimuli and representations in background knowledge. Students engaged in a task are more likely to stay involved with that task, and are more likely to learn during the task (Kounin & Doyle, 1975). While a full discussion of motivation is beyond the scope of this chapter, it is perhaps not surprising that students' intrinsic and extrinsic motivation towards educational tasks can directly impact the likelihood they will participate in and learn from those activities (e.g., Cordova & Lepper, 1996; Ryan & Deci, 2000).
Engagement also implies a level of cognitive activity commensurate with the nature of the task. Evidence for the benefits of focused cognitive activity comes from early work on memory. Levels of processing theory (Craik & Lockhart, 1972; Craik & Tulving, 1975; Hyde & Jenkins, 1969) contends that the quantity and quality of mental activity involved in a study task directly influences the likelihood that information will enter long-term memory. The deeper we process information, the more likely we are to remember that information for later use; the more shallow our processing, the less likely information will be stored in memory. The terms 'deep' and 'shallow' describe endpoints on a continuum ranging from intensive processing (e.g., relating information to prior knowledge) to surface processing (e.g., identifying the font of printed words). When students integrate new information with prior knowledge in memory, they build stronger links that increase the

probability that information will be stored in memory, and subsequently retrieved in the future. Engagement, as defined by increased motivation and a deeper level of cognitive activity, should therefore result in the construction of mental models in memory. Of course, in many cases deeper processing may not always lead to better memory, depending upon the nature of the material or the situations for which the material must be retrieved from memory (e.g., Morris, Bransford, & Franks, 1977; Tulving 1972, 1983). For example, if an individual wishes to memorize particular font types for a typesetting task, deeper processing might be less effective than shallow processing. This suggests that at least in some cases, a mental model may not be necessary for successful performance (e.g., for simple, surface-level tasks). Nevertheless, engagement implies processing that is more consistent with the deeper end of the levels of processing continuum.

Constructivist views of learning depend on notions such as engagement to outline the circumstances requisite for successful knowledge acquisition (Jonassen, 1999; Mayer, 1996; Phillips, 1997). These views contend that students learn best when they actively participate in tasks, building comprehension with relevance to their own interests and prior knowledge. When students construct their own answers, developing personally relevant cues for memory, they are more likely to remember that information. The well-established *generation effect* (Hirshman & Bjork, 1988; Slamecka & Graf, 1978) supports this view. When individuals *develop* their own cues for a list of words, they are more likely to remember words from that list than when *provided* with cues. Linking new information to personally relevant prior knowledge can facilitate later memory retrieval. While mental models have not been specifically invoked in these theories, the implications are fairly clear. Mental models are more likely to be constructed when readers are actively engaged in a learning task.

Interactivity

When students can directly impact the course of a lesson, by changing the pace or topic, or by manipulating characteristics that personalize the material in meaningful ways, the lesson can be described as interactive. Interactive lessons tend to be dynamic, in the sense that they can change in a variety of ways based on the needs of the student and the instructor. Interactivity is important beyond the degree to which it can describe the nature of particular instructor styles and lesson characteristics. Evidence for the success of instructional techniques is often attributed to the degree to which students can interactively direct the lesson (Ferguson & Hegarty, 1995; Wagner, 1997). Many researchers have argued that the positive impact of various types of educational technologies may not be attributable to the instructional features, but rather to the degree of control that students have over the pace or direction of those lessons (see Tversky, Morrison, & Betrancourt, 2002, for such a discussion on animation and learning). Simply, if students have control over the presentation of information, this may result in increased learning. Interactivity also directly relates to the notion of cognitive activity. Interactive instructional technologies make it more likely that students will become actively involved in the situation.

Therefore, interactivity provides students with the opportunity to generate knowledge in an iterative fashion. Current views of education have placed an emphasis on the interactive components of lessons (Scaife & Rogers, 1996). These views have argued against traditional lecture-based views of instruction, for which the common metaphor of 'the student as passive recipient' has been the norm. Interactive educational situations are designed to counter that notion. Interactivity also aligns with views of science education that seek to teach students how to engage in productive scientific thinking. The scientific method relies on interactivity as a means of evaluating the validity of hypotheses. Finally, interactive educational methodologies may facilitate learning as they align with views of mental models as interactive memory representations (e.g., Rapp, Crane, & Taylor, 2003). In a conceptual sense, a mental model is a manipulable memory representation that can be run as a simulation to test possibilities. In that sense, a mental model is also interactive.

Multimedia Learning

Educators have a variety of formats with which to present information. To what degree do these format types influence learning? One view suggests that multimedia presentations should facilitate learning, as redundant sources of information may help establish connections in memory. Early belief in the notion that a combination of visual and verbal information will facilitate learning and memory was offered by Paivio's dual-coding theory (Paivio, 1969, 1971). Words for which we can establish both a mental image and a meaning-based representation (e.g., *pyramid* and *dog*) are easier to remember from studied lists as compared to words for which we can only rely on one type of representation (e.g., *freedom*, *war*). Freedom is an abstract notion that necessitates a verbal description for meaning; a pyramid is a concrete shape that is amenable to the development of both a verbal and visual description. According to dual-coding theory, multiple formats provide multiple cues for encoding and retrieving information from memory. Particular subsystems in working memory process visual and verbal information separately, suggesting that multiple presentation types align with memory functioning (Baddeley, 1992, 2000).

While Paivio demonstrated the utility of his view for educational situations (1991), recent research in multimedia learning has more specifically suggested that multiple presentations may not always be effective. Kalyuga, Chandler, & Sweller (1999) contend that multiple information sources conveying the same idea in different formats can actually lead to interference and confusion. They found that when text was added to picture presentations, the additional verbal information led to memory difficulties as compared to cases in which pictures were presented alone. Kalyuga et al.'s work suggests that facilitation due to multiple media presentations is a function of the degree to which the media align or complement one another. That is, the *quality* of the presentation is more important than the *quantity* of the presentation. Additional evidence for this view has been offered by Mayer and colleagues, who have assessed various combinations of animation, text, and diagrams for science learning (e.g., Mayer, 2001; Mayer & Anderson, 1991; Mayer & Moreno, 1998; Mayer & Sims, 1994; Mayer, Moreno, Boire, & Vagge, 1999). In many cases, added media overloads limited memory resources, leading to interference in memory and comprehension problems.

Thus, while early views suggested that presenting information in multiple forms can facilitate learning, recent research suggests that considerable care must be taken in selecting and developing multimedia learning presentations. Of course, even single format or modality presentations may be ineffective for presenting a concept or idea unless they are designed for a specific learning goal (e.g., Norman, 1988; Tversky, et al., 2002). The take-home message of this research is that instructors must be strategic in their design of instructional presentations. Careful design is more likely to lead to the construction of mental models than simply applying a multimedia presentation. This is not to say that multimedia presentations are entirely ineffective; rather, the point is that multimedia presentations will be useful in a specifically delineated set of circumstances, such as when the intended internal representation is afforded by the design of the external presentation (Tversky et al., 2002).

These three examples provide an indication of some factors that are important for learning situations. Each of these factors is directly related to the notion of mental models, in the sense of shared concerns with respect to the successful acquisition of knowledge, and use of that knowledge outside of classroom settings. To what degree do these factors help us to build expectations for the use of visualizations as educational methodologies? Can visualizations lead to the construction of mental models?

VISUALIZATIONS AND LEARNING

A visualization can be defined as a novel visual presentation of data. Visualizations can include line drawings of data patterns, detailed 3-dimensional mappings of concept spaces, and hypermedia-based environments. In visualization situations, novel presentations are used to provide cohesiveness to concept mappings, explanations for processes, and interface designs for accessing datasets. To provide an example, a visualization of a dataset could take the form of an environmental walkthrough using a modeled terrain space. Individuals might travel the terrain as if they were traveling through an actual environment, accessing information by manipulating components of the virtual world. Another example might involve a chloropleth overlay for GIS-linked data, such as a color-coded map of population change in South America over the previous twenty years. A final example might use detailed animation to illustrate the migration of animal patterns by compressing thousands of years into a two-minute movie, with temporal control of the movie through controls relying on a spatial metaphor. Each of these visualizations is intended to allow the user to easily understand material by relying on previous knowledge, understandable mappings, and familiar metaphors (e.g., moving through environments as moving through layers of data; 'warmer' colors as indicating more densely populated regions in contrast to 'colder' colors; dynamic animated arrows indicating direction of control for viewing an animated sequence or movement data).

Visualizations have become popular in science education as instructors seek new methods for presenting complex concepts and data to students. Visualizations provide one method for describing how particular components in a complex mechanism interact. Much of science involves the explanation of complex, causal relationships in dynamic systems. One line of thinking suggests that only by illustrating these dynamic explanations in a form that captures salient relationships will students understand the complexity (or in some cases, the simplicity) underlying a conceptual theory. Also, as some scientific explanations *cannot* be observed in the everyday world, visualizations can provide experience with these concepts. For example, earth scientists and biologists are interested in developmental processes that take place over thousands of years; observing such change in real time would be impossible. Chemists describe processes that occur at a microscopic level, and these changes may be unobservable in traditional classroom situations. Physicists often rely on theories that describe abstract environments quite different from our own (e.g., frictionless surfaces). Visualizations can allow students to experience these situations in plausible, economically viable ways.

Despite these beliefs, there are several problematic issues to address with respect to the use of visualizations. To begin, a visualization is by no means a panacea for teaching difficult scientific topics. A poorly designed and implemented visualization is no better than a poorly designed lecture, an incoherent textbook explanation, or a rambling conversation. All of these situations can lead to study difficulty in building mental models. This can include a failure to develop a mental model due to information overload, or faulty model building due to incoherent information. To build a mental model that facilitates comprehension, all of these forms of discourse must focus the learner on important concepts and explanations to facilitate encoding in memory. A visualization is not guaranteed to do this. A related problem is that visualizations are often constructed with little thought towards appropriate design. There is no a priori reason to believe that a deep understanding of some content area will transfer to understanding effective visualization design. Researchers in human factors design and media studies have intensively examined the ways in which information can be most directly and successfully conveyed (for example, see Tufte, 2001). Without the use of appropriate cues, colors, designs, and organizations, students are less likely to know what to attend to, and what is being conveyed by the visualization. And lastly, visualizations are often designed without a specific goal. Educational technologies are of limited use if they have not been designed with a specific teaching plan (with particular attention to the topic and student for whom the technology is designed). Of course these concerns are not specific to visualizations; similar arguments have been levied at various types of instructional methods and technologies.

So, to what degree do visualizations facilitate learning? To date there has been mixed evidence towards the use of visualizations. Researchers have argued that visualizations may be successful in helping students as long as they remain consistent with prior notions of effective learning situations (de Vries, 2003; Hegarty, Narayanan, & Freitas, 2002). Without adhering to these notions, there is no a priori reason to believe that visualizations will necessarily lead to enhanced learning beyond traditional teaching techniques. In fact, there is evidence that inappropriately designed multimedia software can actually cause decrements in learning. Distracting visuals and irrelevant details are but two factors that can lead to learning that is poorer than would occur without visualizations (Harp & Mayer, 1998).

SO, WHY VISUALIZATIONS?

Despite concerns over the current utility of visualizations for educational settings, there is little doubt that their popularity and implementation will continue to increase (particularly as a function of lower costs, greater public access to visualization tools, and pressure to implement novel technology). But to what degree can visualizations influence learning in a positive way? We might consider visualizations as holding promise for learning to the degree that they involve cognitive activities demonstrated to enhance learning in other instructional situations. In an earlier section of this chapter, three characteristics of learning were described. These three characteristics were not selected at random - they are specifically relevant to thinking about the design of educationally valid visualizations for science classrooms.

First, visualizations are often quite engaging. An appropriately designed visualization can be quite impressive, using graphics and animation to present information in ways that people have not seen before. However, novelty will only last so long (Clark, 1994). One way to influence the likelihood of learning from visualizations is to design activities that are part of the science curriculum that are also cognitively engaging. Rather than simply having students observe a particular visualization, actively require them to become immersed in hypothesis testing with the visualization. A clear set of tasks, a problem-based learning scenario, or a laboratory session relying on the visualization as an information resource can help focus students on the important characteristics to consider from the presentation. If students are presented with visualizations and asked to think about the material at a deeper level, by connecting it with information in prior knowledge, they are more likely to construct mental models with which they can consider an array of scientific issues. Paired with guided learning and immersive tasks, visualizations are an important component in the instructor's educational toolkit (e.g., Hannafin, Land, & Oliver, 1999; Schank, Berman, & Macpherson, 1999).

Secondly, visualizations can be interactive and as suggested earlier, interactivity can foster learning through the construction of mental models. If students can influence key features of the visualization beyond simple personalization-based manipulations, they can actively engage in hypothesis testing. They can develop inferences, test those inferences, and revise their expectations to evaluate concepts. Feedback can be immediate as a function of the experience. Interactivity runs a continuum from control over a video presentation (allowing students to go back and review information that may not have made sense) to the manipulation of independent variables in a quasi-experimental visualization (e.g., Renshaw, Taylor, & Reynolds, 1998; Taylor, Renshaw, & Jensen, 1997). Control over a visualization can facilitate mental model construction, allowing the user to decide what to evaluate next in line with their interests and preferences. Thus, an important feature in a potentially successful visualization experience will be the degree to which the user can manipulate factors to study specific features and characteristics of intrinsic interest, or to complete task goals.

Finally, learning is fostered by conveying information in a succinct, guided manner that aligns with the nature of mental representations. Mayer (2001) has argued that multimedia learning will be successful to the degree it allows students to encode information in effective format combinations. Thus, the caveats and suggestions offered by psychological research with respect to the presentation of material should be directly applied to the design of visualization experiences. The point is that multimedia presentations will not make learning easier, but that research has suggested sets of circumstances under which multimedia designs can be used to present information in an effective manner, potentially leading to the construction of mental models (e.g., Brunye, Taylor, & Rapp, 2004).

CHALLENGES FOR THE FUTURE

There have been many impressive advances in scientific visualization, with respect to the development of new devices, systems, and software capabilities. However, this focus is problematic in that it neglects both the student and the scientific content in the visualization design. Without including both of these in examinations of visualizations, there is little reason to expect success in the construction of visualization systems for effective educational experiences. I next consider several challenges that are intended to direct the field of science education towards assessing and improving our knowledge of the role of visualizations in classrooms.

A first challenge is to establish ongoing discourse between technologists, science instructors, and cognitive scientists. As in other areas of educational technology, collaboration between these groups has led to a better consideration of how to develop effective educational interventions. Programmers and software designers have the technical know-how to develop complex multimedia environments for the presentation of data. Educators have a wealth of knowledge in topic domains, and notions of the types of material that must be conveyed to students. Cognitive scientists study the mechanisms that underlie thinking by empirically examining how the mind functions. Separately, each area has much to contribute specifically to their respective domains. Together, these groups can interact in an interdisciplinary manner to develop effective technologies that convey educationally relevant material in ways that align with cognitive processes. The field of visualization research needs to continue to foster and support these collaborations.

A second challenge is for the field to engage in more sophisticated and rigorous analyses of the impact of visualizations on learning. Experiments purporting to study educational interventions are often criticized for a variety of reasons, and visualization studies are no exception. First, potential confounds including prior knowledge and expectations can influence the likelihood of knowledge acquisition; these factors are often ignored in traditional studies. Second, control groups may be inappropriate or entirely ignored in the assessment design. For example, is a lecture group an appropriate control for a visualization group? According to Clark (1994), such a control group confounds the media format with the instructional design in the evaluation, affording little insight into the impact of computer-based learning situations. Third, visualization studies often assess short-term effects. If a visualization is effective in helping students to build a mental model for a scientific principle, we would expect students to retain this information beyond a short delay period. An analysis of student knowledge months after the lesson would test whether they encode more stable representations as a function of the learning experience. Fourth, and related to the third point, traditional visualization studies often compare only short-term treatment situations. It is rare to see comparisons of semester or year-long educational interventions. Should we expect significant learning benefits if students are only given a single trial session of the treatment? If we obtain effects, can we eliminate novelty as a potential cause of the effect? Empirical evaluations of visualization must, by necessity, take an evidence-based approach to make claims about the utility of educational visualizations.

A third challenge is for visualization researchers to construct designs that address specific educational concerns. Visualizations can be developed with particular goals, and currently there are many issues in science education to which visualizations could target their focus. For example, considerable research has investigated the ways in which individuals update their mental models as they encounter new information (e.g., Johnson & Seifert, 1998; O'Brien, Rizzella, Albrecht, & Halleran, 1998). However, many researchers have argued that updating is not guaranteed, and instead, individuals tend to rely on prior information often to a fault (Johnson & Seifert, 1994, 1999; van Oostendorp & Bonebakker, 1999). Consider this issue for students with inaccurate mental models. If students possess strongly held but incorrect beliefs about a scientific principle, it may be a challenge for them to correctly revise that information. Can we design visualizations that specifically target these faulty beliefs (common examples include the rotation and shape of the earth, beliefs about the laws of physics, and properties of electricity)? Evidence suggests that previously held beliefs are particularly resistant to change, but this resistance is mitigated if appropriate explanations are provided as to why the faulty belief cannot be correct (e.g., Hynd & Guzzetti, 1998). In some cases, a targeted visualization could provide that explanation. Thus, research should examine the features of visualizations that are necessary to provide target explanations that refute misconceptions. Again, this is only one example of the issues worthy of targeted visualization designs. Existing empirical work on such issues also provides a strong grounding for developing theoretically guided, testable hypotheses on the impact of visualizations for education.

CONCLUSION

Mental models provide a description of the contents of memory for complex concepts, including the understanding of scientific principles and theories. Mental models are mental simulations that can be run, akin to a software program, to apply knowledge in novel ways, in novel settings, for novel problem sets. Science educators have been interested in mental models as a means of evaluating the efficacy of teaching methodologies and to account for student comprehension of lesson material. Visualization researchers have also been interested in mental models, viewing them as a hopeful final product of experience with novel visualizations. While there is ample evidence suggesting that, at least in theory, visualizations may have the capacity to influence learning, there is less evidence that visualizations are being designed in a way that actually fosters the construction of mental models. There has been precious little visualization research that attempts to apply what we know from education sciences, human factors design, and cognitive psychology. Whether research across such diverse areas will converge towards a mutual understanding of mental models and visualization processes remains to be seen. A major goal for the future should be the critical evaluation of visualizations as tools for helping students to build mental models for scientific understanding.

REFERENCES

Baddeley, A. D. (1992). Working memory. *Science, 255,* 556-559.
Baddeley, A. D. (2000). The episodic buffer: A new component of working memory? Trends in Cognitive Sciences, 4, 417-423.
Barsalou, L. W. (1999). Perceptual symbol systems. *Behavioral and Brain Sciences, 22,* 577-660.
Bower, G. H., & Morrow, D. G. (1990). Mental models in narrative comprehension. *Science, 247,* 44-48.
Brunye, T., Rapp, D.N., & Taylor, H.A. (2004). Building mental models of multimedia procedures: Implications for memory structure and content. To appear in *Proceedings of the 26th Annual Meeting of the Cognitive Science Society.*
Carey, S. (1985). *Conceptual change in childhood.* Cambridge, MA: Bradford.
Clark, R. E. (1994). Media will never influence learning. *Educational Technology, Research, and Development, 4,* 21-29.
Craik, F. I. M., & Lockhart, R. S. (1972). Levels of processing: A framework for memory research. *Journal of Verbal Learning and Verbal Behavior, 11,* 671-684.
Craik, F.I.M., & Tulving, E. (1975). Depth of processing and the retention of words in episodic memory. *Journal of Experimental Psychology: General, 104,* 268-294.
Craik, K. (1943). *The nature of exploration.* Cambridge, England: Cambridge University Press.
Cordova, D. I., & Lepper, M. R. (1996). Intrinsic motivation and the process of learning: Beneficial effects of contextualization, personalization, and choice. *Journal of Educational Psychology, 88,* 715-730.
de Vries, E. (2003). Educational technology and multimedia from a cognitive perspective: Knowledge from inside the computer, onto the screen, and into our heads? In H. van Oostendorp (Ed.), *Cognition in a digital world* (pp. 155-174). Hillsdale, NJ: Lawrence Erlbaum Associates.
Diakidoy, I. N., & Kendeou, P. (2001). Facilitating conceptual change in astronomy: A comparison of the effectiveness of two instructional approaches. *Learning and Instruction, 11,* 1-20.
diSessa, A. (1982). Unlearning Aristotelian physics: A study of knowledge-based learning. *Cognitive Science, 6,* 37-75.
Ferguson, E. L. & Hegarty, M. (1995). Learning with real machines or diagrams: Application of knowledge to real-world problems. *Cognition and Instruction, 13,* 129-160.
Franco, C., & Colinvaux, D. (2000). Grasping mental models. In J. K. Gilbert and C. J. Boulter (Eds.), *Developing models in science education* (pp. 93-118). Boston, MA: Kluwer Academic Publishers.
Franklin, N., and Tversky, B. (1990). Searching imagined environments. *Journal of Experimental Psychology: General, 119,* 63-76.
Gentner, D., & Stevens, A. L. (Eds.) (1983). *Mental models.* Hillsdale, NJ: Lawrence Erlbaum Associates.
Gernsbacher, M. A., Hallada, B. M., & Robertson, R. R. W. (1998). How automatically do readers infer fictional characters' emotional states? *Scientific Studies of Reading, 2,* 271-300.
Guzzetti, B. J., Snyder, T. E., Glass, G. V., & Gamas, W. S. (1993). Promoting conceptual change in science: A comparative meta-analysis of instructional interventions from reading education and science education. *Reading Research Quarterly, 28,* 117-159.
Hannafin, M., Land, S., & Oliver, K. (1999). Open learning environments: Foundations, methods, and models. In C.M. Reigeluth (Ed.), *Instructional-design theories and models: A new paradigm of instructional theory* (Volume II, pp. 115-140). Hillsdale, NJ: Lawrence Erlbaum Associates.
Harp, S. F., & Mayer, R. E. (1998). How seductive details do their damage: A theory of cognitive interest in science learning. *Journal of Educational Psychology, 90,* 414-434.
Hegarty, M., Narayanan, N. H., & Freitas, P. (2002). Understanding machines from multimedia and hypermedia presentations. In J. Otero, J. A. Leon, & A. C. Graesser (Eds.), *The psychology of science text comprehension* (pp. 357-384). Hillsdale, NJ: Lawrence Erlbaum Associates.
Hirshman, D. L., & Bjork, R. A. (1988). The generation effect: Support for a two-factor theory. *Journal of Experimental Psychology: Learning, Memory, and Cognition, 14,* 484-494.
Hyde, T. S., & Jenkins, J. J. (1969). Differential effects of incidental tasks on the organization of recall of a list of highly associated words. *Journal of Experimental Psychology, 82,* 472-481.

Hynd, C., & Guzzetti, B. J. (1998). When knowledge contradicts intuition: Conceptual change. In C. Hynd (Ed.), *Learning from text across conceptual domains* (pp. 139-164). Mahwah, NJ: LEA.

Johnson, H. M., & Seifert, C. M. (1998). Updating accounts following a correction of misinformation. *Journal of Experimental Psychology: Learning, Memory, and Cognition, 24*, 1483-1494.

Johnson, H. M., & Seifert, C. M. (1999). Modifying mental representations: Comprehending corrections. In H. van Oostendorp & S. Goldman (Eds.), *The construction of mental representations during reading* (pp 303-318). Mahwah, NJ: Lawrence Erlbaum Associates.

Johnson-Laird, P. N. (1983). *Mental models.* Cambridge, MA: Harvard University Press.

Johnson-Laird, P. N., Legrenzi, P., & Girotto, V. (2004). How we detect logical inconsistencies. *Current Directions in Psychological Science, 13*, 41-45.

Jonassen, D. (1999). Designing constructivist learning environments. In C.M. Reigeluth (Ed.), *Instructional-design theories and models: A new paradigm of instructional theory* (Volume II, pp. 215-239). Hillsdale, NJ: Lawrence Erlbaum Associates.

Kahneman, D., & Tversky, A. (1982). The simulation heuristic. In D. Kahneman, P. Slovic, & A. Tversky (Eds.), *Judgment under uncertainty: Heuristics and biases* (pp. 201-208). New York: Cambridge University Press.

Kalyuga, S., Chandler, P., & Sweller, J. (1999). Managing split-attention and redundancy in multimedia instruction. *Applied Cognitive Psychology, 13*, 351-371.

Kendeou, P., Rapp, D. N., & van den Broek, P. (in press). The influence of readers' prior knowledge on text comprehension and learning from text. In *Progress in Education.* Nova Science Publishers, Inc.

Kintsch, W. (1998). *Comprehension: A paradigm for cognition.* New York: Cambridge University Press.

Kirkby, K. C., Morin, P. J., Finley, F., Rapp, D. N., Kendeou, P., & Johnson, J. (2003). *The role of stereo projection in developing an effective concluding earth science course.* Poster presented at the annual meeting of the American Geophysical Union, San Francisco, California.

Kosslyn, S. M. (1994). *Image and brain: The resolution of the imagery debate.* Cambridge, MA: MIT Press.

Kounin, J. S., & Doyle, P. H. (1975). Degree of continuity of a lesson's signal system and the task involvement of children. *Journal of Educational Psychology, 67*, 159-164.Lewis, E. L. & Linn, M. C. (1994). Heat energy and temperature concepts of adolescents, naïve adults, and experts: Implications for curricular improvements. *Journal of Research in Science Teaching, 31*, 657-677.

Loftus, E. F., Miller, D. G., & Burns, H. J. (1978). Semantic integration of verbal information into a visual memory. *Journal of Experimental Psychology: Human Learning and Memory, 4*, 19-31.

Loftus, E. F., & Palmer, J. C. (1974). Reconstruction of automobile destruction: An example of the interaction between language and memory. *Journal of Verbal Learning and Verbal Behavior, 13*, 585-589.

Mayer, R. E. (1996). Learners as information processors: Legacies and limitations of educational psychology's second metaphor. *Educational Psychologist, 31*, 151-161.

Mayer, R. E. (2001). *Multimedia learning.* Cambridge, United Kingdom: Cambridge University Press.

Mayer, R. E., & Anderson, R. B. (1991). Animations need narrations: An experimental test of a dual-coding hypothesis. *Journal of Educational Psychology, 83*, 484-490.

Mayer, R. E., & Moreno, R. (1998). A split-attention effect in multimedia learning: Evidence for dual processing systems in working memory. *Journal of Educational Psychology, 90*, 1998, 312-320.

Mayer, R. E., Moreno, R., Boire, M., & Vagge, S. (1999). Maximizing constructivist learning from multimedia communications by minimizing cognitive load. *Journal of Educational Psychology, 91*, 638-643.

Mayer, R. E. & Sims, V. K. (1994). For whom is a picture worth a thousand words? Extensions of a dual-coding theory of multimedia learning. *Journal of Educational Psychology, 86*, 389-401.

McCloskey, M. (1983). Naive theories of motion. In D. Gentner & A. L. Stevens (Eds.), *Mental models* (pp. 299–324). Hillsdale, NJ: Lawrence Erlbaum Associates.

Morris, C. D., Bransford, J. D., & Franks, J. (1977). Levels of processing versus transfer appropriate processing. *Journal of Verbal Learning and Verbal Behavior, 15*, 519-533.

Morrow, D. G., Bower, G. H., & Greenspan, S. L. (1989). Updating situation models during narrative comprehension. *Journal of Memory and Language, 28*, 292-312.

Morrow, D. G., Greenspan, S. L., & Bower, G. H. (1987). Accessibility and situation models in narrative comprehension. *Journal of Memory and Language, 26*, 165-187.

Norman, D. A. (1983). Some observations on mental models. In D. Gentner & A. L. Stevens (Eds.), *Mental models* (pp. 7–14). Hillsdale, NJ: Lawrence Erlbaum Associates.

Norman, D. A. (1988). *The design of everyday things.* New York, NY: Doubleday.

Nussbaum, J., & Novak, J. D. (1976). An assessment of children's concepts of the earth utilizing structured interviews. *Science Education, 60*, 535-555.
O'Brien, E. J., Rizzella, M. L., Albrecht, J. E., & Halleran, J. G. (1998). Updating a situation model: A memory-based text processing view. *Journal of Experimental Psychology: Learning, Memory, and Cognition, 24*, 1200-1210.
Osborne, R., & Freyberg, P. (1985). *Learning in science: The implications of children's science.* Hong Kong: Heinemann.
Paivio, A. (1969). Mental imagery in associative learning and memory. *Psychological Review, 76*, 241-263.
Paivio, A. (1971). *Imagery and verbal processes.* New York: Holt, Rinehart & Winston.
Phillips, P. (1997). *The developer's handbook to interactive multimedia.* London: Kogan Page.
Pylyshyn, Z. (2002). Mental imagery: In search of a theory. *Behavioral and Brain Sciences, 25*, 157-238.
Rapp, D. N., & Gerrig, R. J. (2002). Readers' reality-driven and plot-driven analyses in narrative comprehension. *Memory & Cognition, 30*, 779-788.
Rapp, D. N., Gerrig, R. J., & Prentice, D. A. (2001). Readers' trait-based models of characters in narrative comprehension. *Journal of Memory and Language, 45*, 737-750.
Rapp, D. N., & Kendeou, P. (2004). *First impressions: Updating readers' models of characters in narrative comprehension.* Paper presented at the 67th annual meeting of the Midwestern Psychological Association, Chicago, Illinois.
Rapp, D.N., & Kendeou, P. (2003). *Visualizations and mental models - The educational implications of GEOWALL.* Invited paper at the American Geophysical Union Fall Meeting, San Francisco, California.
Rapp, D. N., Taylor, H. A., & Crane, G. R. (2003). The impact of digital libraries on cognitive processes: Psychological issues of hypermedia. *Computers in Human Behavior, 19*, 609-628.
Renshaw, C. E., Taylor, H. A., & Reynolds, C. H. (1998). Impact of computer-assisted instruction in hydrogeology on critical-thinking skills. *Journal of Geoscience Education, 46*, 274-279.
Ryan, R. M., & Deci, E. L. (2000). Intrinsic and extrinsic motivations: Classic definitions and new directions. *Contemporary Educational Psychology, 25*, 54-67.
Schank, R. C., Berman, T. R., & Macpherson, K. A. (1999). Learning by doing. In C.M. Reigeluth (Ed.), *Instructional-design theories and models: A new paradigm of instructional theory* (Volume II, pp. 161-181). Hillsdale, NJ: Lawrence Erlbaum Associates.
Scaife, M., and Rogers, Y. (1996). External cognition: How do graphical representations work? *International Journal of Human-Computer Studies, 45*, 185-213.
Slamecka, N. J., & Graf, P. (1978). The generation effect: Delineation of a phenomenon. *Journal of Experimental Psychology: Human Learning and Memory, 4*, 592-604.
Taylor, H. A., Renshaw, C. E., & Jensen, M. D. (1997). Effects of computer-based role-playing on decision making skills. *Journal of Educational Computing Research, 17*, 147-164.
Tufte, E. R. (2001). *The visual display of quantitative information.* (2nd ed.). Cheshire, CT: Graphics Press.
Tversky, B. (1993). Cognitive maps, cognitive collages, and spatial mental models. In A. U. Frank and I. Campari (Eds.), *Spatial information theory: A theoretical basis for GIS* (pp. 14-24). Berlin: Springer-Verlag.
Tversky, B. (in press). Mental models. In A. E. Kazdin (Ed.), *Encyclopedia of Psychology.* Washington, DC: APA Press.
Tversky, B., Morrison, J. B., & Betrancourt, M. (2002). Animation: Can it facilitate? *International Journal of Human-Computer Studies, 57*, 247-262.
Tulving, E. (1972). Episodic and semantic memory. In E. Tulving & W. Donaldson (Eds.), *Organization of memory* (pp. 381-403). New York: Academic Press.
Tulving, E. (1983). *Elements of episodic memory.* New York: Oxford University Press.
van den Broek, P. (1988). The effects of causal relations and hierarchical position on the importance of story statements. *Journal of Memory and Language, 27*, 1-22.
van den Broek, P., Risden, K., & Husebye-Hartmann, E. (1995). The role of reader's standards of coherence in the generation of inferences during reading. In E. P. Lorch & E. J. O'Brien (Eds.), *Sources of coherence in reading* (pp. 353-374). Hillsdale, NJ: ELA.
van den Broek, P., & Trabasso, T. (1986). Causal networks versus goal hierarchies in summarizing text. *Discourse Processes, 9*, 1-15.
van Dijk, T.A., & Kintsch, W. (1983). *Strategies of discourse comprehension.* New York, NY: Academic Press, Inc.

van Oostendorp, H., & Bonebakker, C. (1999). Difficulties in updating mental representations during reading news reports. In H. van Oostendorp & S. Goldman (Eds.), *The construction of mental representations during reading* (pp. 319-339). Mahwah, NJ: Lawrence Erlbaum Associates.

Vosniadou, S. (2003). Exploring the relationships between conceptual change and intentional learning. In G. M. Sinatra & P. R. Printrich (Eds.), *Intentional conceptual change* (pp. 377-406). Mahwah, NJ: LEA.

Vosniadou, S., & Brewer, W. F. (1992). Mental models of the earth: A study of conceptual change in childhood. *Cognitive Psychology, 24*, 535-585.

Vosniadou, S., & Brewer, W. F. (1994). Mental models of the day/night cycle. *Cognitive Science, 18*, 123-183.

Wagner, E. D. (1997). Interactivity: From Agents to Outcomes. *New Directions for Teaching and Learning, 71*, 19-26.

Zwaan, R. A. (1996). Processing narrative time shifts. *Journal of Experimental Psychology: Learning, Memory, and Cognition, 22*, 1196-1207.

Zwaan, R. A., Magliano, J. P., & Graesser, A. C. (1995). Dimensions of situation model construction in narrative comprehension. *Journal of Experimental Psychology: Learning, Memory, and Cognition, 21*, 386-397.

Zwaan, R. A., & Radvansky, G. A. (1998). Situation models in language comprehension and memory. *Psychological Bulletin, 123*, 162-185.

AUTHOR NOTE

This chapter is based on work supported by a Research and Development in Molecular Visualization mini-grant awarded by the National Science Foundation, a Grant-in-Aid of Research, Artistry, and Scholarship and a Faculty Summer Research Fellowship from the Office of the Dean of the Graduate School of the University of Minnesota, and a Marcia Edwards Fund award from the College of Education and Human Development at the University of Minnesota. I thank Tad Brunye and Panayiota Kendeou for their insightful comments on earlier versions of this chapter. Please address correspondence to David N. Rapp in the Department of Educational Psychology, 206A Burton Hall, 178 Pillsbury Drive SE, University of Minnesota, Minneapolis, MN, 55455 (e-mail: rappx009@umn.edu).

MICHAEL BRIGGS[1] AND GEORGE BODNER[2]

CHAPTER 4

A MODEL OF MOLECULAR VISUALIZATION

[1]*Department of Chemistry, Indiana University-Pennsylvania, Indiana, PA;*
[2]*Department of Chemistry, Purdue University, West Lafayette, IN, USA*

Abstract: We argue that molecular visualization is a process that includes constructing a mental model. Current qualitative research has shown that participants working on a mental molecular visualization/rotation task invoke components of a mental model. Four of the components are static representations: referents, relations, results, and rules/syntax. The fifth component is dynamic: operation. Two examples of operation are visualization and rotation. Participants used the constructed mental models as mental tools to complete the task. This conceptualization of mental model construction constitutes a theory of learning.

ANTECEDENTS

Research from several domains suggests that molecular visualization is a process of mental modeling that uses both representations and operations. To support this claim we draw from both the previous literature and results of our recent research. More than 50 years ago, the social psychologist Morton Deutsch (1951) wrote, "Men think with models." Similar conclusions were reached by Ludwig Wittgenstein (1961), who suggested that people picture (or visualize) facts to themselves and that a model is a picture (or visualization) of reality. In this chapter, we use the terms visualization and representation to refer to specific kinds of mental activity. We will presume that visualization is a mental operation and that representation is the object and result of that operation.

Johnson-Laird (1989) developed the concept of mental modeling as the foundation for model-based reasoning and suggested the basis for a relationship between visualization, as defined above, and mental model construction. Various authors have argued that mental models contain at least one dynamic component that is called an operation (Tuckey & Selveratnam, 1993; Lesh, Hoover, Hole & Kelly, & Post, 2000). We argue that visualization is an operation that produces a one-to-one correspondence between a mental representation and the idea, object or event to which it refers (the so-called referent). Visualization can therefore serve as the dynamic component that supplies the material with which model-based reasoning occurs. This reasoning is a personalized activity, however, because individual reasoners construct their own mental models (Bodner, 1986; von Glassersfeld, 1989). We therefore suggest that reasoning in general can be conceptualized as mental model building but the individual models constructed are idiosyncratic and personal (Kelly, 1955). Current research (Briggs, 2004) supports this conclusion.

METHODOLOGY

A qualitative research methodology allows one to access mental constructs that are difficult or impossible with a quantitative approach (Bodner, 2004). Our chosen methodology was phenomenography (Marton, 1981; Svensson, 1997) because it seeks to answer what we argue is a third-order question. A first-order question might be, "What is the world like?" A second-order question could be, "How do I experience the world?" A third-order question is, "How do I conceptualize my experience of the world?" Our conclusions in this chapter arose from the study of conceptualizations of the experience of mental molecular rotation by undergraduates either taking or preparing to take organic chemistry. In this study, the third-order question being considered was, "How do participants conceptualize and articulate their experience of mental molecular visualization and rotation?"

To develop theory of mental modeling we wanted to study the mental structure and processes involved in visualization and mental model construction. We could do that using the metaphor given by Marton (1981) of "figure" and "ground." In this metaphor, comprehended content is "figure" and the act of comprehension, is "ground." This perspective on process versus content enables one to focus on the cognitive activities rather than the content processed by those activities. The objective was to obtain for study the articulations of the conceptualizations of the experience of mental molecular rotation.

In order to generate a data stream of articulations we chose to use the technique of individual interviews that involved a think–aloud protocol (Simon & Ericsson, 1993). Participants worked on tasks in which they were required to mentally rotate a given two–dimensional molecular representation and produce a drawn artifact of the rotated molecule.

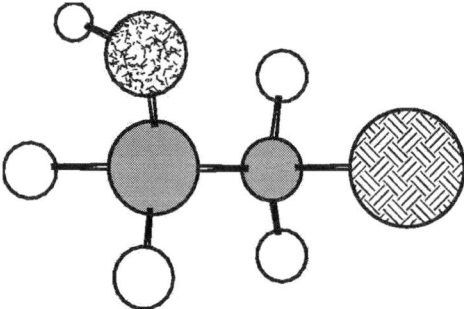

Figure 1. *Example of a task molecule given to participants. The grey filled circles represent carbon atoms; the unfilled circles represent hydrogen atoms; the random filled circle represents an oxygen atom; and the weave filled circle represents a halogen atom*

While working on each task, the participants were encouraged to verbalize what they were thinking at the moment. The various tasks used in this study were classified in order of increasing difficulty on the basis of three criteria: the number of atoms in the molecule, the extent of branching of the carbon chains, and the degree of rotation about each axis. We presented each task as a color–coded representation filling a 10 cm x 20 cm rectangle. The compounds whose molecular representations were used in this study consisted primarily of carbon and hydrogen atoms, but each compound had at least one oxygen or nitrogen atom as a reference for the rotation instruction.

We recorded each participant's articulations on both audio and video tape and collected the drawings they produced during their work as an artifact of the interview. After transcribing the interviews, we coded them in the tradition of grounded theory (Strauss & Corbin, 1998). Analysis of the resulting codes allowed us to develop a theory of mental model construction based on the visualization and mental rotation of molecules.

Multiple participants were used to allow us to compare and contrast the articulations and the conceptualizations of the participants in order to, "… understand in a limited number of qualitatively different ways…" (Walsh, 1993) the answers to the third-order question we asked. The study design was cross–case with thick description. In the following section, we present some of the thick description in the form of interview vignettes as support for claims we make regarding the representations and operations of mental molecular rotation and visualization.

STUDY RESULTS

During the coding of the transcripts, we paid particular attention to articulations that indicated components of a mental model (Lesh, Hoover, Hole & Kelly, & Post, 2000). We felt that evidence of the use of components of mental models by the participants would support a claim for the construction and use of mental models. We found ample evidence of the construction and use of mental models as the following vignettes illustrate. We have arranged this information in accord with the five components of a mental model as described by Lesh, et al. (2000): referent, relation, rules/syntax, operation, and result. Narrative vignettes were chosen from the work of the five participants in this study, who were given the pseudonyms: Oslo, Julia, Chessuana, Pete, and Ryleigh. All of the participants were second–year organic chemistry students at Purdue University.

Referents

Referents are physical objects, labels for objects, and mental representations. The participants acknowledged the use of referents during their work on the tasks. For example, Oslo noted, "So, if I stick with the carbon that we are focusing on …." In this case, Oslo selected a carbon atom in the task molecule and used it as a reference point for determining the relative positions of other atoms in the rotated molecule. Oslo clearly distinguished one atom from the other atoms in the molecule. Making that distinction is critical to completing the task of visualizing and rotating the molecule. In another task, Ryleigh says, "Since this is the one [atom] that they are physically asking me about, that is the one that I'm going to physically rotate around." In a similar statement, Chessuana responds, "And, I would pick, presumably, this corner [atom] right here." In another task, Oslo offers, "Since the line of sight would be drawn from this carbon to this carbon …"

Note that each of the participants indicates a specific atom within a molecule. Failure to distinguish atoms in one of tasks molecule would seem to prevent one from visualizing the task molecule. The point is clear in Julia's articulation, after sitting silently for many seconds, "Because I'm having difficulty with this chain here. Like putting that into space." Careful analysis of the transcript to her interview suggested that Julia had chosen a group of atoms but could not visualize them in space because she was not making distinctions between the atoms in this group and therefore could not place them into proper relation in mental space. The failure to distinguish among individual atoms may indicate a defective or missing visualization operation and would prevent a participant from constructing a useful mental model of the task molecule. We therefore argue that one of the components for a model of visualization of molecules is a defined referent.

Relations

The second component of our model of molecular visualization is relation, which involves the spatial relationship between referents. While working on one of the tasks in our study, for example, Ryleigh reported, "And there were two hydrogens off of that carbon, both going to the back of the page, so now they will come to the front just slightly to the left." In this vignette, Ryleigh focuses on the spatial relation among three atoms. There are three elements of this relation: position in space (109 degrees apart), sequence in bonding (hydrogen-oxygen-hydrogen), and identification of each atom (two hydrogens and one carbon). Ryleigh indicates a significant amount of chemical knowledge in this short statement, including the notion of tetrahedral bond angles, atom valences, and atom electronegativities. Working on another task, Chessuana states, "What I'm trying to do here is I'm trying to take this red oxygen [atom] and then somehow flip this molecule so that these two oxygens are one on top of the other so that I can't see one oxygen atom." In this statement, Chessuana identifies two oxygen atoms in the task molecule and thinks aloud about their relation to one another both before and after the rotation. In his interview, Oslo demonstrated the component of relation by referring to a carbon by position in the task molecule, "… then I'm going to move to the carbon on the left side here, that's with the *t*–butyl group, the number two carbon ..." Several relations are defined in this statement. Oslo identifies a carbon as "number two", indicating sequential relations between the carbon atoms in the backbone of the molecule. Oslo also identifies a connectivity relation between the central carbon of the *t*–butyl group and the number two carbon atom.

On a different task, Julia commented, "… then this one [atom] appears to be even more farther back, farther away, then somehow this has to connect to the blue (atom) ..." Julia's comment was based on a relation between two atoms in terms of "behind" or "in front of". Julia seemed to know specific atoms are related by connectivity in the task molecule, but could not see the relationship in the rotated molecule. She used the word "somehow" to indicate this uncertainty.

Another aspect of the component of relation is between the observer and the observed. Chessuana demonstrated this relation by saying, "I guess when I'm looking at something, I would imagine myself to be ... at the origin of a three-dimensional axis system." This aspect of relation is directed from the observer to a virtual observer rather than the relation between the observer and the observed shown in the preceding vignettes. This relation resembles the notion of perspective taking in the work of Piaget and Inhelder (1956). In an interesting comment, Julia combines these two types of relation, "... the vertical orientation is not going to change just by walking around to the other side of it." The relation of the atoms within the molecule, indicated by the phrase, "the vertical orientation," is a relation within the molecule. Julia then referred to a mental representation of the task molecule that took a perspective as if looking at the molecule from another side; Julia seems to be thinking from the perspective of a virtual observer, a relation external to the referent molecule. .

Rules/syntax

A required component of any mental model is a set of rules and syntax by which to order a mental representation. We define "rule" to be a concept and "syntax" to be the method of implementation of this rule. Thus, the rule and its syntax form a system. Our work suggests that mental imaging is not a random collection of mental "pixels" but rather an ordered visualization operation. This implies that a set of rules and syntax must be operating in order to make sense of the visual input from our eyes. Examples of rules/syntax systems that occurred in our study are given below.

Ryleigh expressed the following while drawing an artifact of a rotated molecule, "And the last methyl group [carbon number five] will just retain its tetrahedral shape [draws the three hydrogen atoms]." Ryleigh indicated an important concept associated with mental model construction: bonds in the task representation must be conserved. This conservation includes the number of bonds, their orientation in space, and the identity of the atoms involved in each bond. This is a sophisticated evolution of visualization because it allows the participant to draw a correct artifact of the rotated molecule. An important form of assessment for the researcher or teacher is available at this point. For years, organic chemists have evaluated the maturity of a student's mental model construction by evaluating the artifacts drawn. Domain acceptable application of the rules/syntax component will produce "correct" artifacts.

While working on another task, Chessuana comments, "This carbon atom is going to come out and its, all of its hydrogens are going to be almost completely in view now [after the rotation]." Chessuana has applied another rule: atoms that are occluded by other atoms still exist and follow the rules of conservation. The task from which this comment was generated happened to have all of the methyl hydrogens visible, but Chessuana comments on their visibility because in other tasks the hydrogen and carbon atoms were sometimes hidden behind other atoms.

Chessuana's articulation indicates that the rule/syntax component of conservation still holds true in the case of occluded atoms.

Another example of the rule/syntax component is comparison against a standard to check for reality or correctness. Julia demonstrated this comparison by saying, "Which also means the third [carbon atom] will kind of hide the carbon that is behind it, too, because there is the same, like plane going through all of those [atoms]." Julia constructed a mental model of a plane, applied it to the visualized molecule, and then used that construct to compare how other atoms were occluded. The comparison process seems to imply another rule/syntax component: atoms in a plane all move in the same manner when the plane rotates. This component must be an integral part of the constructed mental model of a plane or the result would not be useful.

Oslo used a rule/syntax component that noted that rotation conserves conformation of bonds. "But if we say it stays in this conformation then I can draw it [in rotated form]." And on another task, Oslo noted, "Oh, because it is all bonded to the carbon [atom] it would have to be just like this [draws the rotated molecule]," which is another application of the rule of conservation of bonds.

Another rule/syntax component is based on perspective. Visualization requires some method of indicating objects close to the viewer and other objects farther away from the viewer. In chemistry we often use atom size, bond foreshortening, or a system of hashes and wedges to indicate spatial orientation. The rule/syntax component might be: atoms closer to the viewer are larger than atoms farther away from the viewer; or, the end of a bond coming toward the viewer is larger than the end of the bond farther away from the viewer. Another version of this component might be: bonds along the axis of sight from the viewer to an object are foreshortened but bonds perpendicular to that axis are not foreshortened. These cues to spatial position are required for a student to draw a correct artifact of a visualized and mentally rotated molecule. Oslo, for example, visualizes a molecule as, "... my idea usually is, I have that chain and then I have bonds going away into the page and out of the page." In this case Oslo used the plane of the task representation as a reference for visualizing the molecule. Oslo used the hash and wedge system to show perspective when drawing the rotated molecule. An example of a deficiency of a perspective rule/syntax component was shown by Ryleigh, "I can't picture the entire molecule because there are too many bond angles ..." This lack of a useful rule prevented him from completing the task of visualizing the molecule.

The rule/syntax component of a visualization model is an important part of the visualizing ability. Without this component, participants were unable to make sense of the task molecule and could not visualize it. We assumed that participants who could visualize and mentally rotate a task molecule had a well–developed system of rules and their associated syntaxes. We drew this conclusion from the participants' articulations of their conceptualizations of the experience of molecular visualization.

Operation

The foregoing components are static in nature, but the component "operation" is a dynamic entity. We define an operation as the process of transforming a representation from one form to another. This might occur, for example, when a two–dimensional task representation printed on paper is transformed into a mental representation. We have called this operation "visualization" and called the product of the operation a "representation." Another form of this component would involve the transformation of a mental representation by an operation such as rotation to produce a new mental representation. An important operation in our study was the transformation of a mental representation of the rotated molecule to a two–dimensional artifact drawing. Our work suggests that operations are mental activities and components of mental models as the following vignettes show.

Pete demonstrated on operation that involved transformation of a mental representation to a drawn artifact when he commented, "It's just getting it out onto the paper, drawing it on the paper." Julia talked about the operation of visualizing by using a plane, "Let's see, I'm kind of like trying to split it [Julia uses here hands to indicate a plane] and seeing what would be on what side." Chessuana referred to the operation of rotation by saying, "And I would know that this green [atom] would be rotated the farthest from the end and come all the way around and two carbons, I suppose these are, in the middle, would just stay in the middle."

Note the dynamic nature of these examples. There seems to be a fundamental difference between this component and the others we have investigated. Operation is different from the other components in its ability to transform a referent. Each of the other components involve more or less static representations, but operation involves transformation of a referent. We have noticed that this static–dynamic relation plays a role in the structure and processes of mental model construction.

Our participants invented operations to mitigate deficiencies in visualization abilities. Ryleigh illustrates this process by saying, "It's large enough and complex enough that it's too hard to do that [visualize the whole molecule]. So, I come up with sort of a formula, and know if this is the front and this direction [left or right], and I rotate it 180 degrees that way, then it would be in the back now, and so, then all I have [to do] is kind of plug and chug with each bond." Ryleigh breaks the operation of molecular rotation into an operation on each atom in the molecule. This invention worked fairly well on small molecules but broke down on larger ones. Another problem with operation was shown by Pete in this vignette, "My problem right now is actually, when I rotate anything, instead of rotating the molecule as a whole, I'm just sort of twisting it around a bond." Pete realized that a deficiency existed but did not invent a correct operation to replace rotation. Oslo invented a way of assisting the rotation operation by chunking or combining atoms into groups. This was shown by, "… it's pretty easy of you have a grasp on how these methyl groups are oriented, it's pretty easy to visualize a tripod and to flip the whole thing around." If the components known as referents, relations, rules/syntax, and operations are correctly constructed then a participant could construct and use mental models to obtain results.

Results

The component "result" is an important part of model-based reasoning because it is the product of operating on a referent. We use the term "result" to convey both mental and physical products. A mental example would involve visualization producing a mental representation from a conceptualization or a physical object.

That is, visualization as an operation produces a transformed representation: a result. A physical product would occur when the operation of transformation, for example, makes a mental representation into a physical artifact

An example of the component known as result that was found in our study is rotational congruence: an operation on one element of a set will yield the same consequence as the same operation on another element of the set during the mental process of rotating a molecule. Julia demonstrated this when she talked about visualizing a rotated molecule, "… you can't see the ones behind these two methyl groups, because they are going to be hidden just the same as this one …" On another task, Julia said, "… these two carbons and these four hydrogens would all appear to be, like, in the same plane … And I'm looking to see if that appears to be the same thing on this side [of the molecule]…." Julia applies visualization to representations from two different media, a mental representation of the task molecule and the drawn artifact. Chessuana visualized in the same manner. Comparing the task representation to the drawn artifact of a rotated molecule, Chessuana comments, "So I know that up here, I have this red oxygen atom, and that looks to me like these are actually kind of going down a little bit, both this oxygen and this other hydrogen." Chessuana confirmed that the result of the task, the drawn artifact of the rotated molecule, was correct.

In each case, the participants operated on a task molecule and produced a result, an artifact of a rotated molecule. Once a participant produced a result, they could use it as evidence of successful completion of the task, to compare the result to the initial representation of the task to determine correctness, and to learn the conformation of the rotated task molecule. The production of a result seems to be sufficient justification for the exercise of mental model construction. The benefit of building mental models outweighs the costs of using mental energy because we learn from our interactions with the world. This learning then frees us from repetitive problem solving or model building.

In the foregoing paragraphs, we have shown the five components of a mental model. Four of those components are static representations: referents, relations, rules/syntax, and results. The component operation is dynamic. Warrants for the evidence presented in support of our claims come from the work of two researchers, Richard Lesh (Lesh, Hoover, Hole, Kelly, & Post, 2000) and Robbie Case (Case, Okamoto, Stephenson & Bleiker, 1996).

WARRANTS

In *Principles for Developing Thought–Revealing Activities for Students and Teachers in Handbook of Research Design in Mathematics and Science Education*, Lesh, et. al, (2000) discuss mental models and their components. We have made minor changes in some of the terms to allow this work to fit research outside the realm of mathematics. The relationship between the components of a mental model can be stated as follows. An operation on a set of referents produces a result. The referents and results (which can, themselves, become referents) are connected by relations and rules. The concept of a mental model is useful and fruitful. Used as a perspective to analyze our study results, we found that participants use mental models to visualize molecules.

The results of our study of molecular rotation raise several questions, including: "What is the consequence of the dichotomy of static and dynamic components of a mental model of molecular visualization?" And, "How do participants determine the reality and correctness of their visualization results?" Possible answers to these questions can be found by combining the work of Lesh, et al (2000) with the work of Robbie Case, et al, (1996).

In Chapter VIII, Summary and Conclusions in *The Role of Central Conceptual Structures in the Development of Children's Thought*, Case, et. al, (1996) describe a set of structures and processes that account for children's thought. The hypothesized structures differentiate between central conceptual structures and an executive control structure. We have found these structures to be useful ways of conceptualizing mental model construction. We have come to understand the central conceptual structures as the repository of mental operations that are at the core of mental model construction. We conceptualize the executive control structure as a distinct kind of mental model. The central conceptual structure contains mental operations that one uses to construct short–term mental models as required by reasoning in progress. This type of mental model is fleeting and constructed as needed in the intimate problem solving of reasoning.

The executive control structure is different in both duration and content. We argue that this mental structure is a long–term mental component and contains representations of our lifetime experiences. An example is the lesson we learned as young children that the top of a stove can be very hot. We do not have to relearn this lesson because we securely hold it in memory and can access it to warn us of new surfaces that might be hot. Without such a place in memory as the executive control structure, we would have to repeatedly relearn that surfaces can be hot. The executive control structure is a valuable asset because it frees the mind to concentrate on new and important problems instead of the "… buzzing, blooming confusion…" James (1911) we experience as infants. Our minds are freed from repetitious learning, giving us more processing power to apply to new experiences. The executive control structure functions as a standard by which we construe the world. A manner of conceptualizing this mental control structure is as a worldview. The participants in our study continually checked their work to determine if it was correct and used the executive control structure as the standard for comparison.

The work of Lesh, et. al (2000), and Case, et. al (1996), along with our work indicates that the components of mental models, the central conceptual structure, and the executive control structure form a mental system that is the basis for reasoning and learning. This insight has led us to explanations of several diverse fields of study in the domain of science education.

INSIGHTS

The process of visualization seems to precede the operation of rotation in a task of mental molecular rotation. From the moment one gives the participant the instructions for a task, the participant must visualize referents in order to begin a sequence of mental model construction. The task starts with visualizing words and sentences and turning those symbols into meaningful mental models. Then the participant must visualize the task molecule and transform it into a mental representation. We assume that there is significant processing between the eyes and the brain but will not address that in this chapter. Visualization is a dual media operation: it acts upon physical objects in the world around us and acts on mental objects. When told a technical fact we sometimes say, "Oh, I see it now!" This is not just slang but a semiotic response due to the way we conceptualize the action of the operation visualization.

Improper visualization may cause flawed representation and lead to incorrect results. This is an important lesson to teachers. We must be very careful of the manner and precision with which we scaffold our students as they construct their mental models of domain-accepted concepts. A flawed mental model can have an impact on reasoning beyond what one might expect.

The mind has error correction schemes for visualization that we hypothesize arise from the executive control structure. By having a standard by which to compare what one is visualizing, one can make sense of the object of visualization. This can work against us, however. If one builds an incorrect mental model and it becomes the standard by which one compares new mental models, results at variance with domain–accepted concepts may occur. Each of us has experienced the situation in which an optical illusion has tricked us into interpreting a drawing or picture incorrectly. Once we are shown or recognize on our own the correct interpretation, we can usually force ourselves to see the drawing correctly. The same situation with a flawed mental model, however, can cause the holder to reject what the senses are telling them and accept incorrect results, which has happened repeatedly in the history of science. It is for these reasons that teachers must probe for and demand correct construction of mental models of scientific principles.

IMPLICATIONS

The concept of mental modeling and model-based reasoning explains some examples of learning by providing structure and processes. Research into thought processes support the use, by participants, of the components of a mental model. We have found that participants use their constructed mental models as mental tools to navigate a solution path through a task. When a similar task is given several weeks later, the participants recall the former mental model and use it. We hypothesize that the recall consists of reconstructing the mental model when needed. In this sense, mental model building is a process of learning. The components of a mental model have to be constructed before the model is useful for reasoning. As domain- specific concepts become more complex, the model can be reconstructed using new referents, relations, and operations. The construction of the components and the use of mental models is controlled by maturation (Case, et. al,1996). As a result, younger children do not build models as complex as older children and adults, which limits the concepts that students can learn at a given level of maturation.

Learning is a sequential process. Learners must construct basic models before they can construct complex models. In chemistry, for example, students must construct a model of particulate matter in order to construct a more complex model of gas pressure or dissolution. The construction of a mental model of particulate matter requires models of an individual unit, identity, property, and interaction. It takes time to construct mental models and one can often partially construct the models concurrently. This process of interactive mental model construction constitutes our experience as we try to construe the world around us. Some models that are used frequently or successfully in different contexts become part of our executive control structure and are then available for error correction, comparison, and decision making. Use of the executive control structure allows one to modify behavior based on experience and constitutes a definition of learning.

In some ways, misconceptions and conceptual change theory can be thought of in terms of changes to constructed mental models. As infants, we begin to construct mental models of the world around us; for example the cooing of a mother's voice, the discomfort of being wet, and the motion of objects in front of us. Without training in model building, we build the model as best we can and sometimes the relations and rules are not correct. We might label these incorrect models as misconceptions. Changing the misconception is not an easy task, as secondary teachers will attest. We hypothesize that conceptual change is the process of constructing a new, fruitful model to replace (or coexist) with an incorrect one. It may be possible to change mental models but it seems more likely that one builds a new mental model that competes with the old model for usefulness. The model one uses the most or is fruitful in more situations is reinforced and is more likely to be used next time a similar context develops. As warrant, we offer the example of chemical equilibrium. Students with a well developed model of the particulate nature of matter seem to be able to grasp the dynamic nature of chemical equilibrium better than students who do not have such a model or have a model that is not as well developed. As students see chemical equilibrium in action, solve problems involving equilibrium, and reflect on the concept of equilibrium, they build new models with components that make the model useful. This is evidence that students construct mental models as needed to reason about a concept or problem.

In summary, we would like to argue that models and modeling provides a fruitful perspective on learning. In the study of molecular visualization and rotation described, in part, in this chapter, we have found evidence for the construction and use of mental models by the participants in this study.

REFERENCES

Bodner, G. M. (1986). Constructivism: A theory of knowledge. *Journal of Chemical Education, 63*, 873 - 878.
Bodner, G. M. (2004). Twenty years of learning how to do research in chemical education, *Journal of Chemical Education, 81*, 618-628.
Briggs, M. W. (2004). *A cognitive model of second–year organic chemistry students' conceptualizations of mental molecular rotation*, Unpublished Doctoral Thesis, Purdue University, West Lafayette.
Case, R., Okamoto, Y., Stephenson, K. M., & Bleiker, C. (1996). VIII. Summary and conclusions. In Y. Okamoto (Ed.), *The Role of Central Conceptual Structures in the Development of Children's Thought* (Serial No. 246 ed., Vol. 61, pp. 189-263). Chicago: University of Chicago.
Deutsch, K. W. (1951). Mechanism, organism, and society: Some models in natural and social science. *Philosophy of Science*, 18(2), 230-252.
von Glassersfeld, E. (1989). Cognition, construction of knowledge, and teaching. *Synthese, 80*, 121-140.
James, W. (1911). Precept and concept – The import of concepts. In *Some Problems of Philosophy* (p. 50). New York: Longmans Green.

Johnson-Laird, P. N. (1989). Mental Models. In M. I. Posner (Ed.), *Foundations of cognitive science* (pp. 469-499). Cambridge: MIT press.

Kelly, G. A. (1955). Constructive Alternativism. In *The psychology of personal constructs* (Vol. One: A theory of personality, pp. 3-45). New York: W. W. Norton.

Lesh, R., Hoover, M., Hole, B., Kelly, A., & Post, T. (2000). Principles for developing thought-revealing activities for students and teachers. In R. Lesh (Ed.), *Handbook of research design in mathematics and science education* (pp. 591-645). Mahwah: Lawrence Erlbaum.

Marton, F. (1981). Phenomenography – describing conceptions of the world around us. *Instructional science*, 10, 177-200.

Piaget, J., & Inhelder, B. (1956). The rotation and development of surfaces (J. L. Lunzer, Trans.). In *The child's conception of space* (pp. 271-297). London: Routledge & Kegan Paul.

Simon, H. A., & Ericsson, K. A. (1993). *Protocol analysis: Verbal reports as data* (Revised ed.). Cambridge, Mass.: MIT press.

Strauss, A., & Corbin, J. (1998). *Basics of qualitative research: Techniques and perspectives for developing grounded theory* (2nd ed.). Thousand Oaks: Sage.

Svensson, L. (1997). Theoretical foundations of phenomenology. *Higher Education Research & Development*, 16(2), 159-171

Tuckey, H., & Selvaratnam, M. (1993). Studies involving three-dimensional visualization skills in chemistry: A Review.*studies in Science Education, 21*, 99-121.

Walsh, E., Dell'alba, G., Bowden, J., E., M., Marton, F., Masters, G., et al. (1993). Physics students' understanding of relative speed: A phenomenographic study. *Journal of Research in Science Teaching*, 30(9), 1133-1148.

Wittgenstein, L. (1961). *Tractatus logico-philosophicus*. (D. F. Pears and B.F. McGuinness, Trans.) Routledge & Kegan Paul, New York

JANICE D GOBERT

CHAPTER 5

LEVERAGING TECHNOLOGY AND COGNITIVE THEORY ON VISUALIZATION TO PROMOTE STUDENTS' SCIENCE LEARNING AND LITERACY

The Concord Consortium, Concord, MA, USA

Abstract. This chapter defines visualization as it is used in psychology and education. It delineates the role of visualization research in science education as being primarily concerned with external representations and how to best support students while learning with visualizations. In doing so, relevant literature from Cognitive Science is reviewed. Two science education projects, namely, Making Thinking Visible and Modeling Across the Curriculum are then described as exemplars of projects that leverage cognitive theory and technology to support students' science learning and scientific literacy.

WHAT IS MEANT BY VISUALIZATION?

A google search on the word visualization yields over 3 million hits; yet, there are many usages of this term. Three common academic usages of visualization in psychology and educational research are: external visualization, internal visualization, and lastly, visualization as a type of spatial skill.

Visualizations as external representations. External visualizations refer to representations typically used for learning. External visualizations in science are graphics, diagrams, models, simulations, etc., and are, by definition, semantically-rich (Frederiksen & Breuleux, 1988) in that they involve complex, domain-specific symbol systems. As such, semantically-rich external representations are distinguished from iconic visual representations, e.g., a stop sign, and thus, the comprehension of and reasoning with semantically-rich visualizations is much more complex. As I will describe later, different types of external representations have different characteristics, and thus different cognitive affordances for learners and different implications for instruction.

Visualizations as internal representations. The term visualization is also used to describe internal mental constructs, i.e., mental models, thought to be in the mind's eye and used in mental imagery and to solve problems whereby people read off their mental model (Johnson-Laird, 1985). Although there has been considerable disagreement among cognitive psychologists about internal visual representations in terms of their psychological validity, their representational format (propositional/semantic or visual form), and the ways in which they are constructed, stored in memory, and used in reasoning, there is some agreement more recently that internal visual representations may be a bona fide form of mental representation (Kosslyn, 1994).

Visualization as a spatial skill. Lastly, visualization is also used to describe a type of spatial skill. Visualization is "the ability to manipulate or transform the image or spatial patterns into other arrangements" (Ekstrom, French, Harman, & Dermen, 1976) and thought to be important in highly visual domains like architecture (Salthouse, 1991; Salthouse et al, 1990).

It is important to note that these three usages of the term visualization do not imply three mutually exclusive constructs or processes. For example, learning with an external visualization likely requires that one construct an internal mental representation of the object or phenomena under inquiry and that one's spatial visualization skills may play a part in that construction process.

RELEVANCE TO SCIENCE EDUCATION

Each of the three types of visualization described above bear different degrees of relevance to science education. Internal visualization as a type of mental representation is arguably least central to science education and is not empirically tractable via the methodological approaches used in science education research to date. In research programs in which internal visualization is addressed, it is done so as a presupposition to the underlying learning theory employed in research (more detail on model-based teaching and learning will be given later in this chapter). Spatial visualization and its relationship to learning in science is a topic under the rubric of research in science education (cf., Piburn et al, in press). Research in this vein will likely permit us to delineate the role of spatial visualization skills in different scientific disciplines and shed light on the nature of spatial visualization processes as well. At the present state of understanding in the field of science education, a productive approach to advancing the field (in my opinion) is to focus on external visualizations and their role in science education in terms how to best support students' learning with visualizations.

External visualizations and Science Education.

Unpacking the role of scientific visualizations (i.e., external visualizations) in science learning is important, particularly since the onset of the computer generation, because visualizations are now so widely used as knowledge sources for learning at all levels of science education and in the media to convey important scientific information. Despite the proliferation of visualizations as information sources in science and the important role they play in knowledge dissemination, surprisingly little is known about how these information sources are understood, or how their information can be supported. This is greatly contrasted to the comprehension processes of textual and linguistic information which have been very well documented (cf., Kintsch, 1974, 1986; van Dijk & Kintsch, 1983). Thus, in focusing on external representations and their role in science education, the important issues are: unpacking the ways in which external visualizations differ from other types of knowledge media, how we can best support learning with external representations, and how technological advances can be leveraged to enhance students' science learning. The remainder of this chapter is dedicated to these goals.

Forms of Information, Information-Processing & Cognitive Affordances.

Information comes in many forms; in education we primarily deal with information that is textual/linguistic or graphic. Textual and graphic information sources differ in that textual information sources present information in a linear sequence whereas visual information sources provide all information to the learner simultaneously (Thorndyke & Stasz, 1980; Larkin & Simon, 1987). The implications of these differences for learners in terms of information processing are large. The cognitive processing of textual information is directed by the structure of the text; in the case of visual information sources however, the processing of information is directed by the learner. Thus, additional attentional processes for acquiring information from scientific visual information sources are needed. This is particularly important in the case of scientific visualizations because developing a deep understanding of scientific phenomena requires an understanding of causal information, and causal information has a sequential order of some kind. Although the differences between the structure of textual and visual information sources has been acknowledged, what has not been widely acknowledged or studied are the search and knowledge acquisition strategies for acquiring information from complex visualizations in science.

In addition to different search strategies for textual and visual information sources, Larkin and Simon (1987), in their classic paper, "*Why a diagram is (sometimes) worth 10,000 words*", point out two additional ways in which information-processing for the two media differ. These are perceptual cues and inference-making operators. Specifically, in terms of information processing, they state that visual information sources provide perceptual cues that are not present in textual information and thus inference-making on the basis of perceptual cues is different. The three ways in which text and graphic information sources differ outlined by Larkin and Simon (1987) (i.e., search processes, perceptual cues, and inference-making operators) provide a good starting point for unpacking the information-processing and cognitive affordances for textual versus visual forms information.

Extending ideas presented by Larkin and Simon, one way to think about different media (text as well as different types of external visualizations) in terms of their cognitive affordances is to describe them in terms of the degree of visual isomorphism [1] to the object being represented (Gobert, 2003). More specifically:
Textual representations describe in words aspects of science phenomena including the temporal sequence of causal mechanisms. Spatial aspects of scientific objects are difficult to convey in textual form due to the complexity of scientific phenomena.

---

[1] By isomorphic, I mean similar, not identical to the object/process being represented.

As such, text is non-isomorphic to the object or process it represents. In terms of information processing, the sequence of textual information "directs" by its linear nature, students' understanding of temporally-sequenced events.

Diagrams and Models (static) [2] describe in visual format spatial features of phenomena which are static. In this regard, diagrams and models can be isomorphic to the objects they represent. Because of their inherent spatial format, diagrams and models are good media for conveying spatial attributes of scientific object but are poor at conveying causal sequences and processes because they are static. In terms of information-processing on the part of the learner, supports or guidance are needed to direct students' knowledge acquisition because students lack the necessary domain knowledge in order to guide their search processes through diagrams in order to understand the relevant spatial, causal and dynamic information (Lowe, 1993; Head, 1984; Gobert, 1994; Gobert & Clement, 1999). As previously mentioned, constructing a mental model of causal mechanisms is particularly important since their causal processes have a sequential order to them.

Models and **simulations** are dynamic representations and provide a visual explanation of the underlying causal mechanisms and processes underlying scientific phenomena which are not directly observable because of their scale (Gobert, 2000). In terms of information-processing for learners, support and guidance for acquiring this information can promote students' deep learning of complex causal processes (Gobert, 2003; Gobert & Pallant, 2004).

From this continuum of representational formats, it is reasonable to assume that textual representations offer the fewest cognitive affordances for learners and that models and simulations, on the other hand, should offer the greatest number of cognitive affordances for learners. In recognition of the different affordances of text and graphics, in the 80's and 90's a great deal of research was conducted in which diagrams were provided as an adjunct learning source to text (cf., Hegarty & Just, 1989; 1993; Mandl & Levin, 1989; Mayer & Gallini, 1990). It was found that simply providing the two information sources together did not significantly improve students' science learning because it increased cognitive load on learners (Sweller et al, 1990) particularly when the students did not recognize that the object being described by the text and by the diagram was the same object (Gobert, 1994). Additionally, as previously stated, since students lack the necessary domain knowledge in order to guide their search processes to the relevant spatial, causal, dynamic, and temporal information (Lowe, 1993; Head, 1984; Gobert, 1994; Gobert & Clement, 1999), scaffolding is needed. If scaffolding is not provided, the affordances of the representation may be wasted as students become either confused or overwhelmed trying to figure out what is being depicted in the visualization or students may not engage in any deep processing of the visualization (particularly as the novelty of dynamic visualizations lessens). Thus, scaffolding is needed, particularly in the absence of prior knowledge, to support students' knowledge acquisition of and deep understanding of visual representations.

Extrapolating from Larkin and Simon (1987) and other work on visual information processing (Gobert, 1994; Lowe, 1993; Hegarty & Just, 1993) what is

---

[2] Models can be either static or dynamic. In this chapter, I give examples of each type.

needed to better support students' knowledge acquisition from scientific visualizations are: supports to guide search processes for acquiring rich spatial, dynamic, causal, and temporal information from visual representations; supports to make perceptual cues more salient since these can be rich ways to develop students' learning (Gobert, 2000; Clement, Brown, & Zietsman, 1989) and supports for inference-making in order to leverage from perceptual cues afforded by the representation. The use of these types of scaffolds as well as other scaffolds specifically designed to support students' model-building will be described and exemplified in two projects later in this chapter.

Learning from graphics as a constructive process. In thinking about learning processes for graphical information sources, learning is viewed as an active and constructive process. This view of learning is largely due to a seminal paper in memory research, entitled "Levels of processing: A framework for memory research" (Craik & Lockhart, 1972; Lockhart & Craik, 1990) which introduced the notion that the nature of the learner's processing of the stimulus material largely determined the learner's memory representations for that material. The levels of processing framework was originally developed in the context of text materials, but the framework has been subsequently shown to be applicable with visual stimuli including faces (Bower & Karlin, 1974) and cartoon figures (Bower, Karlin, & Dueck, 1975), as well as complex conceptual visual stimuli such as chess (Lane & Robertson, 1979), architecture (Akin, 1979, 1986; Gobert, 1994, 1999), and geology (Gobert & Clement, 1999; Gobert, 2000; Gobert, submitted).

Relationship between initial learning and knowledge representation. Following theories of human cognition and the levels of processing framework (Craik & Lockhart, 1972; Lockhart & Craik, 1990), it is assumed that the memory representation of visual information sources is largely related to the initial learning processes used to acquire information. The body of literature on expertise in visual domains has provided us evidence about the both ways in which processing is directed by prior domain knowledge and how domain-related information is chunked in memory. For example, superior performance exhibited by experts on domain-related tasks in visual domains has been accounted for by the use of domain-specific prior knowledge schemata (Brewer & Nakamura, 1984; Schank & Abelson, 1977; Rumelhart & Norman, 1975) which provide perceptual and cognitive structures that influence the <u>amount</u> and <u>manner</u> in which information is processed and encoded in memory (Chang, Lenzen, & Antes, 1985; Gilhooly et al., 1988; Head, 1984). In electronics, experts were found to represent information using functional units dictated by the type of circuit depicted and their knowledge organization was attributable to the functional units they had identified during their initial learning of the circuit diagram (Egan & Schwartz, 1979). Similarly, in the case of topographic map reading experts were found to have better comprehension of relative heights of the terrain depicted in the map and search strategies identified by eye-tracking showed that they attended to the highest and lowest points depicted (implicitly) in the map in order to fully understand the terrain (Chang et al., 1985). In the case of architecture, experts represent their knowledge in hierarchical structures made up of spatial chunks (Chase & Chi, 1981); again, the nature of the learning processes employed affected the resulting conceptual representations (Akin, 1979; 1986). In another study involving the understanding of a building from its plans, experts were found to employ search strategies that were both more systematic and 3-dimensional when compared to sub-experts; the resulting understanding of the building in both groups reflected their initial knowledge acquisition strategies (Gobert, 1994; 1999).

Important in all of these studies is the finding that experts use knowledge acquisition strategies for learning from visualizations that are highly related to task performance in their domain. Thus, in each case, skills which evolve through experience are especially adapted for performance in the respective domain. These findings highlight the need for scaffolding students' learning from visualizations since they lack necessary domain knowledge to guide their knowledge acquisition and search processes. More detail on scaffolding for learning will be given later in this chapter.

MODEL-BASED TEACHING & LEARNING AS A FRAMEWORK FOR LEARNING WITH VISUALIZATIONS.

Model-based learning and teaching is a synthesis of research in cognitive psychology and science education (Gobert & Buckley, 2000). In model-based teaching and learning, it is assumed that learners construct mental models of phenomena in response to a particular learning task (assuming the task has engaged the learner) by integrating pieces of information about the spatial structure, function/behavior, and causal mechanisms, etc. Reasoning with the model may instantiate evaluation of the model, leading to its revision or elaboration; model revision involves modifying parts of an existing model so that is better explains a given system. Model-based reasoning requires modeling skills to understand representations, generate predictions and explanations, transform knowledge from one representation to another, as well as analyze data and solve problems. It is analogous to hypothesis development and testing seen among scientists (Clement, 1989).

In the two projects to be outlined later in this chapter, model-based learning is the framework employed to guide our research, curriculum development, and scaffolding design. The emphasis on model-based reasoning fits within a current vein of science education which seeks to promote deep and integrated understanding of science by use of model-based tasks. It is believed that having students construct and work with their own models engages them in authentic scientific inquiry, and that such activities promote deep content learning, scientific literacy, and lifelong learning (Linn & Muilenberg, 1996; Gilbert, S., 1991; Sabelli, 1994). Although the need for models and modeling tasks in science education has been well recognized (AAAS, 1993, Giere, 1990; Gilbert, J., 1993; Hesse, 1963; Linn & Muilenberg, 1996; National Research Council, 1996), model-based teaching still does not play a significant role in American science curricula, nor have many research programs addressed how we can scaffold students' model-based learning using technology, a particularly suitable media for rich science learning.

LEVERAGING TECHNOLOGY TO SUPPORT STUDENTS' LEARNING WITH MODELS AND SCIENTIFIC LITERACY.

Conventional definitions of scientific literacy maintain that being scientifically literate includes understanding science content, having scientific process and inquiry skills, and understanding the nature of science, i.e., what is taken as evidence (Perkins, 1986). More recently, it has been argued that epistemic understanding also includes students' epistemologies of the nature and purpose of scientific models (Gobert & Discenna, 1997; Schwarz & White, 1999). Learning and reasoning with visualizations or models, as in the research to be described later in this chapter, fall under scientific process and inquiry skills and, thus also should be considered an integral part of scientific literacy. Briefly stated, the goal of scientific literacy is to have citizens understand science that is relevant to their everyday lives such that knowledge can be integrated across school topics and applied to real world problems (Linn, 1999), e.g., understanding scientific findings described by the media and making decisions that affect their everyday lives.

One very promising approach to achieving scientific literacy is to harness computer technology in service of this goal. Technology has great potential to bring about positive changes in students' learning, and in turn, on scientific literacy, for several reasons. First, technology is a powerful computational medium which can run dynamic simulations and support complex visualizations that were previously not realizable in conventional textbook diagrams (Gobert & Tinker, 2004). Secondly, since computers are becoming more common in schools, issues of access are becoming less problematic (Gobert & Horwitz, 2002). Thirdly, the World Wide Web makes it possible for students to access real-time scientific data and visualizations as part of their science learning, allowing students to see and work with authentic data. The web also allows teachers and students to access these learning environments such as WISE (Linn & Hsi, 2000) as well as other software tools and technological infrastructures which provide rich tools for science learning. Lastly, the interactive component of the web allows students to communicate with teachers and peers about their ideas which are excellent means to promote science learning (Linn & Hsi, 2000). This is not to say that simply the presence of computers will affect change. Rather, it is widely recognized that simply providing schools with computers does not in itself enhance student learning (Cuban, 2001). The greatest likelihood of affecting change on a broad scale will come about if we design learning environments, learning tasks, and pedagogy that are guided by educational research and what we know about student cognition in science.

PROJECTS AS EXEMPLARS TO DEVELOPING STUDENTS' MODEL-BASED REASONING AND SCIENTIFIC LITERACY: 'MAKING THINKING VISIBLE'

The Making Thinking Visible Project[3] (Gobert et al, 2002; Gobert, 2004a; Gobert & Pallant, 2004) harnessed technology in order to promote students' content knowledge, modeling skills (as a type of inquiry), and epistemologies of models in the domain of plate tectonics. Plate Tectonics is a domain in which many students and adults have misconceptions and learning difficulties (Bezzi, 1989; Gobert, 2000; Gobert & Clement, 1999; Jacobi et al, 1996; Ross & Shuell, 1993; Turner et al, 1986), thus, it is an important area for scientific literacy. Also, plate tectonics is an excellent domain in which to study model-based learning because of the plethora of external visualizations used in Geology and the important role that modeling, i.e., as an internal mental activity, plays in understanding geological phenomena.

The project drew on previous research on modeling in continental drift and plate tectonics with middle school students. Briefly studies conducted to date have addressed: the effects of a multimedia environment, CSILE (Scardamalia & Bereiter, 1991), on students' graphical and causal explanations of continental drift (Gobert & Coleman, 1993); the nature of students' pre-instruction models of plate tectonics and learning difficulties encountered in this domain (Gobert, 2000); causal reasoning associated with models of varying levels of integration and the visual/spatial inferences afforded on the basis of models (Gobert, 2000); the benefits of student-generated diagrams versus summaries (Gobert & Clement, 1999) and student-generated diagrams versus explanations on rich learning (Gobert, submitted); and the influence of students' epistemologies of models on learning in this domain (Gobert & Discenna, 1997).

The curricular unit, "What's on your plate?" was designed using the relevant literature on learning in Earth Sciences, namely, misconceptions of plate tectonics of both the inside structure of the earth and of the causal mechanisms underlying plate tectonic-related phenomena (Gobert & Clement, 1999; Gobert, 2000), and findings about students' knowledge integration difficulties in this domain (Gobert & Clement, 1994). Middle and high school students from demographically diverse schools in California and Massachusetts collaborated on-line about plate tectonic activity in their respective location. The students participated in this curriculum using WISE, Web-based Inquiry Science Environment (Linn, 1998), an integrated set of software resources designed to engage students in rich inquiry activities. The 2-week curriculum unit engaged students in many inquiry-oriented, model-based activities including activities in which they constructed and revised models as well as activities in which they learned from and with models. Some of these activities and the scaffolding by which we supported model-based learning are summarized below.

---

[3] Making Thinking visible was funded by a grant from the National Science Foundation (REC-NSF-9980600) awarded to Janice Gobert. Any opinions, etc. are that of the author and do not necessarily reflect the views of the Foundation.

a) Students' Model Building & Explanation of their Models. Students were asked to construct visual models and explanations of plate tectonic-related phenomena in WISE (i.e., mountain formation, volcanic eruption, and mountain building). They were scaffolded to include what happens in each of the layers of the earth in their model and explanation. This prompt was used in previous research (Gobert & Clement, 1999) in order to promote students' understanding and integration of the spatial structure of the earth with the various causal and dynamic mechanisms inside the earth. Once the students had done these two steps, they posted their models and explanations for their learning partners on the opposite coast.

b) Students' Evaluation and Critique of the Learning Partners' Models. Students read two pieces of text in WISE ("What is a Scientific Model? & "How to evaluate a model?") to get some basic knowledge with which to evaluate they're learning partners' models. Students were scaffolded to critique learning partners' models using the following prompts:
1. Are the most important features in terms of what causes this geologic process depicted in this model?
2. Would this model be useful to teach someone who had never studied this geologic process before?
3. What important features are included in this model? Explain why you gave the model this rating.
4. What do you think should be added to this model in order to make it better for someone who had never studied this geologic process before.

c) Students' Model Revision & Justification. Students read the evaluation from their learning partners on the opposite coast. They were then prompted to revise their models based on this as well as the content knowledge they had learned from the unit. They wrote a revised explanation for their new models. Lastly, students justified their changes to their models. Scaffolds here included:
  1. I changed my original model of.... because it did not explain or include....
  2. My model now includes or helps explain....
  3. My model is now more useful for someone to learn from because it now includes....
  4. I revised this on the basis of my learning partners' critique in the following ways....
  5. I revised this on the basis of the activities in these WISE units....

d) On-line field Trips. Students did an on-line field trip and visited multiple USGS websites with current data in order to better understand the differences between the coasts in terms of their mountains, volcanoes, and earthquakes. We chose viewing earthquake maps as the first task since this is the feature for which the difference between the two coasts is most salient. Here we hoped that this saliency would provide a perceptual cue that students would engage students to think deeply about. After each "site visit", students wrote a reflection note about the differences between the two coasts and at the end of the activity we asked students to speculate on the reasons for the differences. The scaffolds were designed to promote students' knowledge building. Students also visited a plate boundaries website in order to speculate about how the location, frequency, and magnitude of geological events (mountains, earthquakes, and volcanoes) are related. Students were scaffolded to reflect and integrate what they have learned the following prompt: "Write one (or two) question(s) you have about plate boundaries or plate movement that will help you better understand why the geologic processes on the West and East coasts are different". Students revisited these questions in a discussion forum later in the unit.

e) Dynamic Models. Students viewed and read about the different types of plate boundaries, namely, collisional, divergent, convergent, and transform boundaries in order to think about how the location of and type of plate boundaries are related to geological occurences on the earth's crust. Students reified their learning by writing reflection notes which were designed to integrate their understanding. Students also visited a model of mantle convection and an accompanying text which scaffolded their understanding of the dynamic and causal features of the model by directing their processing of the causal and dynamic information in the model as it "ran". Students wrote a reflection note to explain how the processes inside the earth relate to plate movement. Lastly, students visited a series of dynamic models which depicted different types of plate convergence, namely, oceanic-oceanic convergence, oceanic-continental convergence, and continental-continental convergence. Again, students' understanding was scaffolded via a text which directed their processing of the causal and dynamic information in each model as it "ran".

The Implementation. The "What's on our plate?" unit was effectively implemented in many middle school and high school classes from 2001-2003 (i.e., 34 classes in 2001 [n=1100 students], 27 classes in 2002 [n=900 students], and 23 classes in 2003 [n=565 students]). The analysis of fifteen middle school classes is briefly described here. Pairs of students from one class on the West coast were partnered with pairs from two classes on the East coast because of the differences in class sizes. Five such sets or "virtual classrooms" (referred to as WISE periods) were created in WISE. Some activities require collaboration of the east-west pairs, some activities did not. Students did a pre- and post-test for content knowledge as well as a pre- and post-test for empistemological understanding. In all of the WISE periods, the students made a significant gain on the post-test compared to the pre-test, meaning that all WISE periods, students acquired content knowledge during the implementation of the "What's on your plate?" unit. We also found a significant change from the pre- to the post-test on students' epistemologies of models; thus, it appears that the unit promoted epistemological change in the students as well. For a fuller description of the content data and some examples of students' models and model revisions, see Gobert et al, (2002). For a fuller description of the epistemological data, see Gobert and Pallant (2004). A full description of the project can be seen at mtv.concord.org.

PROJECTS AS EXEMPLARS TO DEVELOPING STUDENTS' MODEL-BASED REASONING AND SCIENTIFIC LITERACY: 'MODELING ACROSS THE CURRICULUM'

In the Modeling Across the Curriculum project [4] (MAC; http://mac.concord.org), we leverage computer-based technologies to affect science learning and scientific literacy on a broad scale, i.e., in many schools both in the U.S. and internationally. Specifically, we investigate how technology can be used to promote deep learning of **science content**, foster **inquiry skills**, develop understanding of the **nature of science** and **of models**, as well as **improve attitudes toward science learning**. Our interactive curriculum materials for topics in physical science, biology, and chemistry offer students an opportunity to learn by doing while interacting with computer-based hypermodels that are scaffolded to deepen students' understanding of the qualitative causal relationships underlying scientific processes. Our approach combines our domain-general engine, Pedagogica™ (Horwitz et al, 2001; Horwitz, 2002) which collects all students' actions while they manipulate the content models in the curricular activities. Students' interactions with models, stored as log files, are used as an index of their strategies for learning with models and their modeling skills. Pedagogica also reports on all students' actions on embedded and summative assessment tasks and transmits all data back to Concord Consortium's server for subsequent analysis. The results of embedded and summative multiple-choice assessments and the students' open-response answers are made available to teachers through our web portal. Our three domain content engines are:

BioLogica™, a domain content engine for genetics consisting of linked software objects representing pedigrees, organisms, chromosomes, genes, and DNA.

Dynamica™, a domain content engine for classical mechanics that supports representations of objects, such as atoms and walls, as well as graphs and velocity and force vectors.

Connected Chemistry™ (Stieff & Wilensky, 2003), a modeling and simulation package for chemistry in which students observe macro-level chemical phenomena as the result of interactions (on a micro- and submicro-level) of many individual agents.

The underlying theoretical framework for MAC is based on model-based teaching and learning (Gobert & Buckley, 2000). The dynamic, recursive nature of model-based learning in the MAC project is exemplified in the following representation (Gobert, 2004b). Briefly, the diagram depicts that a learner's mental model is based on prior knowledge and the interaction with information acquired from multiple sources (models, text, experiments, experience, etc.). Model reinforcement, revision, or rejection takes place as the learner reasons with his/her model.

---

[4] *Modeling Across the Curriculum project is funded by the Interagency Education Research Initiative (IERI #0115699) from the National Science Foundation and the U.S. Dept. of Education. Any opinions, findings, and conclusions expressed in this paper do not necessarily reflect the views of the Foundation or the U.S. Dept. of Education.*

Scaffolding Students' Interactions with Models.

Using the literature on students' learning with diagrams (cf., Gobert & Clement, 1999; Gobert, 2000; Larkin & Simon, 1987; Lowe, 1993) and model-based learning (Gobert & Buckley, 2000) we developed and refined of a scaffolding framework to support students' model-based learning with our software tools. Some of the types of supports are general such as orienting tasks, advance organizers, and embedded assessments. Others were designed specifically to support model construction, model revision, and model-based reasoning, and transfer. These scaffold types are:

Representational Assistance to guide students' understanding of the representations or the domain-specific conventions in the domain, and to support students in using multiple representations.

Model pieces acquisition to focus students' attention on the perceptual pieces of the representations and support students' knowledge acquisition about one or more aspects of the phenomenon (e.g., spatial, causal, functional, temporal).

Model pieces integration to help students combine model components in order to come to a deeper understanding of how they work together as a causal system.

Model based reasoning to support students' reasoning with models, i.e., inference-making, predictions, and explanations.

Reconstruct, Reify, & Reflect to support students to refer back to what they have learned, reinforce it, and then reflect to move to a deeper level of understanding.

Below is an example of a support for representational assistance from BioLogica. In this example, the students are given two different representations of essentially the same information. On the left, a micro level representation of the cell and the strands of DNA are shown; on the right, the abstracted representation of the information is given. This helps the students link between chromosomes situated in cells and their abstracted representation.

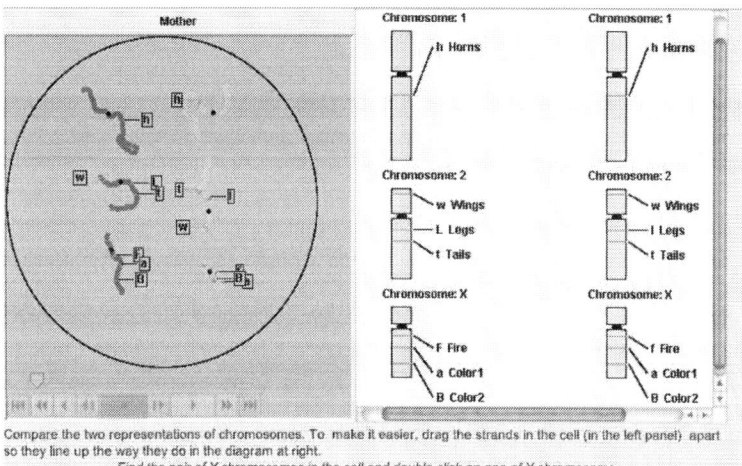

Compare the two representations of chromosomes. To make it easier, drag the strands in the cell (in the left panel) apart so they line up the way they do in the diagram at right.
Find the pair of X chromosomes in the cell and double click on one of X chromosome.

Log file specifications and analysis. Since one of the goals of the MAC project is to promote students' modeling skills, we are leveraging our technology, Pedagogica™ (Horwitz et al, 2001; Horwitz, 2002) to track students' interactions with models as an index of their model-based learning and their modeling skills. We have identified the following screen types across our curricular units, the type of data collected in each and what we can infer from post processing of log files.

Screen/task type	Logged by Pedagogica	Post processing
Telling screens – text with or without models	1. Time in screen	Determining content of screen
Screens with text and an interactive model	1. Time in screen 2. Interaction time with model 3. Inputs to model	Nature of task Tries to success Systematicity (determined from other measures)
Screens with multiple choice questions	1. Time in screen 2. Answer 3. Correct answer	Type of question (scaffolding vs. assessment)
Screens with open-response questions	1. Time in screen 2. Typing time 3. Answer	Type of question (scaffolding vs. assessment) "Quality" of answer
Hint pop-ups	1. Time in Hint 2. Which hint was accessed	

From these log files we extract: **Time in screen**, i.e., time spent with each screen; **interaction time**, i.e., time manipulating models; **inputs to model**, i.e., numbers entered into variables (as in Dynamica or Connected Chemistry, or parents or gametes chosen for breeding (as in BioLogica); **answers** (for auto-scorable answers only); and **typing time**, i.e., time spent typing answers to open response questions. From these data, we determine students' **systematicity**, i.e., whether their approach to learning with models is random, haphazard or mindfully structured (Thorndyke & Stasz, 1980) from a model-based perspective. The variable which is most relevant here is inputs to models. The example below from Dynamica™ is a good illustration of how students' inputs to the model provides an index of how they are approaching the task, i.e., whether their approach is either random or mindfully-structured in a model-based way.

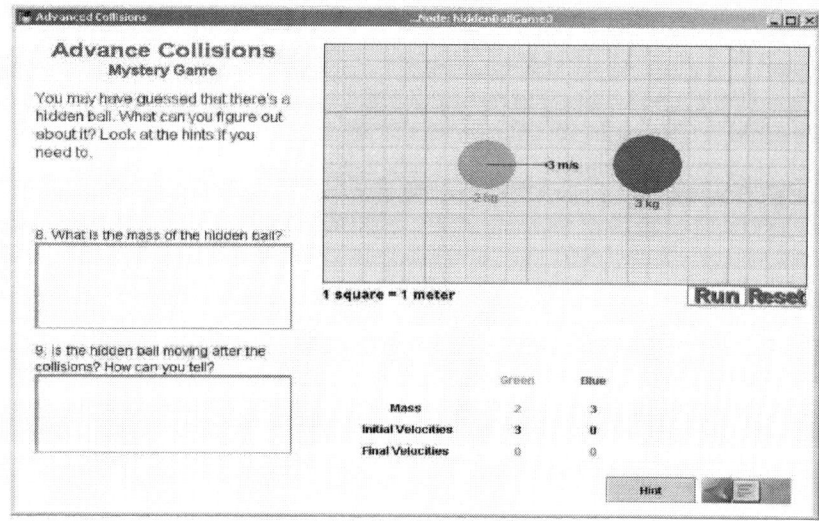

In this activity, there is a hidden ball between the ball on the left and the one on the right. The student's task is to set ball A (on the left) in motion and observe what happens to Ball C (on the right) once it is impacted by the hidden ball (Ball B) then try to figure out the mass of the hidden ball based on how Ball C behaves. In the spring of 2002 (during the activity development phase of Dynamica), we had the opportunity to observe two high school students working on this activity. One student "Sheila" was a straight A science student who likely achieved her excellent grades through rote memorization of formulas and facts; the other student, "Jason" was a low-achieving student who, in his teacher's words "had given up on science" [5]. When encountering the hidden ball activity, Sheila began to apply a brute-force strategy to figuring out the mass of the hidden ball using her calculator. Jason, on the other hand, began with an approximation of the ball's mass and then "fine-tuned" it, thus, he was being much more systematic from a model-based perspective. Both students we able to obtain the correct answer in approximately the same amount of time but Jason's strategy was much more sophisticated from a model-based perspective because he began with an approximation of the ball's mass. Similar estimation strategies are typical of experts whereby it is hypothesized that experts are developing a rich mental model of the task at hand in order to reason

---

[5] The students' names have been changed to protect their identities.

with their model (Paige & Simon, 1966). Although our activities at that time were not logging students' interactions with models, the different strategies used by the two students respectively were not captured via Pedagogica, however, now that our logging capacity is functioning, students' strategies such as these are being logged.

In terms of getting an index of students' modeling skills, we use systematicity in interacting with models, and time on task as two indicators of modeling skills that are automatically generated by Pedagogica ™. We also triangulate these data with predictions, explanations, etc., written by the students. Using data triangulated from many sources (i.e., fine-grained log data, rich data generated to open-response questions, time stamped data, etc.) we are able to get a measure of students' modeling skills both within and across three domains.

Other research questions we address are:
- What is the relationship between students' interaction with models and conceptual understanding as measured by the post-test? Are the log data detailed enough to predict post-test outcomes?
- To what extent could logfile analysis be automated to enable automated tailoring of scaffolding based on students' individual needs in terms of content knowledge and levels of modeling skill?
- Could automated logfile analysis be used to provide formative, diagnostic reports useful for teachers?

SUMMARY & CONCLUSIONS.

In the two projects described we used cognitive theory on learning with visualizations to develop a scaffolding framework in order to support students in learning with models and in model construction and revision. Making Thinking Visible leveraged technology by making use of WISE (Linn & Hsi, 2000), a web-based science learning environment which allows students to access real-time data via the web and interact with peers from geographically distinct locations. In addition to promoting content learning, we sought to promote students' modeling skills and understanding of models as important aspects of scientific literacy. Lastly, the unit was designed and implemented in WISE (Linn & Hsi, 2000) in order to leverage technology to provide students access to real-time data (USGS websites), to run dynamic models and simulations, as well as to support student collaboration from two geographically (and in this case, geologically distinct) locations. Findings from the study suggest that students were able to achieve a deeper understanding of the domain, as well as a deeper understanding of the nature of models.

In the Modeling Across the Curriculum project (Horwitz et al, 2001), we extended our scaffolding framework and also leverage technology in order to provide our materials to many schools across the world (182 as of September, 2004). Important to advancing research in educational technology, we use Pedagogica TM to track students' interactions with models in order to get an index of their model-based learning and modeling skills. Our logging capacity tells us much more about students' learning processes with models than was possible with methodologies used in previous research (i.e., think aloud protocols, Ericsson & Simon, 1980). Future plans include developing adaptive scaffolding which will react to students' needs in terms of content and modeling skills.

In summary, the two projects described herein leverage technology to enhance students' science learning and scientific literacy. The logging capacity developed as part of the MAC project further leverages technology by providing a bird's eye view into the "black box" to develop detailed understanding of students' model-based learning with manipulable models. This is an important contribution to the Learning Sciences. In MAC, we also leverage technology to provide our software and materials to many schools worldwide which is important for scientific literacy, scalability, and sustainability. In the future, we hope to provide individualized, adaptable technology-scaffolded learning that can be faded as student progresses; this will be important for science education, intelligent tutoring, and for educational practice.

REFERENCES

Akin, O. (1979). Models of architectural knowledge: An information-processing view of design. *Dissertation Abstracts International, 41,3*, p. 833 A. (University Microfilms No. 72-8621.)

Akin, O. (1986). Psychology of architectural design. London: Pion.

American Association for the Advancement of Science. (1993). *Benchmarks for Science Literacy*: 1993. New York Oxford: Oxford University Press.

Bezzi, A. (1989). Geology and society: A survey on pupil's ideas as an instance of a broader prospect for educational research in earth science. Paper presented as the 28th International Geological Congress, Washington, D.C.

Bower, G., and Karlin, M.B. (1974). Depth of processing of pictures of faces and recognition memory. *Journal of Experimental Psychology, 103*, 751-757.

Bower, G., and Karlin, M.B., and Dueck, A. (1975). Comprehension and memory for pictures. *Memory & Cognition, 3*, 216-220.

Brewer, W., and Nakamura, G. (1984). The nature and functions of schemas. In R.S. Wyer and T.K. Skull (Eds.), *Handbook of Social Cognition, Volume 1*. Hillsdale, NJ: Lawrence Erlbaum.

Chang, K.-T., Lenzen, T., and Antes, J. (1985). The effect of experience on reading topographic relief information: Analyses of performance and eye movements. *The Cartographic Journal, 22*, 88-94.

Chase, W., and Chi, M. (1981). Cognitive skill: Implications for spatial skill in large-scale environments. In J. Harvey (Ed.), *Cognition, social behavior, and the environment*. Potomac, MD: Lawrence Erlbaum.

Clement, J. (1989). Learning via model construction and criticism: Protocol evidence on sources of creativity in science. In J. Glover, R. Ronning, and C. Reynolds (Eds.), *Handbook of creativity: Assessment, theory, and research*. New York: Plenum Press.

Clement, J., Brown, D., & Zietsman, A. (1989). Not all preconceptions are misconceptions: finding "anchoring conceptions" for grounding instruction on students' intuitions. *International Journal of Science Education, 11*, 554-565.

Craik, F.I.M., and Lockhart, R. (1972). Levels of processing: A framework for memory research. *Journal of Verbal Learning and Verbal Behavior, 11*, 671-684.

Cuban, L. (2001). *Oversold and underused*. San Francisco, CA: Jossey-Bass Publishers.

Egan, D., and Schwartz, B. (1979). Chunking in recall of symbolic drawings. *Memory and Cognition, 7*, 149-158.

Ekstrom, R.B., French, J.W., Harman, H.H., and Dermen, D. (1976). *Manual for Factor Referenced Cognitive Tests*. Princeton, NJ: Educational Testing Service.

Ericsson, K., & Simon, H. (1980). Verbal reports as data. *Psychological Review, 87*, 215-251.

Frederiksen, C., and Breuleux, A. (1988). Monitoring cognitive processing in semantically complex domains. In N. Frederiksen, R. Glaser, A. Lesgold, and M. Shafto (Eds.), *Diagnostic monitoring of skill and knowledge acquisition*. Hillsdale, NJ: Lawrence Erlbaum.

Giere, R. N. (1990). *Explaining Science*. Chicago: University of Chicago Press.

Gilbert, J. K. (Ed.) (1993). *Models and modelling in science education*. Hatfield, Herts: Association for Science Education.

Gilbert, S. (1991). Model building and a definition of science. *Journal of Research in Science Teaching, 28*(1), 73-79.

Gilhooly, K.J., Wood, M., Kinnear, P.R., and Green, C. (1988). Skill in map reading and memory for maps. *The Quarterly Journal of Experimental Psychology, 40A (1)*, 87-107.

Gobert, J. (submitted). The effects of different learning tasks on conceptual understanding in science: teasing out representational modality of diagramming versus explaining. *Journal of Geoscience Education*.

Gobert, J. (2004a). What's on your plate? Exemplary science unit reviewed in *Essential Science for Teachers: Earth and Space Science*, by Harvard-Smithsonian Center for Astrophysics, Science Media Group, Annenberg/CPB Project.

Gobert, J. (April, 2004b). *Teaching model-based reasoning with scaffolded interactive representations: cognitive affordances and learning outcomes.* Presented at the National Association for Science Teaching, April 1-4, Vancouver, B.C.

Gobert, J. (2003). *Students' Collaborative Model-Building and Peer Critique On-line*. Presented at the National Association for Research in Science Teaching, Philadelphia, PA, March 23-26. NB- is this the right ref?

Gobert, J. (2000). A typology of models for plate tectonics: Inferential power and barriers to understanding. *International Journal of Science Education, 22(9),* 937-977.

Gobert, J. (1999). Expertise in the comprehension of architectural plans: Contribution of representation and domain knowledge. In *Visual And Spatial Reasoning In Design '99*, John S. Gero and B. Tversky (Eds.), Key Centre of Design Computing and Cognition, University of Sydney, AU.

Gobert, J. (1994). *Expertise in the comprehension of architectural plans: Contribution of representation and domain knowledge.* Unpublished doctoral dissertation. University of Toronto, Toronto, Ontario.

Gobert, J. & Buckley, B. (2000). Special issue editorial: Introduction to model-based teaching and learning. *International Journal of Science Education, 22(9),* 891-894.

Gobert, J. & Clement, J. (1999). Effects of student-generated diagrams versus student-generated summaries on conceptual understanding of causal and dynamic knowledge in plate tectonics. *Journal of Research in Science Teaching,* 36(1), 39-53.

Gobert, J. & Clement, J. (1994). *Promoting causal model construction in science through student-generated diagrams.* Presented at the Annual Meeting of the American Educational Research Association, April 4-8. New Orleans, LA.

Gobert, J. & Coleman, E.B. (1993). Using diagrammatic representations and causal explanations to investigate children's models of continental drift. *Proceedings of the Society of Research in Child Development.* March 25-28. New Orleans, LA.

Gobert, J. & Discenna, J. (1997). *The Relationship between Students' Epistemologies and Model-Based Reasoning.* Kalamazoo, MI: Western Michigan University, Department of Science Studies. (ERIC Document Reproduction Service No. ED409164).

Gobert, J. & Horwitz, P. (2002). Do modeling tools help students learn science? *In @ Concord*, The Concord Consortium Newsletter, 6(1), p.19.

Gobert, J.D., & Pallant, A., (2004). Fostering students' epistemologies of models via authentic model-based tasks. *Journal of Science Education and Technology.* Vol 13(1), 7-22.

Gobert, J. Slotta, J. & Pallant, A., Nagy, S. & Targum, E. (2002). *A WISE Inquiry Project for Students' East-West Coast Collaboration*, Presented at the Annual Meeting of the American Educational Research Association, New Orleans, LO, April 1-5.

Gobert, J.D., & R. Tinker (2004). Introduction to the Issue. *Journal of Science Education and Technology.* Vol 13(1), 1-6.

Head, C. (1984). The map as natural language: a paradigm for understanding. *Cartographica, 31,* 1-32.

Hegarty, M., and Just, M. (1989). Understanding machines from text and diagrams. In H. Mandl and J.R. Levin (Eds.), *Knowledge Acquisition from Text and Pictures.* North Holland: Elsvier Science Publishers B.V.

Hegarty, M., and Just, M. (1993). Constructing mental models of machines from text and diagrams. *Journal of Memory and Language, 32,* 717-742.

Hesse, M. B. (1963). *Models and analogies in science.* London: Seed and Ward.

Horwitz, P., Tinker, R., Dede, C., Gobert, J., & Wilensky, U. (2001). *Modeling Across the Curriculum*, Intra-Agency Education Research Initiative Grant funded by the National Science Foundation and U.S. Dept. of Education (IERI # 0115699); Awarded September, 2001.

Horwitz, P. (2002, May). *Instructivism: Between CAI and Microworlds.* Presented at EU/US Education Meeting,, Sintra, Portugal.

Jacobi, D., Bergeron, A., & Malvesy, T. (1996). The popularization of plate tectonics: presenting the concepts of dynamics and time, *Public Understanding in Science,* 5, 75-100.

Johnson-Laird, P.N. (1985). Mental models. In A. Aitken, and J. Slack (Eds.), *Issues in cognitive modelling*. London: LEA Publishers.
Kintsch, W. (1974). *The representation of meaning in memory*. Hillsdale, NJ: Lawrence Erlbaum.
Kintsch, W. (1986). Learning from text. *Cognition and Instruction, 3(2)*, 87-108.
Kosslyn, S. (1994). *Image and Brain*. Cambridge, MA: The MIT Press.
Lane, D., and Robertson, L. (1979). The generality of the levels of processing hypothesis: An application to memory for chess positions. *Memory & Cognition, 7(4)*, 253-256.
Larkin, J. H., & Simon, H. A. (1987). Why a Diagram is (Sometimes) Worth Ten Thousand Words. *Cognitive Science, 11*, 65-99.
Linn, M.C. (1998). Supporting teachers and encouraging lifelong learning: A web-based integrated science environment (WISE). Proposal funded by the National Science Foundation.
Linn, M.C. (1999). Designing the knowledge integration environment: The partnership inquiry process. Created for *International Journal of Science Education*.
Linn, M.C., & Hsi, S. (2000). *Computers, Teachers, Peers: Science Learning Partners*. Hillsdale, NJ: Erlbaum.
Linn, M.C., & Muilenberg, L. (1996). Creating lifelong science learners: What models form a firm foundation? *Educational Researcher, 25* (5), 18-24.
Lockhart, R., and Craik, F.I.M. (1990). Levels of processing: A retrospective commentary on a framework for memory research. *Canadian Journal of Psychology, 44(1)*, 87-112.
Lowe, R. (1993). Constructing a mental representation from an abstract technical diagram. *Learning & Instruction, 3*, 157-179.
Mandl, H., and Levin, J. (1989).*Knowledge acquisition from text and pictures*. Amsterdam, The Netherlands: Elsevier Science Publishers B.V.
Mayer, R., and Gallini, J. (1990). When is an illustration worth then thousand words? *Journal of Educational Psychology, Vol. 82*, 715-726.
National Committee on Science Education Standards and Assessment. (1996). *National Science Education Standards: 1996*. Washington, D.C.: National Academy Press.
Paige, J.M., and Simon, H. (1966). Cognitive processes in solving algebra word problems. In B. Kleinmuntz (Ed.), *Problem solving*. New York: Wiley.
Perkins, D. (1986). *Knowledge as design*. Hillsdale, NJ: Erlbaum.
Piburn, M., Reynolds, S., McAuliffe, C., Leedy, D., Birk, J., & Johnson, J. (in press). The role of visualization in learning from computer-based images. To appear in the International Journal of Science Education.
Ross, K. & Shuell, T. (1993). Children's beliefs about earthquakes. *Science Education, 77*, 191-205.
Rumelhart, D., & Norman, D. (1975). The active structural network. In D. Norman, D. Rumelhart, and the LNR Research Group (Eds.), Explorations in cognition. San Francisco, CA: W.H.Freeman.
Sabelli, N. (1994). On using technology for understanding science. *Interactive Learning Environments, 4(3)*, 195-198.
Salthouse, T. (1991). Reasoning and spatial abilities. In F.I.M. Craik and T.A. Salthouse (Eds.), *Handbook of Aging and Cognition*. Hillsdale, NJ: Lawrence Erlbaum.
Salthouse, T., Babcock, R., Mitchell, D., Palmon, R., and Skovronek, E. (1990). Sources of individual differences in spatial visualization ability. *Intelligence, 14*, 187-230.
Scardamalia & Bereiter, (1991). High levels of agency for children in knowledge building: A challenge for the design of new knowledge media. *Journal of the Learning Sciences, 1(1)*, 37-68.
Schank, R., & Abelson, R. (1977). *Scripts, plans, goals, and understanding*. Hillsdale, NJ: Lawrence Erlbaum.
Schwarz, C. & White, B. (1999). *What do seventh grade students understand about scientific modeling from a model-oriented physics curriculum?* Presented at the National Association for Research in Science Teaching, March 28 - 31, Boston, MA.
Stieff, M & Wilensky, U. (2003). The Connected Chemistry Modeling Environment: Incorporating Interactive Simulations into the Chemistry Classroom. *Journal of Science Education and Technology*.
Sweller, J., Chandler, P., Tierney, P., and Cooper, M. (1990). Cognitive load as a factor in the structuring of technical material. *Journal of Experimental Psychology, 119(2)*, 176-192.
Thorndyke, P., & Stasz, C. (1980). Individual differences in procedures for knowledge acquisition from maps. *Cognitive Psychology, 12*, 137-175.
Turner, R. H., Nigg, J. M., & Daz, D. H. (1986). *Waiting for disaster: Earthquake watch in California*. Berkeley, CA: University of California.
van Dijk, T. & Kintsch, W. (1983). *Strategies of discourse comprehension*. New York: Academic Press.

SECTION B

DEVELOPING THE SKILLS OF VISUALIZATION

MIKE STIEFF[1], ROBERT C BATEMAN, JR.[2], DAVID H UTTAL[3]

CHAPTER 6

TEACHING AND LEARNING WITH THREE-DIMENSIONAL REPRESENTATIONS

[1]University of California, Davis; [2]The University of Southern Mississippi; [3]Northwestern University, USA

Abstract. Computer-based visualizations play a profoundly important role in chemistry instruction. In this chapter, we review the role of visualization tools and possible ways in which they may influence thinking about chemistry. There are now several visualization systems available that allow students to manipulate important variables in obtain a solution to a scientific problem. We discuss the fundamental differences between these tools, and we emphasize the use of each within the context of constructivist curricula and pedagogies. We also consider the impact such tools may have on visuo-spatial thinking. We suggest that although visuo-spatial ability may be important in visualization use, its role has at times been overemphasized. We argue for a more nuanced, richer understanding of the many ways in which visuo-spatial reasoning is used in solving chemistry problems. This discussion leads to a set of design principles for the use of visualization tools in teaching chemistry. Finally, we present our work on the Kinemage Authorship Project, a program designed to assist students in understanding spatial structures in complex, biochemical molecules. The Kinemage Authorship Project allows students to construct their own molecular visualizations, and we discuss how this may lead to greater understanding of the spatial properties of molecules. This constructivist program embodies many of the design principles that we present earlier in the chapter.

INTRODUCTION

Visualization tools are among the most important technologies for learning at the high school and undergraduate levels. Educational researchers have devoted considerable effort to the refinement and implementation of visualization tools for science students because of the important role of perceiving, understanding and manipulating three-dimensional spatial relationships for learning and problem solving in many sciences. Although such tools exist for most sciences, chemists and biologists have been among the strongest advocates for visualization tools because their disciplines require the conception of multiple, complex three-dimensional spatial relationships both within and between molecular structures. Repeatedly, instructors find that their students have great difficulty understanding these relationships in these domains, which many novel visualization tools aim to make more comprehensible (Copolo & Hounshell, 1995; Habraken, 1996; Wu, Krajcik, & Soloway, 2001).

In this chapter, we will explore the potential benefits of visualization tools in the teaching and learning of science at the secondary and post-secondary levels. We discuss the advantages of using these tools in conjunction with constructivist pedagogies. We also propose new approaches to investigating their effectiveness as learning aids by focusing on cognitive models of visualization. We detail one novel constructivist visualization-based curriculum that uses *Kinemage* to teach undergraduate biochemistry students concepts of enzyme structure and chemical reactivity. We conclude with a discussion of the opportunities and novel research directions that using a cognitive framework to explore visualization tools can engender.

VISUALIZATION IN THE SCIENCE CLASSROOM

Visualization is universal to the science classroom regardless of domain or level; however, the underlying need for visualization varies with the topic of study. In geology, students must comprehend the spatial relationships between different earth structures that are not perceptually accessible due to both their macroscopic size and their location beneath the crust of the Earth. Visualizing geological structures is complicated because these structures seem ostensibly static, but students must perceive that the structures are actually dynamic objects that move on a time-scale many orders of magnitude greater than that of daily human life. In physics, students perform similar mental feats of visualization that have their own special challenges. Physics students must perceive the spatial relationships between the interaction of forces that result from phenomena such as tension, friction, gravity, and electromagnetism. Because these forces have no visual manifestation that students can perceive, visualization is equally critical for understanding in this domain.

Both instructors and practicing scientists are quick to point out that a critical component to problem solving and comprehension in the sciences lies in a student's ability to visualize the spatial relationships, and transformations of those relationships, among various phenomena or structures. Consequently, many believe that it is critical for a student to generate a mental model of the scientific phenomena under study. In geology, a student might need to visualize how two plates in the Earth's crust slide against one another and what effect the movement has on the surrounding landscape. In physics, a student might imagine mental models of force vectors acting on a vehicle and how manipulating the magnitude of any one vector can alter the vehicle's velocity. Because students have no basis for generating mental models of these phenomena from their everyday experience, instructors provide their students with a plethora of diagrams, models and pictures to assist the visualization process. Unfortunately, the two-dimensional representations commonly used in the science classroom can only approximate the three-dimensional events they represent. This limitation can distort the mental images that students attempt to visualize and hinder learning.

To say that visualization is important for learning in the chemical sciences glosses over the fundamental role that these cognitive processes play at all levels of study in these domains. For example, understanding the nature and importance of spatial relationships is only the most basic component of spatial cognition in the chemical sciences. As students' studies advance, comprehending the particular details of spatial relationships in relevant diagrams, models, and images becomes increasingly important. Not only must students be aware that such relationships exist, but they must also apprehend the particular constraints that molecular structure can have on the macroscopic shape of a substance and the role that these constraints play on molecular interactions and chemical reactivity. While the task may be simpler for small molecules composed of five or fewer atoms, it becomes considerably more difficult when students must deal with biological molecules that can contain thousands of atoms, each of which possesses a unique relationship with the others. For example, when considering a large enzymatic molecule, a student might be required to understand how the position of one particular oxygen atom affects the relationship between thousands of other atoms in the enzyme and how that shape controls the enzyme's chemical activity. To understand this, the student must consider—often simultaneously—properties such as bond angle, bond length, orbital shape, connectivity and chirality, to name a few characteristics.

Particularly in the chemistry classroom, traditional instruction may inhibit or complicate the necessary understanding because it relies on two-dimensional diagrams to represent three-dimensional molecular structures. Although instructors and practitioners are adept at the selection and use of different two-dimensional representations to describe three-dimensional molecular structures, students are rarely successful at interpreting or manipulating the wide variety of representations available (Keig & Rubba, 1993; Kozma, Chin, Russell, & Marx, 2000). For instance, students often report great difficulty understanding that the Fischer projection in Figure 1 represents the same molecule that the ball-and-stick model represents. Although the ball-and-stick model emphasizes the three-dimensional relationships between atoms in the molecule, the Fischer Projection emphasizes the connectivity between the atoms; one must understand the formalisms of the representation to perceive the three-dimensional relationships that are embedded in the two-dimensional diagram. Although the representations in Figure 1 are often restricted to organic chemistry classrooms, the difficulty in perceiving and understanding three-dimensional relationships is ubiquitous among both advanced undergraduate students and novice high school students (Johnstone, 1993).

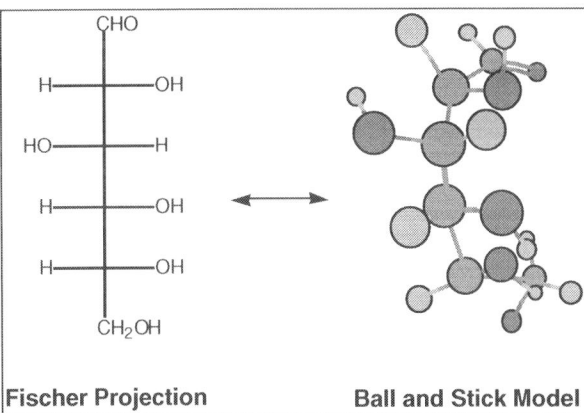

Figure 1. *Unlike ball-and-stick models, Fischer projections obscure three-dimensional spatial relationships and instead highlight atom connectivity, which students find difficult to perceive.*

SUPPORTING VISUALIZATION WITH VISUALIZATION TOOLS

Educational researchers have recently begun to concentrate on the development of a wide variety of visualization tools and novel pedagogies to aid students in science learning at all levels. These tools describe a spectrum of learning environments that support many different types of visualization from concretizing abstract concepts to understanding spatial relationships. Tools are now available that allow students to visualize experimental data sets, simulate experiments, or construct models of imperceptible entities. At their core, each of these tools presents students and instructors with several unique opportunities for teaching and learning science that allow students to visualize complex relationships directly from computer-generated visualizations. Advocates for the use of visualization tools can be found among science teachers at all levels. High-school teachers report that students gain a more robust conceptual understanding of classroom content when their lessons are supported by the use of visualization tools (Copolo & Hounshell, 1995; Wu et al., 2001). Likewise, college teachers have reported similar benefits and support incorporating these visualization tools to enrich traditional pedagogies (Crouch, Holden, & Samet, 1996). In fact, CD-ROMS with visualization tools are packaged with many science textbooks, particular for undergraduate chemistry. Given the compatibility and small disk space required by these programs, a single CD-ROM can support instruction with tools that animate textbook diagrams, model virtual laboratories, and supplement classroom lectures with interactive tutorials.

Visualization tools to enhance teaching and learning in the sciences fall roughly into two groups. The first, which we label *content-specific tools*, accounts for the majority of visualization tools currently in use. Content-specific tools are stand-alone programs that teach specific concepts in a particular science. The other group, which we label *general learning environments,* are general-purpose programs that can be used across a variety of scientific domains. Both categories of visualization tools present unique opportunities to students and instructors in the science classroom. These opportunities are as varied as the tools that have been developed: instructors can use them as simple visual aids to support a lecture or they may fundamentally alter the nature of their teaching by allowing students to explore the tools with little guidance. In essence, each of these tools, regardless of form, attempts to improve learning by visually representing scientific phenomena in conjunction with or in place of text-based descriptions or symbolic notations.

There are, however, fundamental differences between the goals that motivate the two types of visualization tools. Those who favor general learning environments advocate the development of open-ended tools, which instructors can modify easily to tailor the learning experience for many different science classrooms as well as for individual students. These modeling environments emphasize multiple representations of concepts, flexible interfaces, broad functionality, and a strong interactive component (Wilensky, 2001). General learning environments present an alternative to direct instruction by giving students opportunities to explore fully developed models to discover scientific principles. Most general learning environments also allow students to alter given models, or to develop their own, with or without instructor support.

One such environment, *NetLogo,* is based on the assumption that students learn better by connecting the visual representations of microscopic phenomena with macroscopic domain concepts that can often be represented graphically (Wilensky & Resnick, 1999). Educational researchers, in conjunction with instructors and students, have used the NetLogo modeling environment to model and teach phenomena in biology (Centola, Wilensky, & McKenzie, 2000), physics (Wilensky, 1999), and chemistry (Stieff & Wilensky, 2003). Students have used the visual representation of microscopic phenomena, be they molecules, mammalian cells, or gas particles, to deduce fundamental concepts, such as chemical equilibrium, asexual reproduction, or Brownian motion. The NetLogo environment attempts to simulate physical reality with a high degree of fidelity and provides students with direct access to the programming code that controls each model so that they can understand the relationship between the virtual world and the world it represents.

A unique implementation of NetLogo, entitled GasLab, illustrates the goal of this class of visualization tools. GasLab allows students to explore a virtual environment of gas particles in a box to discover the relationship between a visual representation of gas particles in motion, a speed histogram, and a kinetic energy histogram to learn about the cause and effect of the Maxwell-Boltzmann distribution (Wilensky & Resnick, 1999). As the dynamic simulation runs, students are able to observe the visual representation, in which particles constantly collide and transfer energy. Histograms of speed and energy are reported simultaneously, and the student can discern that the total energy of the system maintains an equilibrated distribution. Together, the three representations (illustrated in Figure 2) allow students to develop a richer understanding of the connection between the physical phenomena of particle motion and collisions, the concepts of energy and temperature, and graphical representations of non-normal distributions.

In the geosciences, the development of visualization tools has also focused on providing students with detailed visual representations of domain concepts in the same spirit of tools like NetLogo. Unlike general-purpose modeling environments these tools, such as WorldWatcher (Edelson, Gordin, & Pea, 1999) and Geo3D (Kali & Orion, 1997), have been designed specifically for the purpose of providing students with visual representations of phenomena and concepts in climatology and geology, respectively. Instructors and researchers alike have praised the tools because they show students complex geological structures that are not observable directly due to their macroscopic structure. These tools also allow students to perceive the relationship between different spatial transformations of geological structures as they occur over time. This feature is particular important because study in these domains often require students to comprehend the transformation of the phenomena under study on multiple time scales. For instances, students might need to reason how local albedo or plate tectonics as they transform over a few seconds and over hundreds of years. Reportedly, the dimension of time adds considerable difficulty to students' ability to visualize such phenomena (Kali, 2003). Moreover, verbal descriptions or static diagrams of dynamic relationships can result in significant misconceptions among students, which makes these tools particularly valuable (Kozma & Russell, 1997; Stieff & Wilensky, 2003).

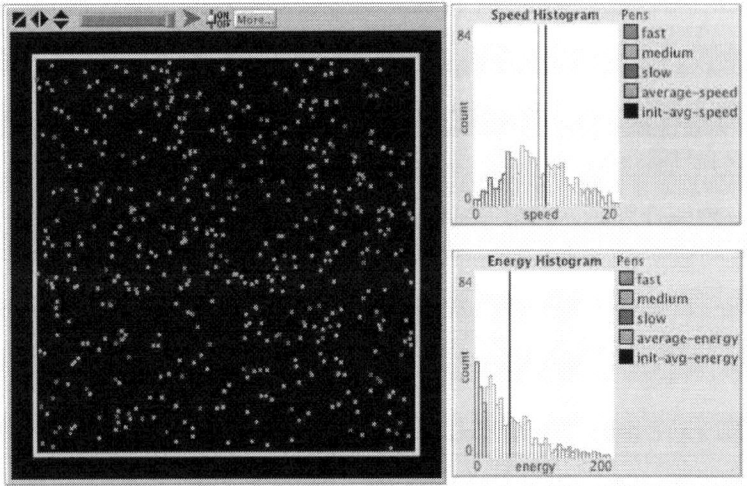

Figure 2. *The NetLogo modeling environment supports the GasLab curriculum that teaches students the relationship between particle motion and the distribution of kinetic energy. A color version of Fig. 2 is given in the Colour Section.*

Many of visualization tools allow students to compare concept maps with visualizations of actual geological data to enrich the learning experience. Figure 3 illustrates how the WorldWatcher interface accomplishes this goal by focusing on the relationship between surface temperature and location on the Earth. With the aid of a visualization of average surface temperatures across the planet, students engage in an investigation to discover how the nature of incoming solar radiation results in different climates. WorldWatcher is equipped with an extensive dataset that allows students to compare temperature data from many different years that climatologists have collected in the field. Using this data, students are able to construct explanations of climate and geography from their own investigations instead of learning them from direct instruction.

In the chemical sciences, educational researchers have focused much of their efforts on the development of visualization tools that allow students to visualize the three-dimensional spatial relationships that are embedded in traditional two-dimensional molecular representations. The tools emphasize how students can learn fundamental domain concepts by understanding the three-dimensional shape and structure of molecules. For example, these tools can help students in molecular biology courses visualize the shape of proteins on the surface of a white blood cell that give the cell the ability to detect and destroy bacteria and viruses. Visualization tools for the chemical sciences often aim at the secondary goal of teaching students the relationships between the different two-dimensional diagrams that domain practitioners most commonly employ. The tools are designed to help students see the relationships among the various two-dimensional diagrams. Students use them to underscore that even though different two-dimensional representations of the same molecule may appear disparate, each represents the same three-dimensional structure.

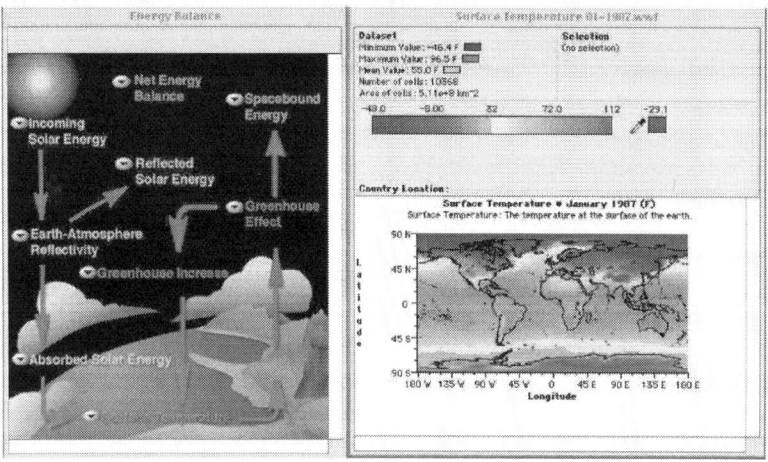

Figure 3. *The WorldWatcher visualization tool allows students to compare concept maps with visualizations of real scientific data in a guided inquiry environment. A color version of Fig. 3 is given in the Colour Section.*

One such visualization tool, eChem (Wu et al., 2001), is designed to allow students to compare two-dimensional diagrams with three-dimensional visualizations to learn organic chemistry. Students can use the tool to construct virtual models of small organic molecules using several common molecular representations (see Figure 4). The tool allows students to compare space-filling models, ball-and-stick models, and wire-frame models vis-à-vis to learn how atomic constituency and connectivity relate to the overall size, shape, and structure of a molecule. The tool also provides students with a database of physical properties for several common molecules. With this information, students can deduce how the size and shape of molecules affect properties such as boiling point, density, hardness, and solubility.

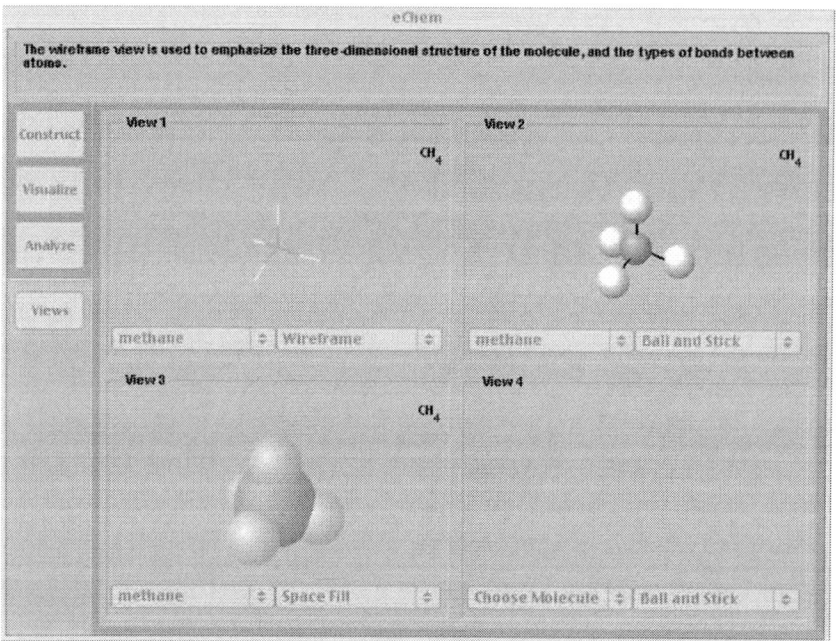

Figure 4. *The eChem visualization tool allows students to compare three-dimensional and two-dimensional representations to better comprehend structure and reactivity.*

TOWARDS A THEORY OF EFFECTIVE VISUALIZATION USE

Although the research community generally agrees that visualization tools are a boon to science education, most of the enthusiasm is based more on anecdotes than on evidence (Horowitz, 2002). In part, the scarcity of data supporting the use of visualization tools is due to the rapid growth in the number and diversity of tools. Many educational researchers and software designers continue to focus their efforts on honing interfaces and constructing curricula that take advantage of visualization tools before turning their attention to gather data on the impact of the tools. In addition, the wide variety of tools available presents a daunting analysis task for the few researchers currently investigating their efficacy. Consequently, relatively little work exists that assess the impact of such tools on both learning outcomes and pedagogy. Despite the paucity of evidence, however, a few studies have aimed to measure the learning outcomes from their use.

The developers of visualization tools have, themselves, conducted the majority of the research on the effectiveness of their tools, and they have reported a wide range of findings on the efficacy of specific pieces of software. Unavoidably, the designers have conducted their studies in conjunction with the ongoing development of a particular tool. Because of this, their research methods have often been hindered by the lack of a fully functional tool. Although many of these studies have been limited to anecdotal reports about student attitudes and understanding, the findings have begun to indicate positive outcomes from each implementation (Crouch et al.,

1996; Wu et al., 2001). Very few studies have attempted to document the impact of their curricula on larger groups of students with more rigorous methods. While these few studies produced some supporting evidence for the anecdotal reports, they have mostly relied on pre- and post-test measures from traditional curricula (e.g., Ealy, 1999; Noh & Scharmann, 1997). In general the limited amount of evidence has been favorable, although some designers have observed declines in problem-solving ability among students who have learned using visualization software (Copolo & Hounshell, 1995). Clearly there is a strong need for research to overcome the empirical limitations regarding the use and effectiveness of visualization tools in science education.

Perhaps more importantly, we lack clear theoretical perspectives to motivate the role of visualization strategies and the effectiveness of visualization tools. There are, of course, suggestions as to why visualization tools do or do not work, but these have not been integrated into a coherent theoretical account of why visualization tools are likely (or unlikely) to help students learn in a given science domain. We need now to integrate these various suggestions into a coherent theoretical account of visualization use and comprehension. New theoretical models of visualization as a cognitive skill for learning and problem solving must replace general assumptions about the critical role of visualization in the sciences. In parallel, new research methods that both determine empirically the role of visualization in learning science and provide detailed descriptions of the quality of learning with novel visualization tools must replace marginal pre- and post-test measures and imprecise anecdotal self-reports. Collaborations between educational researchers, designers, cognitive science, and science practitioners are critical to the development and success of these new theories and methodologies. Members from each of these groups provide valuable input on the role, implementation and value of visualization tools as joint members of the science education community.

Our work ultimately is aimed at these general goals, particularly for visualization tools that teach using three-dimensional representations in the chemical sciences. Perhaps more than any other science domain, the chemical sciences have garnered strong advocates for the use of visualization strategies and visualization tools based on the simple assumption that visualization is critical because the molecular world is three-dimensional (Habraken, 1996). We seek to develop a theoretical framework to replace this assumption regarding the use of visualization tools for the chemical sciences that is motivated by principles and research from cognitive science and the larger science education community. Joint efforts to develop new visualization tools, theoretical models, and research methods have been effectively established to study learning in physics (Sherin, diSessa, & Hammer, 1993) and geology (Keating, Barnett, Barab, & Hay, 2002), but are lacking in the chemical sciences. In the following sections, we present three principles of a theoretical framework that aims to motivate the use of visualization tools and specify the role of visualization as a cognitive strategy in the chemical sciences. Although these principles are not exhaustive, we offer them as a starting point for further research on visualization by the science education community.

Principle 1: Design Visualization Tools for Chemistry to Support Spatial Cognition

Chemistry has always been a very visually oriented science, and novel visualization tools should support students' understanding of the variety of visual representations of chemical compounds and reactions. As mentioned above, the representation of chemical structures at the molecular level is particularly important because of the direct relationship between the structure, the chemical reactivity, and the physical properties of a compound. There are limitless examples of this: the arrangement of individual atoms in a semiconductor is directly related to the semiconductor's ability to conduct electricity, the bonding pattern of carbon atoms in a diamond is responsible for the gem's extreme hardness, the interactions between water molecules in ice results in the shape of snowflakes. All chemists are familiar with these relationships and consider them of primary importance in most courses related to the chemical sciences.

In particular, study in this domain requires that students gain a strong understanding of the concept of stereochemistry. This concept concerns the connectivity and three-dimensional spatial arrangement of atoms within an individual molecule. Even relatively small molecules, such as amino acids, have multiple spatial arrangements that can result in different *stereoisomers*, which are unique molecules that are composed of the same atoms. Understanding the structure and relationship between stereoisomers is crucial to the basic and advanced study of chemistry. In introductory courses, students must know how to name and build each isomer. In advanced biochemistry courses, students must understand how different stereoisomers create different proteins with different functions. For example, only one stereoisomer of amino acids, known as the L isomer, is found in proteins in living systems; incorporation of the R isomer into a protein can a have disastrous impact on biological function. A protein containing a single amino acid of the alternate arrangement could well be dysfunctional and result in systemic disease at the organism-level. Figure 5 illustrates the minor difference in spatial arrangement between the stereoisomers for the amino acid, alanine. Note that although both molecules in the figure contain the same number and type of atoms, they are mirror reflections of one another that cannot be superimposed; consequently, they are unique stereoisomers.

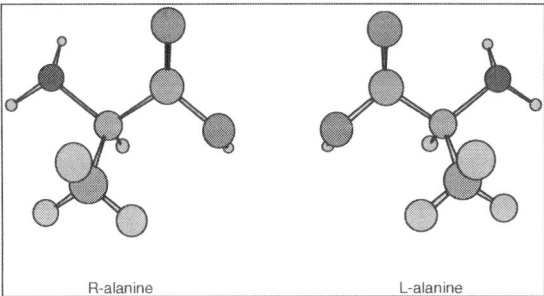

Figure 5. *The two stereoisomers of alanine are composed of the same atoms, but each has a unique spatial arrangements.*

Students must become adept at not only perceiving the spatial relationships between stereoisomers but also at manipulating them as needed to solve problems in the classroom and the laboratory. Two well-known medical examples serve as reminders of the importance of stereochemistry. One unfortunate example concerns a drug, thalidomide, that was given to numerous women in Europe as a treatment for nausea during pregnancy. Although effective, the drug caused severe birth defects. Investigation showed that thalidomide had been synthesized and distributed as a pair of stereoisomers. Only one stereoisomer was responsible for the therapeutic effects of the drug; the other stereoisomer caused the serious side effects. Thalidomide was quickly banned as an antinauseant banned because of its side effects, but today medicinal chemists are now testing the therapeutic stereoisomer as an anticancer agent. The second example reveals the importance of understanding stereochemistry for diagnosing disease. Sickle cell anemia is caused by single amino acid replacement mutation in the protein hemoglobin. This mutation places an incorrect amino acid in a critical position on the surface of hemoglobin that causes the protein molecules to stick together and form long, insoluble fibers. These insoluble fibers distort the red blood cells and give them the characteristic "sickle" shape after which the disease is named. Understanding the relationship between the structure of the mutated protein and the depleted function of the diseased red blood cells has provided new avenues of research toward treatments for the disease.

It is important, then, for chemists and biochemists to be able to visualize the spatial arrangement and connectivity between atoms in molecules both small and large. Chemists have traditionally used molecular models to help students visualize the proper mental image of molecules and their interactions. Typically, these models are the plastic ball-and-stick modeling kits that are familiar to anyone who has taken a chemistry course. Other chemists have used balloons to illustrate atomic orbitals or inverted music stands to show displacement reactions. Unfortunately, these models are static in nature and do not allow students to see dynamic interactions between molecular structures. Indeed, some students can come to believe that molecular bonds exist as rods that hold nuclei together in the same way that the plastic pieces connect the pieces of the modeling kit. These types of models work well for small molecules and give the student a tactile way to approach their formulation of a mental model, but many students often fail to understand the relationship between the models, diagrams, and real molecules (Keig & Rubba, 1993; Wu et al., 2001).

Visualization tools stand to aid students' perception and mental manipulation of three-dimensional relationships by providing virtual models of actual molecular structures that maintain a high degree of fidelity with real molecules. Such tools give students direct access to authentic representations of the phenomena under study in the chemical sciences. Designers of educational software for chemistry have already realized the potential of visualization software to provide rich details about shape and structure of the molecules for advanced research, and we believe this same potential applies to the chemistry classroom. These tools provide students with the ability to compare different stereoisomers, to learn techniques for translating one molecular representation into another, and to construct models of molecule interactions and subsequent chemical reactions. Moreover, we suggest that there the use of visualizations may benefit students in ways that extend beyond the benefits that have been assumed.

As we continue to develop visualization tools, the generic goal of using them to help students appreciate the overall size, shape and structure of particular molecules is insufficient. Instead, we must focus our designs in such away that students learn to use such tools to directly support visualization strategies on appropriate tasks. That is to say, dynamic visualization tools allow students to manipulate virtual molecules in much the same way they might manipulate a mental model of a chemical reaction. Therefore, productive tools should offer users the capability to rotate molecules to gain new perspectives or viewing angles of molecular structures in isolation or interaction. Additional features that allow students to rotate individual bonds within a molecule to see how the overall shape of a structure changes as a consequence provide even more versatility and authenticity. For example, students might use them as researchers do to learn about DNA structure by constructing a visualization to illustrate how replacing one DNA base-pair with another can distort the double helix. Each of these capabilities of visualization tools provides an advanced external representation with which students can perceive and manipulate spatial transformations during problem solving. These transformations are nontrivial and even the most adept students stand to benefit from the use of a tool to assist in visualizing them.

Principle 2: Investigate the Role and Efficacy of Visualization Tools with Cognitive Models

The prior section suggested that a mental visualization is perhaps the key to success in chemistry and hence should be the cornerstone for the development of visualization tools. Many authors have taken this approach, suggesting that visualization alone is the most important predictor of success in chemistry. Traditionally, many researchers have cited a student's inability to visualize the three-dimensional structure embedded in two-dimensional representations as the primary barrier to learning. Although some have suggested that the barrier results from the total amount of information that each particular representation contains (Keig & Rubba, 1993; Kozma et al., 2000; Stieff & Wilensky, 2003), most researchers have repeatedly emphasized that it is the spatial information, in particular, embedded within each representation that confuses students (Bodner & Guay, 1997; Brownlow & Miderski, 2001; Carter, LaRussa, & Bodner, 1987; Coleman & Gotch, 1998). The latter group of researchers emphasizes that the three-dimensional structure of a molecule plays a large role in the molecule's properties and reactivity, and they suggest that students must stay mindful of these features in order to understand and complete most chemistry tasks. These claims have been made largely independent of the level of instruction or the topic of discussion: chemistry education researchers have made the case for the role of visualization in learning chemistry equally for both secondary students of general chemistry and for undergraduate students of organic chemistry.

We agree that the abilities to perceive and understand three-dimensional spatial relationships are important for many science domains, particularly chemistry, but the precise role of such abilities in these domains is relatively unclear. Assumptions that postulate a central role of visualization in the sciences on the simple observation that these domains concern imperceptible three-dimensional objects fail to consider the complex interaction between visualization, spatial reasoning, and external representations of spatial information. Perhaps more problematic is that such assumptions disregard the unique details of individual tasks, which may be essentially unrelated to the representations or the spatial information present in the task. For example, consider a task from an organic chemistry classroom that concerns the reaction between two large, spatially-complex molecules that produces two new molecules. At first, it may appear that students with well-honed visuo-spatial abilities would be predisposed to excel on this task. However, further inspection of the task reveals that it simply requires the student to determine the relative ratio of each molecule in solution after the reaction concludes. Therefore, despite the presence of molecules with complex and detailed spatial information, the task is fundamentally a math problem contextualized in the chemistry classroom.

As we develop novel visualization tools and curricula for the science classroom, we must develop stronger cognitive models regarding the role of visualization as an aspect of spatial cognition that includes a principled foundation for the use of visualization tools. Instead of simple assumptions, we must base these models and frameworks on research and theoretical principles from cognitive science regarding spatial cognition as a fundamental cognitive strategy. Some earlier research programs attempted to characterize visualization as a form spatial cognition in the chemical sciences by basing their work on the assumption that individual differences in visuo-spatial abilities were an underlying cause of variations in performance. Indeed, those studies found several moderate correlations between measures of visuo-spatial ability and performance; however, the results of such studies were not definitive, and several are quite contradictory (Carter et al., 1987; Ealy, 1999; Keig & Rubba, 1993). Paradoxically, findings from that line of research have revealed that the strongest correlations exist between standardized measures of visuo-spatial ability and performance for chemistry tasks that do not include any spatial information. Although well established, the focus on individual differences has led to the basic claim that students who perform well on standardized measures of visuo-spatial ability also perform well in chemistry and other sciences. This line of research has generated little information regarding the type of cognition underlying learning and problem solving in the chemical sciences.

A more rigorous cognitive model that explicates the form and function of spatial cognition for scientific problem solving can provide a productive theoretical framework for both the design, implementation, and assessment of visualization tools. We advocate a theoretical framework that acknowledges that spatial cognition is neither simple nor uniform. We posit that the role of spatial cognition in problem solving can vary as a function of the demands of a task and the features of the external representations present in the task. At the least, a cognitive model of the role of visualization tools for learning must address two particular features of spatial cognition: the manner in which students encode spatial information and the cognitive strategies with which students manipulate that information.

How Do Students Perceive Spatial Information?

New cognitive models of spatial cognition and visualization tools must first account for research from cognitive science regarding information processing and memory modality. Such work has established that individuals can encode information independent of its modality; that is, individuals do not always visualize mental images of physical objects simply because they are three-dimensional. Instead, individuals can encode information that is superficially spatial in nature (e.g. the shape of an object or the distance between two objects) with both analog and propositional mental representations selectively during problem solving based on problem context (Kosslyn, 1994; Markman, 1999; Pylyshyn, 1981). Information that is encoded analogically, as a mental image, is manipulated and stored in a memory system devoted to the processing of spatial information. This system is distinct from a verbal memory system, which stores information encoded as propositional statements (den Heyer & Barrett, 1971; Finke & Pinker, 1982; Pickering, 2001). To further complicate the issue, a reciprocal relation exists between the two systems. For instance, non-spatial information in the form of words or numbers, can cue the visualization of mental images, such as when individuals compare size and shape given only object names (Shepard & Chipman, 1970). Because of these issues, experiments in spatial cognition are careful to limit the amount of non-spatial information that may facilitate cognitive processes other than spatial cognition (Just & Carpenter, 1987; Shepard & Metzler, 1971). Although many representations contain both spatial and non-spatial features that likely involve both spatial and verbal cognition, chemistry educators have yet to consider the role that each system plays.

This latter finding is particularly relevant to investigations concerning the impact of visualization tools in science learning. Although some chemistry education researchers have suggested the spatial memory system described by psychologists is quite obviously the primary cognitive system required for chemistry thinking (Bodner & McMillen, 1987; Coleman & Gotch, 1998; Habraken, 1996), there are two reasons to suspect that visualization plays a more complex role in the cognition underlying scientific problem solving. First, the use of visualization strategies to encode information and to solve problems decreases with development. By the age of eight, most individuals begin to rely less on the perceptual features of objects and more on the verbal labels that they have given objects (Pickering, 2001). For example, when one considers what groceries one must purchase on a shopping trip, they are more likely to think of the necessary items as a list of food names, than as a list of mental images of each item. Despite their predisposition to encode information verbally, adults are better able to recall items encoded as visual images than as verbal labels, which has been coined the *picture-superiorty effect* (Madigan, 1983; Paivio, 1986). Because of the increased dependence on verbal labels for problem solving among adults, the extent to which visualization plays a role in problem solving in chemistry, or any other discipline, becomes an empirical question.

The fact that many molecular representations contain both spatial and verbal information suggests that they require a cooperative effort between both verbal and visual memory. Adolescents' and adults' preference for verbal encoding may increase the difficulty of problem solving in a science domain that requires the student to visualize to effectively problem solve. If this is the case, then the use of visualization tools to reinforce the encoding and manipulation of a mental image of molecular representations could provide substantial support to the science student.

The precise role of the tool to support visualization in this way requires further investigation.

How Do Students Manipulate Spatial Information?

A novel cognitive model must not only determine how visualization tools help students encode visually represented information, they must also determine how the tool affects the cognitive strategies that students use to manipulate that information. Research in cognitive science has shown that alternative strategies are available based on the nature of visually represented information, but the impact of these strategies for problem solving in science remains unknown. Early work in this area established that the visualization of mental images was a crucial strategy for encoding and comparing different representations of three-dimensional objects (Shepard & Metzler, 1971). In those early studies, participants were required to view pairs of three-dimensional shapes (see Figure 6) and asked to determine if the shapes were identical to one another or mirror images. The results of the study revealed that participants required more time to compare the shapes, on average, as the angular disparity between the shapes increased. The linear relationship has repeatedly been interpreted as evidence that individuals visualize and manipulate mental images to solve most tasks that involve three-dimensional spatial information. The nature of the task in Shepard and Metzler's classical experiment is nearly identical to the stereochemistry tasks required in advanced chemistry studies. Indeed, a replication of Shepard and Metzler's original experimental design using molecular representations revealed the same linear relationship between response time and angular disparity (Stieff, 2004). The finding strongly suggests that students routinely attempt to visualize molecular structures for mental rotation when completing tasks regarding stereochemistry.

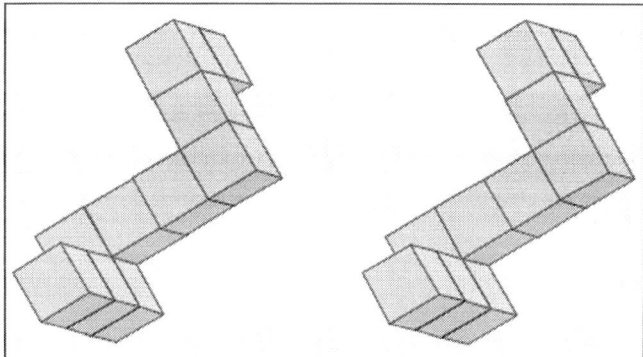

Figure 6. *Individuals are quick to invoke mental imagery strategies to visualize representations of three-dimensional objects*

Although some have suggested the similarity between the laboratory task and stereochemistry problems establishes the primacy of visualization strategies for problem solving in chemistry (Habraken, 1996), additional findings suggest a more nuanced role of visualization strategies in solving stereochemistry problems. Experiments conducted after Shepard and Metzler's original work revealed that mental rotation is a more complicated process than previously assumed. For example, people do not use mental rotation to solve all problems that involve mental comparison of spatial figures. In fact, they seem to rely on mental rotation only for objects that are not rich in information. That is, when comparing generic three-dimensional objects, individuals are quick to use mental imagery and mental rotation to compare objects such as those in Figure 6. However, Just and Carpenter (1987) showed that a very different process is used for the same objects investigated by Shepard and Metzler, if those objects contain additional information, as simple as alphanumeric characters as shown in Figure 7. When presented with these stimuli, participants in the laboratory experiment abandoned a mental rotation strategy and instead made direct comparisons of the stimuli to problem solve. For example, when comparing the two objects in Figure 7, participants did not attempt to visualize and each shape and perform a Gestalt mental rotation; instead they simply noted if the letter 'F' on each object was identical.

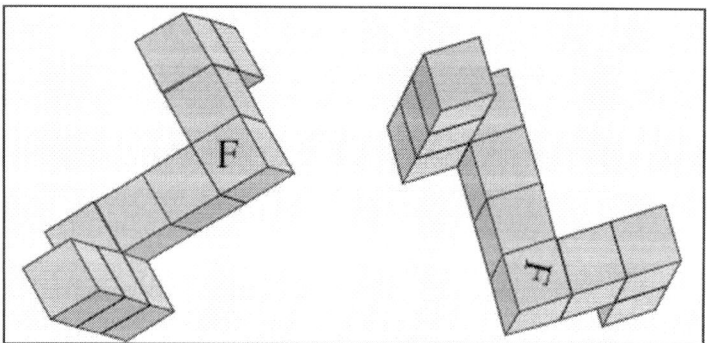

Figure 7. *When comparing "information rich" three-dimensional shapes individuals can avoid visualization and mental rotation.*

Similarly, educational researchers and cognitive scientists have begun to document the use of cognitive strategies alternative to the visualization of mental imagery in science domains even when it is ostensibly required. In mechanical engineering, students have revealed that they can use either mental imagery or invoke abstract rules that obviate visualization to solve tasks regarding gear trains (Hegarty, 1992; Schwartz & Black, 1996). Instead of visualizing a mental animation of a gear train to determine the direction each gear rotates, experienced students can instead invoke a parity rule that states every other gear turns in the same direction. Similarly, science educators have revealed that students use similar abstract rules for solving problems in chemistry. This work has established that students are able to make decisions about stereochemistry by directly inspecting molecular representations without generating a mental image (Stieff, 2004). Instead, the students simply look for planes of symmetry within a molecule to make their decisions. Interestingly, students only apply the rule in limited cases and often resort to visualization when the rule fails to provide an immediate solution. Thus, only when they must treat the molecules as generic objects do students resort to the visualization and mental rotation strategies that Shepard and Metzler described originally.

These findings suggest that more detailed investigations of the use of visualization in the chemical sciences is needed, especially when visualization tools are employed. In particular, the interaction between the use of a visualization strategy and the form of an external representation suggests a nuanced role for visualization tools in the chemistry classroom. We suggest that novel research programs involving the use of visualization tools investigate the varied use of visualization strategies among students without assuming that it is the central cognitive strategy. It must also be made clear how visualization tools alter the form and function of visualization and mental rotation strategies for problem solving in the sciences. The theoretical frameworks and empirical findings from cognitive science suggest that visualization tools have the capability to both enhance and inhibit several aspects of spatial cognition. The extent to which this affects learning and problem solving ability must remain a central feature of novel research programs that hope to use visualization tools more extensively than visual aids.

Principle 3: Implement Tools with Constructivist and Constructionist Pedagogies

Although a variety of novel visualization has developed rapidly in the past decade, concomitant curricula and pedagogies that take advantage of these tools have evolved at a much slower pace. Much to the dismay of the software designers, instructors often fail to incorporate the tools successfully into their classrooms even though many of the tools are available at no cost. In many cases, visualization tools packaged on CD-ROMs and bundled with standard textbooks quickly find their way into the wastebasket instead of the curriculum (Butler & Sellbom, 2002). Educational researchers have cited many reasons for this including a lack of sufficient instructional support, the absence of detailed operating instructions, and a general reluctance of instructors to adopt new tools (Butler & Sellbom, 2002; Duhaney, 2001; Krueger, Hansen, & Smaldino, 2000). Despite the barriers that these factors present to the successful implementation of new visualization tools, both instructors and designer remain committed to overcoming them and to seeking new avenues for successful adoptions.

The design community has advocated that novel science curricula can take full advantage of each of the opportunities that a visualization tool affords by adopting a constructivist approach to teaching and learning (Edelson et al., 1999; Wilensky & Resnick, 1999). Constructivist theories of learning advocate that each student comes to understand scientific concepts based on their individual experiences and their own interpretation of information (Collins, 1996; Papert, 1991; Piaget & Inhelder, 1967). This position stands opposed to the information transmission theories of old that postulate learning takes place when information possessed by the expert instructor is successfully transferred in its entirety to the attentive student. Constructivism acknowledges that individual students can possess different understandings of the same principle that are unique. Although students' individual understandings may differ from the understanding of their instructors, the theory posits that each student can still problem solve and communicate with equal ease. It is important to note that acknowledging alternative understandings does not suggest that all understandings are correct understandings, just that individuals vary in how they apprehend particular concepts.

With regard to visualization tools, a constructivist theory of learning advocates that students learn best when given opportunities to tie together their own knowledge with classroom material to discover or deduce desired principles for themselves (Perkins, 1991). The pedagogy suggested by this tenet of constructivism sharply diverges from didactic or Socratic methods in which instructors lecture to large groups of students about what relevant information must be noted and remembered. In contrast, constructivism takes the position that the latter pedagogies result in poor engagement, low retention, and "inert knowledge" (Brown, Collins, & Duguid, 1989). That is, under traditional instruction students remain passive participants with a tendency to memorize information that they can apply to only the most routine tasks. As such, pedagogical approaches that define the use of visualization tools as simple visual aids to support traditional lectures have little efficacy from the perspective of a constructivist.

Instead, constructivists advocate that students learn through active and varied engagement with visualization tools to maximize their learning benefits. This position takes several different forms, the most popular of which is that of *guided inquiry*. As a pedagogical approach in the science classroom, guided inquiry environments are based on a curriculum in which the student places the role of a research scientist (Edelson, 2001). In the role of a team of scientists, the students receive an authentic scientific problem, which they must solve by designing a research study, collecting and analyzing data with a visualization tool, and presenting their findings to other students. The entire activity is performed under the guidance of the classroom instructor, who helps students make decisions about their study and models the practices of professional scientists. The BGuILE curriculum provides one example of such a guided inquiry environment with the use of data visualization tools, such as *Animal Landlord* and *The Galapagos Finches* (Reiser et al., 2001). BGuILE casts students in the role of field biologists who are attempting to explain such phenomena as food webs and extinction by collecting and analyzing data on climate, animal behavior, and reproduction.

A second pedagogical approach to the use of new visualization tools, which is gaining some precedence in the science education community, is that of *constructionism*. This approach posits that students are better able to apprehend complicated concepts through the act of building their own models of the concepts with the aid of computer-based technologies (Papert, 1991). The constructionist approach specifically supports the use of visualization tools to provide students with opportunities to build models of scientific phenomena in order to increase or improve both learning and problem solving. It is clear that the constructionist approach has particular benefits for courses in the chemistry sciences. In accordance with the tenets of constructionism, students stand best to learn from activities in which they use visualization tools to build models of molecular-level objects and explore their interactions and properties. We suggest that new pedagogical approaches adopt such a pedagogy to provide students with more interactive experiences instead of using visualization tools as a supplemental example of representing the molecular world.

We perceive the benefits from allowing students to build their own models of the molecular world as threefold. First, constructing a model allows students to concretize their understanding with a visual representation that provides them with an opportunity to reflect on their own learning as they build. The embedded features of a tool can help guide the student to such reflections; for instance, eChem highlights the number of bonds that are possible for individual atoms when students construct molecules, which supports learning about the concept of molecular hybridization and three-dimensional structure. Second, visualization tools provide immediate visual feedback to the student about the viability of the models that they construct. In this way, students have the opportunity to evaluate what they are building and compare it with other concepts learned in the classroom. The Connected Chemistry modeling environment achieves this by offering students the opportunity to simulate chemical interactions. If a student constructs a model in which the collision of an acid molecule and a base molecule raises the pH of a solution, a pH monitor in the visualization tool can warn students that the model's behavior violates accepted concepts of pH. Finally, constructed models can provide feedback to both the teacher and the entire classroom about one student's understanding. By viewing the model that a student constructs with a visualization tool, the instructor has a tangible measure of the student's progress to more effectively provide critical support and guidance. In much the same way, the entire class can view individual student work to share ideas and learn from each other about alternative perspectives about an idea or concept.

We advocate that novel visualization tools should be based on the theoretical tenets of constructivism and constructionism. Such environments allow students the widest range of opportunities to learn scientific concepts based on direct experience with visualization tools. This is not to say that the tools should not be used in any other manner. We believe that instructors can also employ them as effective visual aids when necessary, but such practices do not take full advantage of the capabilities of these novel tools. Activities in which students are able to build their own models of imperceptible objects adhere closest to the tenets of this principle. New tools should be embedded in curricula that allow great flexibility and interactivity beyond that of simple animations aimed at direct instruction. In the next section, we provide an example of one such novel curriculum, the Kinemage Authorship Project, that employs a constructivist pedagogy and adheres to both our design and cognitive principles.

THE KINEMAGE AUTHORSHIP PROJECT: TEACHING AND LEARNING WITH THREE-DIMENSIONAL REPRESENTATIONS

Teaching students to build ball-and-stick models allows students to not only manipulate and visualize the molecular models, but to build a new structure from a simple chemical formula. In this way, the student must provide the extra information needed to place the atoms in proper orientation. Building such a model is like thinking aloud. It is an expressive way to construct a mental model. Such a constructionist approach promotes deep understanding and forces the student to work through misconceptions that can arise with a simple lecture or textbook illustration. Indeed, individual construction projects do not have the time pressure of lecture where a brief look at a model can be misleading and students can adjust memory of the image to fit their preconceptions or generate new misconceptions.

Today, most chemists and biochemists work with more complex structures than students mighty typically see in introductory courses. Because physical models are limited in their ability to portray such structures, advanced research scientists in these domains have become dependent on computer graphics modeling programs to aid in the visualization of macromolecules. In fact, there tremendous progress over the last ten years in the development and implementation of these programs, which continue to become more widespread for research and industrial applications as well as publication (Voith, 2003). In the classroom, however, these tools have not yet enjoyed such versatility. Instructors still rely on textbooks to show illustrations captured from the programs and student interaction is often limited to a brief tutorial that uses pre-constructed renderings of molecules or molecular interactions.

In the Kinemage Authorship Project, we have developed a pedagogy for teaching with molecular graphics that follows the constructionist approach mentioned above that makes molecular modeling a more central component of the curriculum. In our courses, students choose a particular macromolecule to research and write a "molecular story" about the structure and function of that molecule. Each story includes a series of interactive graphic images along with extensive annotation via a hyperlinked text window. The student authors deliberately construct the graphic images to illustrate concepts and features of importance. The intent is to use a variety of approaches such as alternate renderings of the same structures and simple animation between alternate conformations to illustrate molecular motion to convey multilayered information embedded in the macromolecule. Overall, our guiding philosophy of the approach is to remove the clutter by taking away the "irrelevant" features of each image so that the student's attention is drawn to the subject of interest (Richardson & Richardson, 1994, 2002; White, Kim, Sherman, & Weber, 2002). This disembedding of information from a complex structure allows the student to focus on one thing at a time to promote learning and problem solving.

Because the graphic images are completely interactive, even editable, an author's original story can be changed at any time over the course of the project. New features can be added or original features can be altered or removed altogether. The entire project is saved as a plain text file called a *kinemage*, which is generated using the menu-based utility PREKIN and viewed using the program MAGE or its progeny program KING. The kinemage, then, is deliberately designed to be a means of communication of three dimensional concepts (four dimensional when one includes animation of motion). A kinemage can be likened to a children's museum where the displays are labeled, annotated, and interactive. In our curriculum, however, the students are the designers of each displayed kinemage. As much as children enjoy and learn from these museums, how much more could they learn if they actually helped construct the museum displays?

The kinemage format was chosen for this project because kinemages were originally developed as an electronic accompaniment for the journal *Protein Science* (Richardson & Richardson, 1992). Journal authors publishing a structure-based article could include a kinemage which would convey in three dimensions what the article described in two dimensions. The kinemage format was unique when developed because other authoring tools that used molecular graphics programs were not designed in this manner. Although other such programs, such as Rasmol and DeepView, contain scripting capability, they do not have the capability to annotate the scripted image through text windows, caption windows, and graphic labels. This multilayered annotation provides a direct linkage between the spatial and verbal information conveyed to the viewer. In addition to the inherent communication capabilities of the kinemage, the graphics also have features not found in the other graphic programs such as excellent depth cueing, simulation of molecular motion through animation between alternate conformations, and the ability to move features independently in order to dock or overlay features. Each of these features was designed so that the viewer might use the kinemage to support the visualization of the relevant structures instead of relying solely on an imagined mental image.

Kinemage construction is time consuming and is best done as an out-of-class assignment stretched over the course of a semester. Of course, teachers must embed such a construction project within the overall curriculum so that students can work up to it and so that it makes sense in the context of the discipline. Students must be able to draw connections between the individual project and what they are learning in lectures where instructors models the use of these three-dimensional images with both visual images and verbal explanations. This is an opportunity for the instructor to correlate the graphic images with physical models as well as both words and images from the textbook. Lecture is also an opportunity for students to see alternate renderings of a molecule (ball and stick, backbone, ribbon, spacefilling, etc.) and understand that these are images of the same molecule. As with the construction project, it is important to keep images used in lecture simple so that students are not confused by too much complexity, too soon.

In addition to introducing the students to the graphic images in lecture, it is important to also introduce them to the software and point out some of the features they will want to learn to move forward in their construction project. These include basic manipulation such as rotation, zooming, atom identification, and distance measurements. The use of motion is important to emphasize the illusion of three dimensionality of the graphic image. Motion, particularly a gentle rocking motion, is an effective way for even people without effective stereo vision to see the three dimensional aspects of a structure.

Today's students are quite experienced with observing complex, multidimensional computer graphics in computer and video games. This does not mean, however, that they see the same things that the instructor sees when they look at the projected image. An important part of using graphic images in lecture, then, is both to ask the students what they see and to draw the student's attention to visible features by asking questions to make them look more carefully at the image. The particular benefits of the visual display apply not only to student-teacher interactions but also to student-student interactions. For example, two students, John and Jane, may initially believe that there are no similarities between a protein:DNA complex that John constructed and a protein:protein complex constructed by Jane. However, when the instructor gives John and Jane the opportunity to examine the graphics annotations of each other's kinemage, they are able to put the verbal and spatial cues together to discover the common strategies nature uses to form interactions within and between molecules. Proper annotation of the kinemage associates the verbal information in the label with the spatial information in the visualization to generate a more complete understanding for both students.

In addition to exposure to kinemages in lecture, students must manipulate kinemages as part of their homework during the first half of the semester. These guided inquiry homework assignments are designed to become progressively more demanding as students gain both content knowledge and familiarity with the software. For example, students may be asked very simple questions about a kinemage at the beginning, then later be asked to generate a Ramachandran plot by measuring dihedral angles along pieces of a protein, and still later they may be asked to find charge interactions between a protein and a piece of DNA. Most standard biochemistry courses would stop at this point and assume that manipulation and interaction with molecular images assures a sufficient level of familiarity with the concepts of macromolecular structure.

To achieve a deeper level of understanding, we have students go the next step and ask them to make changes in an image and analyze the result. An example of this is found in the kinemage construction tutorial originally written by Jane and David Richardson (1994). The tutorial focuses on the construction of a variety of renderings of the caster bean biotoxin ricin. The active site of ricin contains the amino acid glutamate, which is critical for the toxicity of this poisonous protein. However, mutants of ricin which contain a different amino acid in this position are still active. Students must find out how the mutant biotoxin can remain active by using the graphic image to show how nearby glutamates can rotate into the same position as the mutated glutamate and "rescue" the activity of the toxin. To accomplish this, students must be able to visualize the 3D spatial arrangement of all of the amino acid sidechains in the vicinity as well as have a prior knowledge of the chemistry of the various functional groups on each of these sidechains. The kinemage assists students in applying their fundamental knowledge of protein sidechains by providing the external visualization of the protein so that students have more than their own mental images of the structure to rely on.

After completing the homework assignments, the software tutorial, and a literature search on their chosen topic, students then begin the actual construction of their kinemage. This requires a series of decisions about what to show and how to show it. These decisions force the student to construct an initial mental model of the molecule before they can begin construction on the graphic model. The process of construction requires an ongoing refinement of both mental and graphic models. Because the students are working with experimentally determined coordinate files as their source of molecular data, the data is in front of them and there is little room for "getting it wrong". Students are encouraged to incorporate "new" features into their kinemages. For instance, students can show or propose things that the original scientist that solved the structure failed to consider (or at least report). This experience of constructing and annotating a series of molecular images requires much more of the student in terms of time, mental effort, and initiative than the manipulation and exploration of a preconstructed image. Consequently, we expect the mental model generated by kinemage construction to be more detailed, accurate, and long lasting than one generated from guided exploration of a preconstructed molecular structure.

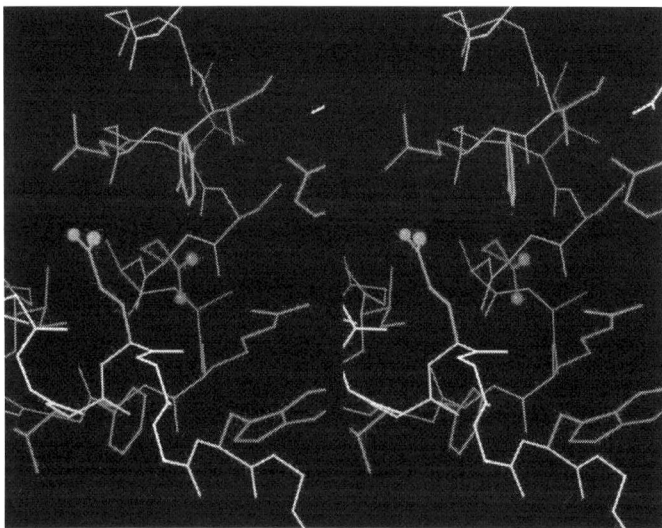

Figure 8. *Stereo kinemage of the active site of ricin showing the critical glutamate (red) and the nearby "rescue" glutamate (green). (A colour version of this diagram is included in the Colour Section)*

Our attitude assessments using surveys and interviews show that the students themselves believe this is the case (Bateman, Booth, Sirochman, Richardson, & Richardson, 2002) and anecdotal evidence long after the semester ends suggests a long lasting effect of this form of learning. For example, student authors believe they have learned more about all levels of protein structure, particularly "higher order" structure, than the control group. The more motivated students in the class particularly liked the project because it provided them an open-ended opportunity to explore as much as they like. Interestingly, we have seen no evidence of gender differences in either the quantitative or qualitative assessments despite the long-standing assumption regarding individual differences in visualization and science achievement. This is not surprising since the construction project provides lengthy training and prior studies (Roberts & Bell, 2000; Scali, Brownlow, & Hicks, 2000) have shown that gender differences in spatial ability are largely due to experience and that training can eliminate gender bias. Our project facilitates long-term experience because it extends beyond a unit or lesson as a self-paced, out-of-class assignment.

To be honest, our qualitative interviews with students have uncovered some student complaints about the kinemage construction project. The project is obviously is very time consuming and some students felt that time spent on the project took away from course fundamentals. Other students complained that the project required learning new, unfamiliar software and that this was asking too much of students. Another complaint was that the project was too focused and that students had trouble generalizing or transferring the knowledge gained about their topic to a new molecular structure. This problem of transference can be addressed in the lecture where the instructor explains connections between one image and another. This responses from students are not surprising as they are common perceptions shared by science instructors who attempt to adopt novel visualization tools (Butler & Sellbom, 2002). As we continue with our iterative development of the curricula, we hope to overcome these criticisms and maximize the positive feedback from students.

CONCLUSION

Throughout this chapter we have advocated for the use of visualization tools in the science classroom. Visualization tools are now available to aid students in the visualization of raw data, abstract concepts, and imperceptible entities. We believe that these tools hold much promise for the teaching and learning of science by providing virtual three-dimensional representations of such objects, events, and phenomena, which are often directly imperceptible to students. Although most current claims regarding the effectiveness of these tools remain based on anecdotal evidence, we believe that more substantiated claims are forthcoming. We consider one fruitful path to such claims to lie in collaborations between designers of visualization tools, science instructors, and cognitive scientists. At present, we suggest that such collaborations should aim to develop strong theoretical frameworks that motivate and explicate the role of visualization tools in the science classroom and the use of visualization as a cognitive strategy in science learning and problem solving.

To that end, we have offered an initial theoretical framework that combines cognitive theories of visualization with constructivist and constructionist theories of pedagogy. Our framework posits three principles for the design and implementation of visualization software in the sciences, particularly chemistry. Our first principle stipulates that visualization tools should be designed to support spatial cognition in learning and problem solving. The general goal that visualization tools allow students to appreciate imperceptible entities is insufficient to achieve significant learning gains. Instead, we believe it is more productive to identify the specific manner in which a visualization tool assist students both to perceive complicated or imperceptible three-dimensional relationships and to generate more accurate mental representations of concepts and phenomena.

Second, we suggest that a strong cognitive model of the role of visualization in science learning and problem solving is needed to specify and assess the role of visualization tools in students' learning. Too many designs have been motivated by the assumption that visualization is important simply because science domains often deal with three-dimensional objects interacting in an imperceptible three-dimensional space. We consider visualization to be an essential and task-specific cognitive strategy in science based on research from cognitive science regarding the encoding and manipulation of spatial information. That research has shown that individuals can perceive and encode spatial information in multiple modalities as a function of the form in which that information is displayed. Moreover, similar lines of research in science education community have indicated that students can alternate between the use of visualization strategies and non-imagistic heuristics as their experience grows. We suggest that productive research agendas lie in investigating the interaction between these two strategies and how visualization tools may mediate that interaction.

Finally, we believe that students and teachers stand to gain the most positive learning outcomes from implementing visualization strategies according to constructivist and constructionist philosophies of learning and teaching. That is, visualization tools are least effective when implemented as a supplemental aid to traditional lessons in the science classroom. A more productive use of these tools is achieved when they become a central feature of novel pedagogies. According to these philosophies students benefit the most from opportunities to build their own visual representations of domain concepts and phenomena in activities like those in the Kinemage Authorship Project. Activities such as these provide students and teachers with direct visual feedback on learning and understanding and they promote recursive self-reflection and collaboration among students.

The three principles of our theoretical framework present unique questions for the science education community to probe. For example, clarifying the role of visualization as a cognitive strategy in science problem solving suggests new agendas to the study of learning and performance. Although previous research agendas have focused specifically on determining the impact of individual differences in visuo-spatial ability in science learning, our framework suggests that more pointed questions are necessary. At the least, we must ask when and how students employ visualization with both traditional problems and with the use of visualization tools. Such an investigation moves beyond determining simple correlations and asks whether visualization tools support poor visuo-spatial abilities or illicit unique problem solving strategies that obviate visualization altogether. Consequently, the findings from such studies can implicate new methods for assessing the impact of visualization tools on learning. By first targeting the role of visualization strategies during problem solving, we can begin to clarify the contradictory evidence for learning gains produced to date.

More broadly, as the effectiveness of using visualization tools in the science classrooms is revealed, we can begin to ask fundamental questions about the implications such tools have for future science curricula. Currently, most instructors employ visualization tools sporadically in curricula that remain mostly traditional in nature. Alternatively, the Kinemage Authorship Project provides one example in which the visualization tool significantly alters the type of learning in which students engage. Instead of simply using the tool as a visual aid, students use kinemages to conduct research investigations by constructing models of proteins and enzymes to probe function. If the end result of such an implementation proves to instill deeper conceptual knowledge and active engagement, the continued use of traditional modes of instruction must be scrutinized. As visualization tools are optimized through iterative development and strong theoretical frameworks emerge to situate the role of these tools in scientific cognition, new models of science curricula must adapt to take advantage of them. These changes will be facilitated by the kinds of interdisciplinary collaborations between chemists, cognitive scientists, and educators that are a characteristic of the chapters in this volume.

REFERENCES

Bateman, R. C., Jr., Booth, D., Sirochman, R., Richardson, J. S., & Richardson, D. C. (2002). Teaching and assessing three-dimensional molecular literacy in undergraduate biochemistry. *Journal of Chemical Education, 79*(5), 551-552.

Bodner, G. M., & Guay, R. B. (1997). The Purdue visualization of rotations test. *The Chemical Educator, 2*(4), 1-18.

Bodner, G. M., & McMillen, T. L. B. (1987). Cognitive restructuring as an early stage in problem solving. *Journal of Research in Science Teaching, 23*(8), 727-737.

Brown, J. S., Collins, A., & Duguid, P. (1989). Situated cognition and the culture of learning. *Educational Researcher, 18*(1), 32-42.

Brownlow, S., & Miderski, C. A. (2001). *How gender and college chemistry experience influence mental rotation ability.* Paper presented at the Annual Meeting of the Southeastern Psychological Association, Atlanta, GA.

Butler, D., & Sellbom, M. (2002). Barriers to adopting technology for teaching and learning. *EDUCAUSE Quarterly, 25*(2), 22-28.

Carter, C. S., LaRussa, M. A., & Bodner, G. M. (1987). A study of two measures of spatial ability as predictors of success in different levels of general chemistry. *Journal of Research in Science Teaching, 24*(7), 645-657.

Centola, D., Wilensky, U., & McKenzie, E. (2000). A hands-on modeling approach to evolution: Learning about the evolution of cooperation and altruism through multi-agent modeling-The EACH Project. In B. Fishman & S. O'Connor-Divelbiss (Eds.), *Fourth International Conference of the Learning Sciences* (pp. 166-173). Mahwah, NJ: Erlbaum.

Coleman, S. L., & Gotch, A. J. (1998). Spatial perception skills of chemistry students. *Journal of Chemical Education, 75*(2), 206-209.

Collins, A. (1996). Design issues for learning environments. In S. Vosniadou, E. D. Corte, R. Glaser & H. Mandl (Eds.), *International perspectives on the design of technology-supported learning environments* (pp. 347-361).

Copolo, C. F., & Hounshell, P. B. (1995). Using three-dimensional models to teach molecular structures in high school chemistry. *Journal of Science Education and Technology, 4*(4), 295-305.

Crouch, R. D., Holden, M. S., & Samet, C. (1996). CAChe molecular modeling: A visualization tool early in the undergraduate chemistry curriculum. *Journal of Chemical Education, 73*(10), 916-918.

den Heyer, K., & Barrett, B. (1971). Selective loss of visual and verbal information in STM by means of visual and verbal interpolated tasks. *Psychometric Science, 25*(2), 100-102.

Duhaney, D. C. (2001). Teacher education: preparing teachers to integrate technology. *International Journal of Instructional Media, 28*(1), 23-30.

Ealy, J. B. (1999). A student evaluation of molecular modeling in first year college chemistry. *Journal of Science Education and Technology, 8*(4), 309-321.

Edelson, D. C. (2001). Learning-for-use: A framework for integrating content and process learning in the design of inquiry activities. *The Journal of Research in Science Teaching, 38*(3), 355-385.

Edelson, D. C., Gordin, D. N., & Pea, R. D. (1999). Addressing the challenges of inquiry-based learning through technology and curriculum design. *Journal of the Learning Sciences, 8*(3/4), 391-450.

Finke, R., & Pinker, S. (1982). Spontaneous imagery scanning in mental extrapolation. *Journal of Experimental Psychology: Learning, Memory & Cognition, 8*(2), 142-147.

Habraken, C. L. (1996). Perceptions of chemistry: Why is the common perception of chemistry, the most visual of sciences, so distorted? *Journal of Science Education and Technology, 5*(3), 193-201.

Hegarty, M. (1992). Mental animation: Inferring motion from static diagrams of mechanical systems. *Journal of Experimental Psychology: Learning, Memory & Cognition, 18*(5), 1084-1102.

Horowitz, P. (2002). *Simulations and visualization: Issues for REC.* Retrieved May 21, 2004, from http://prospectassoc.com/NSF/simvis.htm#3

Johnstone, A. H. (1993). The development of chemistry teaching. *Journal of Chemical Education, 70*(9), 701-705.

Just, M. A., & Carpenter, T. P. (1987). *The psychology of reading and language comprehension.* Boston: Allyn and Bacon.

Kali, Y. (2003). A virtual journey within the rock-cycle: A software kit for the development of systems-thinking in the context of the Earth's crust. *Journal of Geoscience Education, 51*(2), 165-170.

Kali, Y., & Orion, N. (1997). Software for assisting high-school students in the spatial perception of geological structures. *Journal of Geoscience Education, 45*, 10-21.

Keating, T., Barnett, M., Barab, S. A., & Hay, K. E. (2002). Developing conceptual understanding of scientific concepts through building three-dimensional computational models. *Journal of Science Education and Technology, 11*(3), 261-275.

Keig, P. F., & Rubba, P. A. (1993). Translation of representation of the structure of matter and its relationship to reasoning, gender, spatial reasoning, and specific prior knowledge. *Journal of Research in Science Teaching, 30*(8), 883-903.

Kosslyn, S. M. (1994). *Image and brain: The resolution of the imagery debate.* Cambridge: MIT Press.

Kozma, R., Chin, E., Russell, J., & Marx, N. (2000). The role of representations and tools in the chemistry laboratory and their implications for chemistry learning. *Journal of the Learning Sciences, 9*(2), 105-143.

Kozma, R., & Russell, J. (1997). Multimedia and understanding: Expert and novice responses to different representations of chemical phenomena. *Journal of Research in Science Teaching, 34*(9), 949-968.

Krueger, K., Hansen, L., & Smaldino, S. E. (2000). Preservice teacher technology competencies. *TechTrends, 44*(2), 47-50.

Madigan, S. (1983). Picture memory. In J. C. Yuille (Ed.), *Imagery, memory, and cognition: Essays in honour of Allan Paivio* (pp. 65-89). Hillsdale, NJ: Erlbaum.

Markman, A. B. (1999). *Knowledge representation.* Mahwah, NJ: Lawrence Erlbaum Associates.

Noh, T., & Scharmann, L. C. (1997). Instructional influence of a molecular-level pictorial presentation of matter on students' conceptions and problem-solving ability. *Journal of Research in Science Teaching, 34*(2), 199-217.

Paivio, A. (1986). *Mental representations: A dual-coding approach.* Oxford: Oxford University Press.

Papert, S. (1991). Situating constructionism. In I. Harel & S. Papert (Eds.), *Constructionism* (pp. 1-11). Norwood, NJ: Ablex.

Perkins, D. (1991). Technology meets constructivism: Do they make a marriage? *Educational Technology, 31*(5), 18-23.

Piaget, J., & Inhelder, B. (1967). *The child's conception of space.* New York: W. W. Norton & Co.

Pickering, S. (2001). The development of visuo-spatial working memory. *Memory, 9*(4/5/6), 423-432.

Pylyshyn, Z. W. (1981). The imagery debate: Analogue media versus tacit knowledge. *Psychological Review, 88*, 16-45.

Reiser, B. J., Tabak, I., Sandoval, W. A., Smith, B., Steinmuller, F., & Leone, T. J. (2001). BGuILE: Strategic and conceptual scaffolds for scientific inquiry in biology classrooms. In S. M. Carver & D. Klahr (Eds.), *Cognition and instruction: twenty-five years of progress* (pp. 263-305). Mahwah, NJ: Erlbaum.

Richardson, D. C., & Richardson, J. S. (1992). The kinemage: a tool for scientific communication. *Protein Science, 1*, 3-9.

Richardson, D. C., & Richardson, J. S. (1994). Kinemages-simple macromolecular graphics for interactive teaching and publication. *Trends in Biochemical Sciences, 19*, 135-138.

Richardson, D. C., & Richardson, J. S. (2002). Teaching molecular 3-D literacy. *Biochemistry and Molecular Biology Education, 30*(1), 21-26.

Roberts, J. E., & Bell, M. A. (2000). Sex differences on a computerized mental rotation task disappear with computer familiarization. *Perceptual and Motor Skills, 91*, 1027-1034.

Scali, R. M., Brownlow, S., & Hicks, J. L. (2000). Gender differences in spatial task performance as a function of speed or accuracy orientation. *Sex Roles, 43*(5-6), 359-376.

Schwartz, D. L., & Black, J. (1996). Shuttling between depictive models and abstract rules: Induction and fallback. *Cognitive Science, 20*(4), 457-497.

Shepard, R. N., & Chipman, S. (1970). Second-order isomorphism of internal representations: Shapes of states. *Cognitive Psychology, 1*(1), 1-17.

Shepard, R. N., & Metzler, J. (1971). Mental rotation of three-dimensional objects. *Science, 171*, 701-703.

Sherin, B. L., diSessa, A. A., & Hammer, D. (1993). Dynaturtle revisited: Learning physics through the collaborative design of a computer model. *Interactive Learning Environment, 3*(2), 91-118.

Stieff, M. (2004). *A localized model of spatial cognition in chemistry.* Unpublished doctoral dissertation, Northwestern University, Evanston, IL.

Stieff, M., & Wilensky, U. (2003). Connected Chemistry - Incorporating interactive simulations into the chemistry classroom. *Journal of Science Education and Technology, 12*(3), 285-302.

Voith, M. (2003). Make your own 3-D Models. *Chemical & Engineering News, 81*(49), 36-38.

White, B., Kim, S., Sherman, K., & Weber, N. (2002). Evaluation of molecular visualization software for teaching protein structure. *Biochemistry and Molecular Biology Education, 30*(2), 130-136.

Wilensky, U. (1999). GasLab-an extensible modeling toolkit for connecting micro- and macro-proerties of gases. In N. Roberts, W. Feurzig & B. Hunter (Eds.), *Modeling and simulation in science and mathematics education* (pp. 151-178). New York: Springer.

Wilensky, U. (2001). *Modeling nature's emergent patterns with multi-agent languages.* Paper presented at the EuroLogo 2001, Linz, Austria.

Wilensky, U., & Resnick, M. (1999). Thinking in levels: A dynamic systems perspective to making sense of the world. *Journal of Science Education and Technology, 8*(1), 3-18.

Wu, H.-k., Krajcik, J. S., & Soloway, E. (2001). Promoting conceptual understanding of chemical representations: Students' use of a visualization tool in the classroom. *Journal of Research in Science Teaching, 38*(7), 821-842.

AUTHOR NOTE

Color versions of all figures included in this chapter are given in the accompanying CD-ROM.

ROBERT KOZMA[1] AND JOEL RUSSELL[2]

CHAPTER 7

STUDENTS BECOMING CHEMISTS: DEVELOPING REPRESENTATIONAL COMPETENCE

[1]*Center for Technology in Learning, SRI International, Menlo Park, CA*
[2]*Department of Chemistry, Oakland University, Rochester, MI; USA*

Abstract: This chapter examines the role that representations and visualizations can play in the chemical curriculum. Two types of curricular goals are examined: students' acquisition of important chemical concepts and principles and students' participation in the investigative practices of chemistry—"students becoming chemists." Literature in learning theory and research support these two goals and this literature is reviewed. The first goal relates to cognitive theory and the way that representations and visualizations can support student understanding of concepts related to molecular entities and processes that are not otherwise available for direct perception. The second goal relates to situative theory and the role that representations and visualizations play in development of representational competence and the social and physical processes of collaboratively constructing an understanding of chemical processes in the laboratory. We analyze research on computer-based molecular modeling, simulations, and animations from these two perspectives and make recommendations for instruction and future research.

INTRODUCTION

Over the past decade or so, there has been a significant effort to change the curricular goals of science education in the U.S.A. At the postsecondary level, science agencies (American Association for the Advancement of Science, 2001; National Research Council, 1996b; National Science Foundation, 1996) have promoted literacy in science, mathematics, and technology by direct experience with the methods and processes of inquiry. At the primary and secondary levels, national science standards emphasize the learning of science both as a process of investigation, as well as a body of significant concepts (American Association for the Advancement of Science, 1993). For example, the National Research Council (1996a) states that a new vision of science standards

> "requires that students combine processes and scientific knowledge as they use scientific reasoning and critical thinking to develop their understanding of science. . . Science as inquiry is basic to science education and a controlling principle in the ultimate organization and selection of students' activities" (p. 103).

In this chapter, we examine new visualization technologies in chemistry and consider the implications that these technologies and the above stated goals have for helping students become chemists. Clearly, very few high school and college students actually go on to major in a science and even fewer go on to major in chemistry. Nonetheless, a strong case can be made for using the investigative practices of scientists as the orienting themes for the science curriculum (O'Neill & Polman, 2004). Traditionally, the primary focus of the chemistry curriculum has been students' acquisition of significant chemical concepts, such as bonding, structure, reactivity, equilibrium, oxidation, acidity, and so on. An overemphasis on concepts has lead to a curriculum in the U.S.A. that is sometimes characterized as a "mile wide and an inch deep" (Vogel, 1996, p 335). The new goals expressed above do not replace the acquisition of concepts, although they may compete for limited time in the curriculum. Rather, they shift the emphasis to the acquisition and use of concepts within the context of scientific investigation in which students pose research questions, design investigations and plan experiments, construct apparatus and carry out procedures, analyze data and draw conclusions, and present findings (Krajcik, et al., 1998). Within this context, students have a deeper understanding of (perhaps fewer) concepts because they are the focus of intensive investigation. But perhaps more important, with these goals students also come to understand the basis for scientific claims and theories related to these concepts (O'Neill & Polman, 2004).

- In the following sections, we look at the role of visualizations for both kinds of goals for chemical education: those related to the acquisition of significant concepts and those related to scientific investigation. However, our emphasis is on the use of visualizations to support students' investigative practices and to develop a set of skills that we call "representational competence." Specifically the goals of this chapter are:
- Examine the representational skills and practices that chemists use while conducting scientific investigations.
- Compare these representational skills and practices to those of chemistry students.
- Review related theories of learning.
- Consider the implications of these research findings and theory for the new goals of chemical education and the development of students' representational competence.
- Review research related to several kinds of visualization technologies that can promote learning in chemistry.
- Propose recommendations for changes in the chemistry curriculum and for future research in chemical education, based on these theories and research results.

To achieve our goals, the chapter is divided into six major sections: chemists' uses of visualizations; students' uses of visualizations; theories of learning; new goals for chemical education; research on visualizations in chemical education, including molecular modeling, simulations, and animations; and implications for the chemistry curriculum and future research on visualization in chemical education. In the subsequent, companion chapter (Russell & Kozma, this volume) we describe a number of specific visualization software environments and assessment systems.

CHEMISTS' USES OF VISUALIZATIONS

Eminent chemist and former President of the National Science Board, Richard Zare (2002, p. 1290) characterizes chemists as ". . . highly visual people who want to 'see' chemistry and to picture molecules and how chemical transformations happen." There are two types of representations that chemists use to understand chemical phenomena – those that are internal, mental representations and those that are external symbolic expressions. All chemists have developed the ability to "see" chemistry in their minds in terms of images of molecules and their transformations. In this chapter we refer to such internal representations as concepts, principles, or "mental models" that encompass the state of chemical understanding of the individual. Chemists also construct, transform, and use a range of external representations—symbolic expressions, such as drawings, equations, and graphs. In conversations between chemists they spontaneously draw equations and structural diagrams to visually depict components of their mental models and the composition and structure of the compounds that are the object of their work. Indeed, Nobel chemist Roald Hoffmann contends that, "In an important sense, chemistry is the skillful study of symbolic transformations applied to graphic objects" (Hoffmann & Laszlo, 1991, p.11). In this chapter, we refer to such symbolic expressions as "visualizations" or merely "representations" (Kozma, Chin, Russell, and Marx, 2000). Others refer to them as "inscriptions" (see Roth & McGinn, 1998.) Thus visualizations are perceptible, symbolic images and objects in the physical world that are used to represent aspects of chemical phenomena, much of which can not be seen. In this sense the representations that Zare "sees" within his mind are mental models while the figures he draws on paper or constructs on a computer screen are visualizations.

An Historical Analysis

The role of representations and visualization technologies are central to the development of chemical understanding—not just the understanding of students as they struggle to learn about aperceptual chemical entities and processes but central to the understanding of chemists and, indeed, to the development of the field. There has always been a strong relationship between chemists' understanding of chemical phenomena and the external representations that they use to represent them. In a short historical review of representation in chemistry (Kozma, Chin, Russell, & Marx, 2000), we found that in the development of the discipline, new representations and tools have corresponded to new approaches to the study of chemical substances and new ways of thinking about the aperceptual chemical entities and processes that underlie and account for the material quantities of these physical substances. Within a community of shared goals, knowledge, activities, and discourse, chemists came to use these new representations and tools and they, in turn, shaped the way chemists thought about and conducted their work.

This process is most dramatically illustrated in the late 18th century when chemistry became a modern science. What distinguished pre- from modern chemistry was the transformation from the practice of conducting qualitative experiments (such as noting that a red mineral changed to a silver liquid upon heating) to quantitative experiments (such as noting that 100 g of the red solid produced 86.2 g of the silver liquid). This quantitative approach facilitated a change in the way chemistry was represented, a change in the most fundamental way that any field is represented—its language. Prior to the work of 18th century chemist Anton Lavoisier and his contemporaries, chemicals were named based upon their physical properties. The red solid, mercury(II) sulfide, was called vermilion and the silver liquid, mercury, quicksilver. By the late 18th century Lavoisier and others began to establish analytic techniques needed to find the underlying composition of substances as the focus of chemistry. At the same time, Lavoisier, Guyton de Morveau and colleagues developed a nomenclature system based upon elemental composition rather than physical properties (Anderson, 1984). For Lavoisier the connection between the development of the language and new activities within the discipline was explicit. Specifically, the naming of a substance required its experimental decomposition so that the component elements could be identified. The evolution of the chemical formula allowed chemists to display how molecules decomposed and combined and these symbolic expressions corresponded to the experimental procedures used in the laboratory to decompose and combine physical substances. Thus the language and symbol system were structured such that operating on symbols would be analogous to operating on substances. By embedding in chemical nomenclature and symbol systems a shift in focus from physical surface features to aperceptual elemental composition, Lavoisier created a new way of thinking about chemistry, a new set of practices, and a new chemical community. Lavoisier stated that:

> A well formed language, a language in which one will have captured the successive and natural order of ideas, will bring about a necessary and even prompt revolution in the manner of teaching. It will not allow those who profess chemistry to diverge from the march of nature. They will either reject the nomenclature, or else to follow irresistibly the route that it will have marked out. (as cited in Anderson, 1984, p. 176).

Developments in chemistry have continued to be shaped by developments in the way chemical phenomena are represented or visualized. In the 20th century chemical synthesis—the design and formation of molecules—became a major emphasis of the field. In parallel with this new activity was the development of structural formulas or diagrams. Structural formulae show both the composition and the bonding pattern of atoms in molecules and such visualizations aid the analysis of sites within a compound that will react to form new molecules.

Between the 1930s and mid-1960s, chemists developed physical 3-D structural models composed of elemental components (sometimes balls) and (sometimes) sticks representing bonds between elements (Francoeur, 1997, 2002). These structures made the dimensional arrangement of elements more explicit and allowed for rotation and inspection of the molecular model. With the advent of sophisticated computers and molecular modeling software beginning in the 1960s, it was very easy to construct ball and stick, space filling, and electron density models for even very large molecules. Such interactive molecular graphics have come to replace physical models. The advantage of the computer modeling programs is that they support additional analyses, such as measuring bond lengths and angles. At each stage in the development of more refined and meaningful models the new representations afforded new chemical practices and new ways of thinking about chemistry, such as considerations of chirality, steric factors, and electrophilic and nucleophilic species. The outputs of new tools (chromatographs, infrared spectra, NMR spectra, mass spectra) also provided graphical visualizations of physical properties of the aperceptual molecules that supported the synthesis of new compounds.

An Ethnographic Study of Chemists

These historical examples illustrate that the design and use of visualization systems is a key cultural activity that affects the composition of a community, the activities of its members, and the understanding its members share about its domain of knowledge. In our ethnographic study of chemists we described how this finding is realized in everyday chemical practice (Kozma, Chin, Russell, & Marx, 2000). In this study, we investigated the interplay between various forms of visualization and the practices and discourse of chemists at an academic and a pharmaceutical laboratory, both focusing on the synthesis of new compounds.

In the academic lab we observed two graduate students, a postdoctoral fellow, and a faculty mentor as they worked and in the pharmaceutical lab six chemists including one project director were observed working at the bench and participating in group problem-solving meetings. For all these synthetic chemists, in every conversation, and written on laboratory hood doors and on white boards in conference and coffee rooms, chemical equations using structural formulae were omnipresent. The next most common forms of visualizations used by chemists in both settings were nuclear magnetic resonance (NMR) and mass spectra. These visualizations served two important, interrelated roles in the practices of chemists: material and social. First, visualizations provided a material representation of otherwise aperceptible entities and processes, representations that could be perceived and manipulated. Second, visualizations served to support the social discourse of chemists as they worked to synthesize intended compounds.

There were several patterns in representational practices that we noticed in our observations. First, chemists used visualizations to express their research goals in terms of molecular structures of compounds. That is, chemists used structural diagrams to describe the composition and geometry of the compounds that they were trying to synthesize. Second, they used visualizations to reason about the chemical and physical processes needed to synthesize these compounds. They used diagrams and equations to think though the possible reaction mechanisms by which reagents would be transformed to desired products and the physical procedures in the laboratory that would support these transformations. Third, chemists used visualizations to verify the composition and structure of the synthesized compounds. They would analyze instrumental displays and printouts to determine if the products were the compounds that they were trying to synthesize. Fourth, they used visualizations in a rhetorical sense to convince the scientific community that the compound they synthesized was the one intended. In this regard, specific features of the printouts were used as warrants for claims that the products were indeed structured as intended. Finally, chemists' use of various visualizations confirmed their membership in the scientific community. By using various representations in these ways, graduate students and postdoctoral fellows developed the chemical skills and practices that integrated them into the community of chemists.

Students' Uses of Visualizations

The various ways that chemists visualize chemical entities and processes differ significantly from the ways chemistry students use representations. They differ both in their laboratory practices and in their ability to use and understand various forms of visualization.

An Ethnographic Study of Students

In an observational study of an organic chemistry course, we examined the laboratory practices of college students (Kozma, 2000b, 2003). Pairs of students were audio and video recorded as they interacted with each other while synthesizing a compound in the laboratory. We analyzed their actions and interactions during one laboratory session and found several patterns. First, in sharp contrast to academic and professional laboratory practice, we found infrequent use of visualizations by students during their wet lab experiments. As a consequence, the laboratory practices and discussion of chemistry students was focused on the physical aspects of their experiments—the laboratory equipment and the physical properties of reagents. The discussion among student lab partners was focused primarily on setting up equipment, trouble shooting procedural problems, and interacting with the physical properties of the reagents they were using (e.g., was their crystalline product washed enough or dry enough). Second and most important, students rarely discussed the molecular nature of the reactions that they were running on the lab bench. Unlike the discourse of chemists, we observed very little discussion among students about the molecular properties of the compounds they were synthesizing or the reaction mechanism that might be taking place during their experiments. The lack of representational use even shaped the discourse of the laboratory instructors. In these discussions between students and their instructors, the mention of molecular properties and processes was absent.

An Expert-Novice Study of Visualization

These differences between chemists and chemistry students in laboratory practices correspond to significant differences in chemists' and students' understanding of various forms of chemical visualization. In an earlier study (Kozma & Russell, 1997), we examined what expert chemists and novice students understand when they see and use various chemical visualizations such as graphs, equations, and animations of molecular phenomena. We also wanted to know if subjects saw connections between different chemical visualizations corresponding to the same phenomena or if they understood something different for each type of visualization. The novice pool consisted of 11 first semester chemistry students from a Midwestern research university. The expert pool consisted of five doctoral students from this university, five chemists from a pharmaceutical company, and one community college instructor. All subjects were videotaped as they worked at a computer to complete two tasks.

Subjects were shown 14 dynamic and still images corresponding to several chemical reactions. The representations included videos of the experiments, animations of the molecular events, dynamic graphs of a physical property of the system, and chemical equations or formulae. Subjects were then given a card corresponding to each representation, with its dynamic image being represented by a single still frame. They were asked to sort these cards into logical subsets, give a name to each subset, and explain the meaning of the name. In this sorting experiment we observed that both experts and novices created chemically meaningful groups. However, novices formed their meaningful groups from a small number of cards often from the same media type (e.g., all graphs, all equations, etc.) while experts used larger groups composed of multiple media forms. Also the reasons for forming particular groups that experts gave were largely conceptual while novices' reasons often were based upon surface features.

In the second task subjects were shown various visualizations—chemical equations, videos of experiments, dynamic graphs of a physical property of some experiment, and animations of the molecular events of an experiment. They were asked to describe what they saw and then to transform the visualization into various other forms, as specified. (Digitized versions of these tasks are included on the CD with this book under the chemistry section, entitled sorting tasks and transformation tasks.) From this transformation experiment we found that experts were much better than novices at providing verbal descriptions, due to their deeper understanding of chemical principles and concepts. Experts were only slightly better than novices with transformations requiring a choice between answers (e.g., given an equation and matching it to one of several video segments), since surface features could be utilized to make a choice. However, they were much better than novices for transformations that required a constructed response such as drawing a graph or writing a chemical equation. They were particularly better in giving verbal descriptions as transformations.

Although this study used a variety of visualizations not used in prior expert-novice studies (Chi, Feltovich & Glaser, 1981; Glaser & Chi, 1988; Larkin, McDermott, Simon & Simon, 1980), its conclusions were consistent with these earlier studies. Experts in our study were able to use their deep conceptual knowledge and mental models of chemical phenomena to construct the meaning of these visualizations while novices based meaning on the surface features of the visualizations or the physical properties of the reactions. Experts as members of the chemical community not only have a more extensive knowledge base but they also were experienced in using diverse forms of visualization as they communicate with others.

LEARNING THEORY AND VISUALIZATIONS IN CHEMICAL EDUCATION

Learning theory can play an important role in guiding research and curriculum development related to the use of visualizations. There have been important developments in learning theory over the past several decades that have implications for work in the area of visualization. The focus of learning theory in the 70's and

80's was on the mind of the individual learner—cognitive structures and processes. Beginning in the late 80's and early 90's, theorists focused on how learners interacted with the social and physical resources in their environment and how they integrated into communities of practice—what has come to be called "situative theory." Together these theories address the two types of goals for the chemistry curriculum. Cognitive theory has significant implications for how instruction could be designed to support the acquisition of concepts and problem solving procedures. Situative theory has implications for goals related to investigative practices of students "becoming chemists." Both have implications for the use of representations and visualizations in classrooms and for research in this area.

Learning Concepts

An important goal for chemical education is student acquisition of key concepts and principles, such as bonding, structure, reactivity, equilibrium, acidity, and so on. High school and even college students have significant difficulty understanding these concepts and principles (Gable, 1998; Krajcik, 1991; Nakhleh, 1992), in large part because they related to phenomena that are not available for direct inspection.

Mayer (2001, 2002, 2003) describes a cognitive theory of multimedia learning and proposes principles for the design of effective multimedia instruction to address the learning of concepts and principles. This important work was summarized for the chemical education community by Robinson (2004). Mayer uses three assumptions from cognitive psychology: dual channel, limited capacity, and active processing. The dual channel assumption states that the human brain has separate channels for processing auditory-verbal and visual-pictorial inputs (Baddeley, 1999; Paivio, 1986). The limited capacity assumption is that there is a relatively small limit to the part of the human brain called working memory that processes and manipulates inputs from the auditory-verbal and visual-pictorial channels (Baddeley, 1999; Johnstone, 1997). There are both limits to the amount of information that can be processed from each channel as well as a non-additive overall limit. Learning occurs when the student actively selects, organizes, and integrates information from auditory and/or visual inputs (Mayer, 2001). Mayer (2002) uses this theory and research findings from numerous studies to develop eight principles of multimedia learning that use both channels, address limited cognitive capacity, support students' active processing, and result in deeper learning—that is learning that results in understanding of difficult concepts and principles which can then be used to solve novel problems. These are:

1. *Multimedia Principle.* Deeper learning occurs when both words and pictures are used than from the use of words alone.
2. *Contiguity Principle.* Deeper learning results from presenting words and pictures simultaneously rather than successively.
3. *Coherence Principle.* Deeper learning occurs when extraneous words, sounds, or pictures are excluded rather than included.
4. *Modality Principle.* Deeper learning occurs when words are presented as narration rather than as on-screen text.
5. *Redundancy Principle.* Deeper learning occurs when words are presented as narration rather than as both narration and on-screen text.

6. *Interactivity Principle*. Deeper learning occurs when learners are allowed to control the presentation rate than when they are not.
7. *Signaling Principle*. Deeper learning occurs when key steps in the narration are signaled rather than non-signaled.
8. *Personalization Principle*. Deeper learning occurs when words are presented in conversational style rather than formal style.

These eight principles have a direct bearing on how visualizations can be used in chemical education to support the learning of difficult chemistry concepts and principles. For example, the multimedia, contiguity, and modality principles encourage the simultaneous use of visualizations along with oral instruction to provide a synergistic learning effect by supporting the active processing of both visual and audio inputs in working memory. Multimedia software would be more effective if it used audio narration rather than printed text. The coherence and redundancy principles suggest that educators and instructional designers use the simplest possible visual and audio components to achieve the specific learning objective. Visualizations should reduce unnecessary details, rather than be as realistic as possible, in order to achieve a specific learning goal.

The cognitive theory and design principles described by Mayer (2001, 2002, 2003) and Robinson (2004) assume an educational model in which learning occurs as a result of the student's engagement with some instructional presentation, be it software or lecture. The focus is on learning difficult concepts and principles. When students interact with information that is represented in both pictures and words they are more likely to learn difficult concepts and principles, retain what they learn longer, and use what they learn to solve problems, than if the information is presented in words alone.

To illustrate the application of this theory to chemical education, a common organic chemistry convention is to represent unreactive portions of molecules as "wiggle lines," "picket fences," or "R;" this convention would be useful for instructional purposes, as well, because it reduces extraneous information (*Coherence Principle*). Similarly, in a molecular animation, it is likely to be more effective to show the equilibrium reaction $N_2O_4 \rightleftharpoons 2NO_2$ by using single and double spheres of different colors rather than space filling models with unnecessary detail. The Personalization and Signaling Principles provide guidelines for writing and reading audio narration that accompanies visual representations of molecular processes. The Interactivity Principle supports numerous studies that favor active learning and software design with at least a minimum amount of user control over the rate of movement through the lesson. In a later section of this chapter and in another chapter (Russell & Kozma, this volume), these principles will be used in reviewing specific visualization studies and products.

Learning Investigative Practices

If the educational goal is to engage students in chemical practices of laboratory investigation then visualizations need to be used in a different way than that proposed by Mayer. Correspondingly, a broader theoretical approach is needed to account for the use of these various representations to visualize chemical phenomena in the laboratory. In this regard, we draw on situative theory to complement Mayer's cognitive theory.

In brief, situative theory posits that the physical and social characteristics of a specific setting shape the interpersonal and psychological processes that occur within that setting, including those processes related to understanding and learning (Greeno, 1998; Roth, 1998, 2001; Lave & Wenger, 1991). Characteristics of physical objects in a specific setting enable (i.e., afford) or constrain the way those in that setting talk, think, understand, etc. For example, the availability of chemical reagents in a laboratory affords certain kinds of discussions about the physical properties—color, viscosity, smell, etc.—of the compounds. On the other hand, the physical properties of chemical reagents do not as such facilitate discussions about the composition or structure of the molecules or the processes by which these molecules are transformed, as observed in our ethnographic study of chemistry students (Kozma, 2000b).

Situative theory also emphasizes interactions among individuals as they are engaged in activities within an organized social system. Regular patterns of activities within these systems are characterized as practices of a community. From this perspective, learning occurs as individuals become attuned to the constraints and affordances of the patterned physical and social situations of a community and become more centrally involved in the community's practices. Through these social interactions within and outside the group, identities and affiliations are formed which influence the motivation to participate in the community's practices and adopt community standards and norms.

Representations play a particularly important role within the situative theoretical perspective since representations allow the consideration and discussion of objects and processes that are not present or otherwise apparent in a specific situation. Representations—such as written or drawn symbols, iconic gestures or diagrams—"stand for" or "refer to" other objects or situations and thus they become part of the material resources of the current context. It is by the use of these representations that chemists are able to visualize, discuss, and understand the molecules and chemical processes that account for the more perceivable reagents and phenomena they observe in the laboratory. However, the meaning of a representation is not embedded in the representation itself but is assigned to the representation through its use in practice—in the case of chemistry, the use of various chemical symbol systems in the context of laboratory investigations. A crucial part of any social system or community is the convention of interpreting meanings of representations. As individuals become integrated into a community of practice, they progressively use its representational systems in meaning-making activities. And in turn, representations come to be useful tools for constructing and communicating understanding.

From the situative perspective (Greeno, 1998), classrooms that support investigative practices are ones that encourage students to participate in activities in which representations are used in the formulation and evaluation of conjectures, examples, applications, hypotheses, evidence, conclusions, and arguments. In the subsequent section, we derive implications from our study of chemists' use of representations (Kozma, et al., 2000) for the design of educational experiences in chemistry, if the educational goal is to promote learning by laboratory investigation

(Krajcik, et al., 1998). In the section following that, we use both the cognitive and situative perspectives to analyze certain technology applications and studies which examine the use of various visualizations to support students' learning of difficult concepts and principles and their laboratory investigations in chemistry.

NEW GOALS AND RESOURCES FOR CHEMICAL EDUCATION

Our comparisons between expert chemists and chemistry students and the learning theories described above have significant implications for chemical curricula and pedagogy. The laboratory practices of chemists suggest that a curriculum based on investigation should not only emphasize the experimental aspects of chemistry but the representational and social aspects, as well. That is, the process of chemical investigation depends not only on experimental procedures but on the social discourse by which the purposes and results of laboratory investigations are represented and understood.

Krajcik and his colleagues (1998) propose a project-based approach to science learning in which students work in pairs or groups to conduct extended investigations in which they pose scientific questions, plan and design investigations and procedures, construct apparatus, carry out their experiments, interpret data, draw conclusions, and present their findings. Discourse is an essential component of such an investigative approach for it is by "talking science" that students come to understand the phenomena that they are investigating (Lemke, 1990). Our analysis of the practices of chemists supports this position. It also supports the central role that visualizations play in scientific meaning making. We have drawn on the findings of our own research in chemistry (Kozma, 2000a, 2000b, 2000c; Kozma, Chin, Russell, & Marx, 2000; Kozma & Russell, 1997) and that of others (Amman & Knorr Cetina, 1990; Chi, Feltovich, & Glaser, 1981; diSessa, et al., 1991; Dunbar, 1997; Glaser & Chi, 1988; Goodwin, 1995; Larkin, 1983; Larkin, McDermott, Simon, & Simon, 1980; Roth, 1998; Woolgar, 1990) to develop a notion of "representational competence" as an important set of skills and practices to be included in the chemistry curriculum (Kozma 2000a).

Representational Competence

Representational competence is a term we use to describe a set of skills and practices that allow a person to reflectively use a variety of representations or visualizations, singly and together, to think about, communicate, and act on chemical phenomena in terms of underlying, aperceptual physical entities and processes (Kozma, 2000a, 2000b; Kozma & Russell, 1997). The act of using representations to successfully construct chemical understanding at once constitutes the meaningfulness of the representation and confirms the user's ability to participate in this representational, meaning-making activity. While those with little representational competence in a domain rely primarily on the surface features of representations to derive meaning (Chi, Feltovich, & Glaser, 1981; diSessa, et al. 1991; Kozma & Russell, 1997) or the mechanical application of symbolic rules (Krajcik, 1991), those with more skill have come to use a variety of formal and informal representations together to explain phenomena, support claims, solve problems, or make predictions within a community of practice (Amman & Knorr Cetina, 1990; Dunbar, 1997; Goodwin, 1995; Kozma, et al. 2000; Kozma & Russell, 1997; Roth, 1998; Woolgar, 1990).

The performance of chemists in our studies suggests that the following skills might constitute the core of a substantive curriculum of representational competence in chemistry:

- The ability to use representations to describe observable chemical phenomena in terms of underlying molecular entities and processes.
- The ability to generate or select a representation and explain why it is appropriate for a particular purpose.
- The ability to use words to identify and analyze features of a particular representation (such as a peak on a coordinate graph) and patterns of features (such as the behavior of molecules in an animation).
- The ability to describe how different representations might say the same thing in different ways and explain how one representation might say something different or something that cannot be said with another.
- The ability to make connections across different representations, to map features of one type of representation onto those of another (such as mapping a peak of a graph onto a structural diagram), and to explain the relationship between them.
- The ability to take the epistemological position that representations correspond to but are distinct from the phenomena that are observed.
- The ability to use representations and their features in social situations as evidence to support claims, draw inferences, and make predictions about observable chemical phenomena.

We propose a conceptual structure of these skills that organizes them into characteristic patterns of representational use at five stages or levels (see Table 1). This structure corresponds to a developmental trajectory that generally moves from the use of surface features to define phenomena which is characteristic of novices within a domain (Chi, Feltovich, & Glaser, 1981; Glaser & Chi, 1988; Kozma & Russell, 1997; Larkin, 1983; Larkin, McDermont, Simon, & Simon, 1980)—to the rhetorical use of representations, which is characteristic of expert behavior (Amman & Knorr Cetina, 1990; Goodwin, 1995; Dunbar, 1997; Kozma et al., 2000; Woolgar, 1990).

There are a number of assumptions embedded in the representational competence structure. First, we assume that the acquisition of these skills follows a developmental trajectory. However, we do not feel that this trajectory is a stage-like, Piagetian progression of personal development (Piaget, 1972). Rather, we ascribe to the Vygotskian notion that development depends on the person's development as well as the physical, symbolic, and social situation (Vygotski, 1980, 1986). Vygotski describes a "zone of proximal development" in which an individual's personal development is supplemented by interactions with the material and social resources in the environment. This position is quite sympathetic with situative theory, with its emphasis on the progressive use of representational conventions to become more centrally involved in a community of practice.

Table 1. *Summary of Representational Competence Levels*

LEVEL 1: REPRESENTATION AS DEPICTION When asked to represent a physical phenomenon, the person generates representations of the phenomenon based only on its physical features. That is, the representation is an isomorphic, iconic depiction of the phenomenon at a point in time.
LEVEL 2: EARLY SYMBOLIC SKILLS When asked to represent a physical phenomenon, the person generates representations of the phenomenon based on its physical features but also includes some symbolic elements to accommodate the limitations of the medium (e.g., use of symbolic elements such as arrows to represent dynamic notions, such as time or motion or an observable cause, in a static medium, such as paper). The person may be familiar with a formal representation system but its use is merely a literal reading of a representation's surface features without regard to syntax and semantics.
LEVEL 3: SYNTACTIC USE OF FORMAL REPRESENTATIONS When asked to represent a physical phenomenon, the person generates representations of the phenomenon based on both observed physical features and unobserved, underlying entities or processes (such as an unobserved cause), even though the representational system may be invented and idiosyncratic and the represented entities or processes may not be scientifically accurate. The person is able to correctly use formal representations but focuses on the syntax of use, rather than the meaning of the representation. Similarly, the person makes connections across two different representations of the same phenomenon based only on syntactic rules or shared surface features, rather than the shared, underlying meaning of the different representations and their features.
LEVEL 4: SEMANTIC USE OF FORMAL REPRESENTATIONS When asked to represent a physical phenomenon, the person correctly uses a formal symbol system to represent underlying, non-observable entities and processes. The person is able to use a formal representation system based on both syntactic rules and meaning relative to some physical phenomenon that it represents. The person is able to make connections across two different representations or transform one representation to another based on the shared meaning of the different representations and their features. The person can provide a common underlying meaning for several kinds of superficially different representations and transform any given representation into an equivalent representation in another form. The person spontaneously uses representations to explain a phenomenon, solve a problem, or make a prediction.
LEVEL 5: REFLECTIVE, RHETORICAL USE OF REPRESENTATIONS When asked to explain a physical phenomenon, the person uses one or more representations to explain the relationship between physical properties and underlying entities and processes. The person can use specific features of the representation to warrant claims within a social, rhetorical context. He or she can select or construct the representation most appropriate for a particular situation and explain why that representation is more appropriate than another. The person is able to take the epistemological position that we are not able to directly experience certain phenomena and these can be understood only through their representations. Consequently, this understanding is open to interpretation and confidence in an interpretation is increased to the extent that representations can be made to correspond to each other in important ways and these arguments are compelling to others within the community.

We make no assumption that the development is automatic or uniform. Quite possibly, a person may display behaviors associated with a higher level (say Level 3) in one context and others that are coded at a lower level (Level 2) in another context. Or a person may display a higher level skill on one occasion and a lower level on the same skill at a later time. Also, different levels of development may be displayed with different formal symbol systems or representations. That is, a person may be more competent and show higher level skills with a particular system (say chemical equations) than another (for example, graphs). However, over time and given appropriate sets of physical, symbolic, and social situations, a student will increasingly display more-advanced representational skills, come to internalize these, and integrate these into regular practice.

Pedagogical Activities that Support Representational Competence

The findings of our research and that of others (Roth, 1998; Roth & McGinn, 1998; Bowen, Roth, & McGinn, 1999) suggest that representational skills can best be developed and used within the context of student discourse and scientific investigations. The use of language and representations during the investigative process is also more likely to lead to a deeper understanding of chemical phenomena. Krajcik and his colleagues (Krajcik, Blumenfeld, Marx, Bass, Fredricks, & Soloway, 1998) propose a five-phase approach to structuring student investigation and collaboration. Language and representations can be embedded in each of these phases, as students pose questions, plan, execute, analyze, and discuss their investigations.

- Ask questions: In this phase, students draw on their personal experiences and background knowledge to ask questions, make predictions, and judge the worthiness and feasibility of their investigation. Questions and predictions can not only be in verbal form but can use other representations (e.g., "When the reaction reaches equilibrium, the lines of the graph will flatten."). In our ethnographic study (Kozma, et al., 2000), we often saw chemists use molecular diagrams to think through the hypotheses for their experiments.
- Design investigations and plan procedures: Here students design their investigation, decide on what variables to use, design measurements to make, and how they will manage data collection. Language and representations can be used to help students think how the physical experiments that they design can help them understand the chemical processes that underlie the phenomena they observe.
- Construct apparatus and carry out investigations: With this step, students select or build apparatus, make observations, take measurements, and record data. An issue embedded in this phase is how the data should be represented (e.g., As numbers, as a graph, as a diagram? What do you gain or lose with one form or the other?).
- Analyze data and draw conclusions: Students transform, analyze, and interpret data, and use them to draw explanations, arguments, and conclusions. This can involve assigning meaning to a specific feature of a representation (e.g., a peak on a graph) and the coordination of meaning across representations (e.g., that

the peak on a graph corresponds to a particular observable event, like a change in color) to help them understand the underlying chemistry.
- Present findings: Finally, students exchange information, share and clarify ideas, give and receive assistance and feedback, create artifacts, and present findings. Again, representation is key here. How do students present their findings in a convincing way that helps others understand the underlying chemistry?

Various visualization technologies can play a key role in supporting investigation and representational practices. As students conduct their investigations, the need to acquire relevant concepts and principles takes on greater salience for them. Representations and visualization technologies can also play a role in this regard, as we saw in our review of Mayer's (2001, 2002, 2003) cognitive theory of multimedia.

VISUALIZATIONS IN CHEMICAL EDUCATION

What implications do these findings have for visualizations in chemical education? In the following sections we examine how visualizations fit into the two complementary theories of learning and understanding. We draw implications from these theories for the use of visualizations in chemical education and use the implications to consider two principal types of chemical visualizations for education: molecular models and simulations and animations.

Molecular Modeling

Gilbert and Boulter (1998) discuss the learning of science through models and modeling. These authors use the term "model" broadly to include representations of ideas, objects, events, processes, or systems. For historical reasons, we use the term here more narrowly. In chemistry, "model" has come to refer to a physical or computational representation of the composition and structure of a molecule and that is how we use the term in this section. In the subsequent section, we refer to representations of systems of molecules as "simulations" or "animations".

Historically, chemical equations were the earliest representation of molecules and molecular systems. Chemical equations identify the reactants and products of a chemical reaction but they do not show the bonding patterns or geometric structure of molecules. They show conservation of atoms and thus mass in a reaction but do not show the mechanism for the reaction. Structural diagrams are better at showing the geometric arrangement of atoms in the molecule but they do not allow for manipulation of the structure. As mentioned earlier, the development of molecular models—first as physical structures, then as computer representations—allowed chemists to explicitly display and manipulate the geometric structure of molecules (Francoeur, 1997, 2002).

Molecular modeling software allows the chemist to generate and visually display electron density, electrostatic potential surfaces, and images of the highest occupied molecular orbitals (HOMOs) and lowest unoccupied molecular orbitals (LUMOs). This visualization technology is now an essential supplement to the chemical equation for most chemical applications including research, textbooks, and instructional software. Many modeling programs allow the users to construct the molecules from atoms, find the lowest energy geometric structure, measure bond lengths and angles for this structure and manipulate the visualization. Typically, users can rotate the model and look at it from different angles. This supports the laboratory practice of synthesis by allowing chemists to look for reactive sites within molecules and speculate on reaction mechanisms.

Several universities have integrated computational chemistry with extensive molecular modeling experiences throughout their undergraduate curricula (Jones, 2001; Kantardjieff, Hardinger, & Van Willis, 1999; Martin, 1998; Paselk & Zoellner, 2002). Molecular modeling exercises using Spartan or HyperChem have been used as laboratory supplements in introductory chemistry (Cody & Wiser, 2003), organic chemistry (Hessley, 2000), and inorganic chemistry (Montgomery, 2001) or as the basis for entire courses such as structural biochemistry (Dabrowiak, Hatala, and McPike, 2000). Applications of molecular modeling are most common in organic chemistry from simple textbook figures, to Chime-based molecule databases, to animations of reaction mechanisms using ball and stick, space filling, and ball and stick with superimposed HOMOs and LUMOs (Fleming, Hart, and Savage, 2000).

Some authors suggest that molecular modeling visualizations in the form of electron density plots and electrostatic potential surfaces are superior to orbital models for introducing students to molecular structure and bonding (Matta & Gillespie, 2002; Shusterman & Shusterman, 1997). The electron density, $\rho(x,y,z)$, represents the probability of finding any one of the electrons in a molecule in a small volume element around point (x,y,z). Electron density is often visualized as a cloud of negative charge that varies in density throughout the molecule. Isodensity surfaces of electron density can be displayed to show bonding types and as electrostatic potential surfaces to probe charge variations over the molecule (see figures in Shusterman & Shusterman, 1997). Electrostatic potential models are useful for investigating intermolecular interactions, identifying reactions sites for electrophilic and nucleophilic reactions, and relative acidities of acids. Both sets of authors suggest that the concept of electron density is easier for students to grasp than the concept of orbitals since orbitals are pure mathematical constructs that cannot be determined by experiments. A further advantage of using electron density models to introduce chemical bonding is its sound theoretical foundation with measurable outputs that do not require modification or replacement in subsequent courses, as students' understanding deepens.

Dori and Barak (2001) showed that high school students also can benefit from the use of molecular models. Israeli high school students who had performed a series of inquiry-based learning tasks using both physical models and computerized molecular modeling had a better understanding of organic compounds than did a control group of students studying the same topics without the modeling learning tasks. The experimental students were better at explaining concepts such as isomerism and most used sketches of ball-and-stick or space-filling models in their explanations. These modeling tasks had developed students' abilities to communicate chemical concepts using appropriate visualizations showing expert-like skills. Use of the modeling tasks also eliminated the gap between students with high and low pre-course knowledge to provide explanations. The control group students were frequently unable to provide explanations with their answers to questions about molecular structures.

With university students Dori, Barak, and Adir (2003) investigated the effect of a single extra credit molecular modeling assignment on performance on the final exam and a special test of chemical visualization and reasoning skills. Of the 215 students who completed the special test as a first week pretest and fourteenth week posttest, 95 volunteered to do the molecular modeling assignment. These students reported spending 5-10 hours drawing structural formulas, building models for an assigned compound such as vitamin A or DDT, displaying and manipulating the models as wire-frame, ball-and-stick, and space-filling forms, and finding the hybridization and electron charge density of each atom. The experimental group had significantly higher scores than the control group on the post-test ($\rho < 0.001$) and final exam ($\rho < 0.02$). Each third of the groups based upon pretest scores showed improvements on the posttest with differences in gains between the experimental and control groups significant for the middle and high thirds. The authors did note that it may be possible that students volunteering for the extra credit assignment were those that worked more throughout the course.

Not only can molecular models facilitate the acquisition of chemical concepts, they can support laboratory practice. In our ethnographic study of organic chemistry students (Kozma, 2000b, 2003), we saw how the use of molecular modeling software can dramatically change students thinking and talking in a laboratory course. In addition to observing students as they worked with lab equipment and reagents, we also observed and coded the discourse and behavior of the same students as they used a molecular modeling package (Spartan) in the computer laboratory to construct and analyze the same compound that they had previously synthesized in the wet lab. In contrast to their discourse during the wet lab session, students using the computer modeling software exhibited far more references to chemical concepts, such as atoms, bonds, electronegativity, dipole moment, and so on. Student discourse while using modeling software was much more like that of laboratory chemists in our study (Kozma, et al., 2000) than was their discourse in the wet lab. Furthermore, the laboratory instructors were more likely to discuss chemical concepts with students than they were in the wet lab.

Computer Simulations and Animations

While we use the term molecular model to refer to visualizations of individual molecules, we use the terms "simulation" or "animation" to refer to representations of dynamic chemical processes or systems. Simulations allow users to select values for input variables from within suitable ranges and observe the results on output variables. With chemical simulations, users might change pressures in a gaseous system or concentrations of regents in a solution system and observe the impact of these changes on the species in the system. Simulations can be used to explore chemical systems or processes in order to derive or test possible underlying explanations or theoretical models. Gilbert and Boulter (1998) refer to these uses as "exploratory" applications. Three examples of chemical simulations involve concepts related to kinetics (Allendoerfer, 2002), equilibrium (Paiva, Gil, & Correia, 2002) and acid-base titrations (Papadopoulos & Limniou, 2003). All three simulations can be used to not only explore fundamental chemical concepts but to supplement laboratory experiments either as pre-laboratory exercises or as extensions to hands-on laboratories. All these simulations, along with several others, are included on the General Chemistry Collection, 7th Edition, CD-ROM (Holmes & Gettys, 2003) available from *Journal of Chemical Education Software*.

Chemical animations are dynamic, molecular-level (sometimes referred to as "nano-level" or "nanoscopic-level") representations of chemical systems as they change and their species react. These displays often consist of individual balls or dots, or clusters of these, that represent individual molecules that move about, collide, and react. The results or outputs of simulations are often displayed as animations. However, other simulations (for example, Schoenfeld-Tacher, 2000) represent the simulated experiments with icons that correspond to the experimental apparatus, reagents, instrument readouts, and observable results of experiments, much as they would appear in the wet lab. While animations can represent the outputs of simulations, they can also be mere "canned" movies that illustrate a particular concept at a molecular level without allowing users to manipulate variables.

Most research studies that have been conducted in this area have used animations as movies that illustrate specific concepts, rather than as the output of simulations of chemical systems that students explore. For example, Yang, Greenbowe, and Andre (2004) used molecular-level animations of oxidation-reduction reactions and the movement of ions and electrons in dry cell batteries in flashlight circuits in a tutorial designed to teach concepts about electrochemical cells, such as the movement of electrons between electrodes within the cell, and about electric circuits, in order to overcome such common misconceptions as a decrease in current as electrons flow through a light bulb. An experimental group of volunteers attended an instructor-guided tutorial session rather than the lecture on electrochemical cells. These students, guided by the instructor and worksheets, used the animation software to study the chemical and physical processes occurring in a common flashlight. These animations allowed users to zoom-in to view the processes at each battery electrode separately and to pause, continue, and to replay each animation. Students using the tutorial rather than attending the lecture were more likely (80% versus 57%) to identify the carbon electrode as an inert electrode on the posttest. This study shows the effectiveness of well-designed animations enhancing student understanding of chemical concepts, although the study design confounded the use of animations with the use of tutorials.

In another study, animation and instructional approach were disentangled. Sanger and Greenbowe (2000) examined the impact of animations and conceptual change strategies on students' conceptions of current flow in electrolyte solutions. In this 2 X 2 design, university students in one group received a lecture along with an animation accompanied by lecturer's narration showing the electrochemical processes occurring in a galvanic cell at the nanoscopic level with a focus on the chemical half-reactions occurring at each electrode and the transfer of ions through a salt bridge. Students in the other lecture section saw only static chalkboard drawings of the process. Half the students in these groups also received a conceptual change strategy that confronted typical misconceptions while the other half did not. Students were tested using algorithmic, visual, and verbal test items on both an immediate and a delayed post test. There were no significant differences between any of the treatment and control groups on the algorithmic or visual test items. There was, however, a main effect in favor of the conceptual change strategy (but not one for the use of animations) on the verbal items on both the immediate and delayed post-tests. There was also an interaction between the animation and conceptual change treatments for these items on the immediate post-test such that students who did not receive the conceptual change strategies scored higher with the animations than those students who received the conceptual change strategies but not the animations.

In an experiment by Sanger, Phelps, and Fienhold (2000), students in the treatment group viewed an animation accompanied by narration that showed representations at both macroscopic (i.e., a can being heated, sealed, and crushed upon cooling) and nanoscopic levels (i.e., numbers of 2-D balls representing molecules moving around both inside and outside the can, more quickly or slowly). Students in both the treatment group and the control group received a lecture on gas behavior and observed a physical demonstration of the can-crushing experiment. Students in the treatment group scored significantly higher on measures related to understanding a similar demonstration. In another experimental study (Sanger & Badger, 2001), students in the treatment group received instruction on and demonstrations of miscibility of polar, non-polar, and ionic compounds supplemented by both molecular-level animations of inter-molecular attractions, as well as shaded 2-D pictures of electron density plots ("eplots") that used color to show the polar (i.e., charged) regions of various molecules. Students in the control group received only instruction and demonstrations. Students in the treatment group scored significantly higher on an experimenter-constructed test in which students were asked to identify polar molecules not discussed in the instruction, make predictions about the miscibility of compounds, and explain their answers.

Williamson and Abraham (1995) investigated the use of animations of properties of states of matter and chemical reactions with a quasi-experimental, posttest control-group design. They compared a control group of college students who attended lectures using only static visualizations for these topics with two experimental groups that also used a total of 13 animations of nanoscale properties each with one to two minute durations in lectures. One of the experimental groups also used the animations in their recitation sections held in a computer lab. All groups received the same series of 10 lectures on concepts related to gases, liquids, solids, and solution reactions. Conceptual understanding was measured using the Particulate Nature of Matter Evaluation Test (PNMET) (Williamson, 1992). Students in the experimental groups had mean scores one half a standard deviation above those in the control group on the PNMET. All three groups were equivalent on course achievement, as measured by an hour exam and final exam. Williamson and Abraham suggest that use of animations may increase conceptual understanding by prompting formation of dynamic mental models of the phenomena.

Also at the college level, we (Kozma, Russell, Jones, Marx, & Davis, 1996) examined the impact of a software environment described in Russell & Kozma (1994) and in the companion chapter (Russell & Kozma, this volume) in which students conducted simulated experiments using multiple representations. In this study, students changed variables related to pressure, concentration, and temperature and examined the effect of these changes on the relative concentrations of species in gas-phase and solution equilibrium systems, as represented in videos of the experiments, dynamic graphs, and molecular level animations. In this quasi-experimental design, students significantly increased their understanding of concepts related to equilibrium and reduced the number of common misconceptions.

Animations and simulations can also be used to support laboratory practice. Schoenfeld-Tacher, Jones, and Persichitte (2001) used simulations of a virtual laboratory experience as part of a goal-based scenario (GBS) exercise in biochemistry in a study to determine the relationships of cognitive and demographic variables on learning outcomes. This was a macro-level simulation of actual laboratory activities rather than a nanoscale simulation, as those discussed above. Schank, Fano, Bell, and Jona (1993, p. 304) state, "Goal-Based Scenarios are problems in the domain of a student's interest that present definable goals and encourage learning in service of achieving those goals. A GBS is a type of learn-by-doing task with very specific constraints on the selection of material to be taught, the goals the student will pursue, the environment in which the student will work, the task the student will perform, and the resources that are made available to the student." Students in the Schoenfeld-Tacher, Jones, and Persichitte study (2001) used *Whodunnit?* (Schoenfeld-Tacher, 2000), a virtual crime lab that uses DNA fingerprinting techniques to identify a murderer. A pre- and posttest design was used with an experimental group of 458 students in seven sections of biochemistry at three universities and a community college. Results were compared with a 37 person pseudo-control group at one university. The entire *Whodunnit?* GBS was completed in 1-1.5 hours, although there was no indication of how long the simulated laboratory experience took. The control group performed a standard DNA laboratory activity rather than the GBS. A paired comparison t test was used to demonstrate a significant difference between pre- and posttest scores for the experimental group on a multiple choice test measuring knowledge of material presented in the simulation. There were no pre-post differences for the control group. Four demographic variables were studied (gender, race, final course score, and prior science coursework). Only the number or prior science courses in high school showed a significant correlation with learning outcomes. Logical thinking ability but not spatial ability or disembedding ability was related to learning outcomes. This GBS experience with its simulated laboratory appeared to be equally effective for all types of students.

In an application of visualization technology that Gilbert and Boulter (1998) refer to as "expressive," Schank and Kozma (2002) describe a software package called *ChemSense* that allows students to construct molecular-level animations of chemical systems (see also the Russell & Kozma chapter in this volume for more description and an example screen shot). While animations were not among the representations that we saw in our observation of laboratory chemists (Kozma, et al., 2000), the assumption behind *ChemSense* is that by enabling students to represent the dynamic, molecular nature of the reactions that they are investigating, the software will support their investigative practices and their chemical discourse about the physical phenomena they observe in the laboratory. In a study of high school students (Michalchik, Rosenquist, Kozma, Schank, & Coppola, 2004), we observed that students used *ChemSense* in the wet lab to move their discourse from a focus on the physical characteristics of their experiments to an analysis of the molecular entities and processes that accounted for physical characteristics and to think more deeply about chemical concepts related to chemical geometry and connectivity. Students came to use the *ChemSense* representations that they constructed as rhetorical artifacts in their discussions with each other and with the teacher. In addition, we found that over the two week-long unit, students significantly improved their scores on measures of both chemical understanding and representational competence (see a discussion of this measure in the Russell & Kozma chapter).

CONCLUSIONS AND RECOMMENDATIONS

For Instruction

Molecular models, simulations, and animations can support both the traditional concept building goals of the chemistry curriculum, as well as new investigative practice goals; although far more research has been done on the former than the latter. Studies have shown that students can use models to understand concepts related to bonding and structure. They can also support students' laboratory practices, shifting their discussions from the physical aspects of the experiments that they conduct to the chemical entities that underlie these physical phenomena. We saw that animations can help students understand difficult concepts related to equilibrium, electrochemistry, and solution chemistry. They can also be used by students to support their laboratory investigations by helping students discuss their experiments in terms of molecular entities and dynamic processes.

We recommend the widespread use of these visualization resources in chemical education. We particularly recommend the expressive use of visualizations to help students acquire the representational competencies and laboratory practices of chemists.

For Research

Much more research is needed on the impact of visualizations on student learning. Classroom research of the sort we examined is important in order to establish the external validity of the findings on visualizations—that they work in real classroom situations. But there is also a need for carefully controlled experiments in the cognitive laboratory that supplement these classroom studies. For example, some of Mayer's design principles were not tested in several studies because the narration was not included as part of the software presentation and verbal information came from the lecturers. Carefully controlled experiments are needed to confirm the effectiveness of the Mayer principles in chemistry.

Additional classroom studies could examine the effectiveness of these visualizations for different topics or different student groups. For example, more classroom research needs to be done on the use of students' manipulation of simulations of chemical systems. Additional work needs to be done on animations. Are they best for concepts that involve dynamic processes, such as acid-base reactions and oxidation? Are they at all useful for less-dynamic concepts such as molecular structure or is this concept best taught with molecular models? Are animations or simulations useful for analyzing aggregation effects in systems of large numbers?

There is also a significant need for research that focuses on the process of laboratory investigation and how it is that the use of various symbol systems and multimedia environments facilitates investigative practices. There is a need to extend research beyond the lecture hall or computer lab into the chemistry laboratory. New assessments must be designed and used that measure investigation practices and related skills, such as visualization skills or representational competence as discussed in Russell & Koza (this volume). Finally, in order to integrate traditional and new curricular goals in chemistry, research studies must examine the relationship between learning investigative practices and the development of chemical understanding and how visualization technologies can facilitate this relationship.

ACKNOWLEDGEMENTS

Robert Kozma's participation in this chapter was made possible by the National Science Foundation under Grant No. 0125726. Any opinions, findings, and conclusions or recommendations expressed in this material are those of the authors and do not necessarily reflect the views of the National Science Foundation.

REFERENCES

American Association for the Advancement of Science, 1993, *Benchmarks for science literacy*, Oxford University Press, New York.

American Association for the Advancement of Science, 2001, *Atlas of science literacy*, American Association for the Advancement of Science, Washington, D.C.

Amman, K., and Knorr Cetina, K., 1990, The fixation of (visual) evidence, in: *Representation in scientific practice*, M. Lynch, and S. Woolgar, eds., MIT Press, Cambridge, MA, pp. 85-122.

Allendoerfer, R., 2002, KinSimXP, a chemical kinetics simulation, *Journal of Chemical Education*, 79(5): 638-639.

Anderson,W., 1984, *Between the library and the laboratory: The language of chemistry in eigthteenth-century France*, Johns Hopkins University Press, Baltimore, MD.

Baddeley, A., 1999, *Human Memory*, Allyn & Bacon, Needham, MA.

Chi, M., Feltovich, P., and Glaser, R., 1981, Categorization and representation of physics problems by experts and novices, *Cognitive Science*, 5: 121-152.

Cody, J., and Wiser, D., 2003, Laboratory sequence in computational methods for introductory chemistry, *Journal of Chemical Education*, 80(7): 793-795.

Dabrowiak, J., Hatala, P., and McPike, J., 2000, A molecular modeling program for teaching structural biochemistry, *Journal of Chemical Education*, 77(3): 397-400.

diSessa, A., Hammer, D., Sherin, B., and Kolpakowski, T., 1991, Inventing graphing: Meta-representational expertise in children, *Journal of Mathematical Behavior*, 10(2): 117-160.

Dori, Y., and Barak, M., 2001, Virtual and physical molecular modeling: Fostering model perception and spatial understanding, *Educational Technology & Society*, 4(1): 61-74.

Dori, Y., Barak, M., and Adir, N., 2003, A web-based chemistry course as a means to foster freshmen learning, *Journal of Chemical Education*, 80(9): 1089-1092.

Dunbar, K., 1997, How scientists really reason: Scientific reasoning in real-world laboratories, in: *The nature of insight*, R. Sternberg and J. Davidson, eds., MIT Press, Cambridge, MA, pp. 365-396.

Flemming, S., Hart, G., and Savage, P., 2000, Molecular orbital animations for organic chemistry, *Journal of Chemical Education,* 77(6): 790-793.

Francoeur, E., 1997, The forgotten tool: The use and development of molecular models, *Social Studies of Science, 27*: 7-40.

Francoeur, E., 2002, Cyrus Levinthal, the Kluge, and the origins of interactive molecular graphics, *Endeavour, 26*(4): 127-131.

Gable, D., 1998, The complexity of chemistry and implications for teaching, in: *International handbook of science education,* B. Fraser, and K. Tobin, eds., Kluwer Academic Publishers, Dordrecht/Boston/London, pp. 233-248.

Glaser, R., and Chi, M., 1988, Overview, in: *The nature of expertise,* M. Chi, R. Glaser, and M. Farr, eds., Lawrence Erlbaum Associates, Hillsdale, NJ, pp.xv-xxviii.

Gilbert, J., and Boulter, C., 1998, Learning science through models and modeling, in: *International Handbook of Science Education,* B. Fraser, and K. Tobin, eds., Kluwer Academic Publishers, Dordrecht/Boston/London, pp. 53-66.

Goodwin, C., 1995, Seeing in depth, *Social Studies of Science, 25*: 237-274.

Greeno, J., 1998, The situativity of knowing, learning, and research, *American Psychologist, 53*(1): 5-26.

Hessley, R., 2000, Computational investigations for undergraduate organic chemistry: predicting the mechanism of the Ritter reaction, *Journal of Chemical Education,* 77(2): 202.

Hoffman, R., and Laszlo, P., 1991, Representation in Chemistry, *Angewandte Chemie,* 30(1): 1-16.

Holmes, J., and Gettys, N., 2003, General chemistry collection, 7th Edition abstract of special issue 16, 7th Edition, a CD-ROM for students, *Journal of Chemical Education,* 80(6): 709-710.

Jones, M., 2001, Molecular modeling in the undergraduate chemistry curriculum, *Journal of Chemical Education,* 78(7): 867-868.

Johnstone, A., 1997, Chemistry teaching – science or alchemy?, *Journal of Chemical Education,* 74(3): 262-268.

Kantardjieff, K., Hardinger, S., and Van Willis, W., 1999, Introducing computers early in the undergraduate chemistry curriculum, *Journal of Chemical Education,* 76(5): 694-697.

Kozma, R., 2000a, *Representation and language: The case for representational competence in the chemistry curriculum.* Paper presented at the Biennial Conference on Chemical Education, Ann Arbor, MI.

Kozma, R., 2000b, Students collaborating with computer models and physical experiments, in: *Proceedings of the Conference on Computer-Supported Collaborative Learning 1999,* J. Roschelle, and C. Hoadley, eds., Erlbaum, Mahwah, NJ.

Kozma, R., 2000c, The use of multiple representations and the social construction of understanding in chemistry, in: *Innovations in science and mathematics education: Advanced designs for technologies of learning,* M. Jacobson, and R. Kozma, eds., Erlbaum, Mahwah, NJ, pp. 11-45.

Kozma, R., 2003, Material and social affordances of multiple representations for science understanding, *Learning and Instruction, 13*(2): 205-226.

Kozma, R, Chin, E., Russell, J., and Marx, N., 2000, The role of representations and tools in the chemistry laboratory and their implications for chemistry learning, *Journal of the Learning Sciences,* 9(2): 105-143.

Kozma, R., and Russell, J., 1997, Multimedia and Understanding: Expert and Novice Responses to Different Representations of Chemical Phenomena, *Journal of Research in Science Teaching,* 43(9): 949-968.

Krajcik, J. S., 1991, Developing students' understanding of chemical concepts, in: *The psychology of learning science,* S. Glynn, R. Yeany, and B. Britton, eds., Erlbaun, Hillsdale, NJ, pp.117-147.

Krajcik, J., Blumenfeld, P., Marx, R., Bass, K., Fredricks, J., and Soloway, E., 1998, Inquiry in project-based science classrooms: Initial attempts by middle school students, *Journal of the Learning Sciences, 7*(3&4): 313-351.

Larkin, J., 1983, The role of problem representation in physics, in *Mental models*, D. Gentner, and A. Stevens, eds., Erlbaum, Hillsdale, NJ, pp. 75-98.

Larkin, J., McDermott, J., Simon, D., and Simon, H., 1980, Expert and novice performance in solving physics problems, *Science*, 208: 1335-1342.

Lave, J. and Wenger, E., 1991, *Situated learning*, Cambridge University Press, New York.

Lemke, J., 1990, *Talking science: Language, learning, and values*, Ablex, Norwood, NJ.

Martin, N.,1998, Integration of computational chemistry into the chemistry curriculum, *Journal of Chemical Education*, 75(2), 241-243.

Matta, C., and Gillespie, R.,2002, Understanding and interpreting molecular electron density distributions, *Journal of Chemical Education*, 79(9): 1141-1152.

Mayer, R., 2001, *Multimedia Learning*, Cambridge University Press, Cambridge, UK.

Mayer, R., 2002, Cognitive theory and the design of multimedia instruction: An example of the two-way street between cognition and instruction, *New Directions for Teaching and Learning*, 89: 55-71.

Mayer, R., 2003, The promise of multimedia learning: Using the same instructional design methods across different media, *Learning and Instruction, 13*(2): 125-139.

Michalchik, V., Rosenquist, A., Kozma, R., Schank, P., and Coppola, B., 2004, *Representational resources for constructing shared understandings in the high school chemistry classroom* [technical report], SRI International, Menlo Park, CA.

Montgomery, C., 2001, Integrating molecular modeling into the inorganic chemistry laboratory, *Journal of Chemical Education*, 78(6): 840-844.

Nakhleh, M. B., 1992, Why some students don't learn chemistry: Chemical misconceptions, *Journal of Chemical Education, 69*: 191-196.

National Research Council, 1996a, *The National Science Education Standards*, National Academy Press, Washington, D.C.

National Research Council, 1996b, *From analysis to action: Undergraduate education in science, mathematics, engineering, and technology,* National Academy Press, Washington, D.C.

National Science Foundation, 1996, *Shaping the future: New expectations for undergraduate education in science, mathematics, engineering, and technology*, National Science Foundation, Washington, DC.

O'Neill, D., and Polman, J., 2004, Why educate "little scientists?": Examining the potential of practice-based scientific literacy, *Journal of Research in Science Teaching, 41*(3): 234-266.

Paiva, J, Gil, V., and Correia, C., 2002, LeChat: Simulation in chemical equilibrium, *Journal of Chemical Education*, 79(5): 640.

Paivio, A., 1986, *Mental Representations: A Dual Coding Approach*, Oxford University Press, New York.

Papadopoulos, N. and Limniou, 2003, pH Titration Simulator, *Journal of Chemical Education*, 80(9): 709-710.

Paselk, R., and Zoellner, R., 2002, Molecular modeling and computational chemistry at Humboldt State University, *Journal of Chemical Education*, 79(10): 1192-1194.

Piaget, J., 1972, *The psychology of the child*, Basic Books, New York.

Robinson, W., 2000, A view of the science education research literature: Scientific discovery learning with computer simulations, *Journal of Chemical Education*, 77(1): 17-18.

Robinson, W., 2004, Cognitive theory and the design of multimedia instruction, *Journal of Chemical Education*, 81(1): 10-13.

Roth, W-M., and Bowen, G., 1999, Complexities of graphical representations during lectures: A phenomenological approach, *Learning and Instruction*, 9: 235-255.

Roth, W-M., and McGinn, M., 1998, Inscriptions: a social practice approach to representations, *Review of Educational Research, 68*: 35-59.

Russell, J., and Kozma, R., 1994, 4M:Chem – multimedia and mental models in chemistry, *Journal of Chemical Education*, 71(8) 669-670.

Sanger, M., and Greenbowe, T., 2000, Addressing student misconceptions concerning electron flow in aqueous solutions with instruction including computer animations and conceptual change strategies, *International Journal of Science Education, 22*(5): 521-537.

Sanger, M., Phelps, A., and Fienhold, J., 2000, Using a computer animation to improve students' conceptual understanding of a can-crushing demonstration, *Journal of Chemical Education, 77*(11): 517-1520.

Schoenfeld-Tacher, R., 2000, *Relation of Student Characteristics to Learning of Basic Biochemistry Concepts from a Multimedia Goal-Based Scenario*, PhD dissertation, University of Northern Colorado.

Schoenfeld-Tacher, R., Jones. L., and Persichitte, K., 2001, Differential effects of a multimedia goal-based scenario to teach introductory biochemistry – who benefits most?, *Journal of Science Education and Technology*, 10(4): 305-317.

Schank, P. and Kozma, R., 2002, Learning chemistry through the use of a representation-based knowledge-building environment, *Journal of Computers in Mathematics and Science Teaching, 21*(3): 253-279.

Schank, R., Fano, A., Bell, B., and Jona, M., 1993, The design of goal-based scenarios, *Journal of the Learning Sciences*, 3, 305-345.

Shusterman, G., and Shusterman, A., 1997, Teaching chemistry with electron density models, *Journal of Chemical Education*, 74(7): 771-776.

Williamson, V., 1992, *The effects of computer animation emphasizing the particulate nature of matter on the understandings and misconceptions of college general chemistry student,* Unpublished doctoral dissertation, University of Oklahoma.

Williamson, V., and Abraham, M., 1995, The effects of computer animation on the particulate mental models of college chemistry students, *Journal of Research in Science Teaching, 32*(5): 521-534.

Woolgar, S., 1990, Time and documents in researcher interaction: Some ways of making out what is happening in experimental science, in: *Representation in scientific practice*, M. Lynch, and S. Woolgar, eds., MIT Press, Cambridge, MA, pp. 123-152.

Yang, E., Greenbowe, T., and Andre, T., 2004, The effective use of interactive software program to reduce students' misconceptions about batteries, *Journal of Chemical Education*, 81(4): 587-595.

Vogel, G. (1996). Science education: Global review faults U.S. curricula, *Science, 274*, 335.

Vygotsky, L., 1986, *Thought and language*, MIT Press, Boston.

Vygotsky, L., and Vygotsky, S., 1980. *Mind in society: The development of higher psychological processes*, Harvard University Press, Cambridge, MA.

Zare, R., 2002, Visualizing Chemistry, *Journal of Chemical Education*, 79(11): 1290-1291.

GALIT BOTZER AND MIRIAM REINER

CHAPTER 8
IMAGERY IN PHYSICS LEARNING - FROM PHYSICISTS' PRACTICE TO NAIVE STUDENTS' UNDERSTANDING

Department of Education in Science and Technology Technion, Haifa, Israel

Abstract: The main issue in this chapter is imagery in physics learning. Three epistemological resources are used to address this issue: Imagery in the history of physics, cognitive science aspects of imagery, and educational research on physics learning with pictorial representations. A double analysis is used. The first analysis is focused on imagery in classical test cases in the history of physics, such as Faraday's work on magnetism and Einstein's thought experiments described in the 1905 papers. The categories identified in the first analysis were used for the second: analysis of imagery in naive students' reasoning. In particular we describe a learning experiment, which examined naive students' representations of magnetic phenomena, during hands-on activities in the physics laboratory. We show that naive students use imagery in making sense of the physical phenomena; that modes of naive students' imagery resemble, on several levels cognitive mechanisms identified in physicists' imagery strategies; and that the product of imagery, pictorial representations, mirror processes of changes in conceptual understanding. We conclude with suggestions and implications for physics learning.

INTRODUCTION

Physics practice often involves cognitive processes such as mental simulations (Clement and Monagham 1999; Clement 1994) mental animations (Hegarty 1992) and thought experiments (Reiner 2000; Gilbert and Reiner 2000; Reiner and Gilbert 2000; Reiner 1998). All of these, require a form of 'seeing with the mind's eye', visualizing an event, mentally exploring a diagram, or comparing pictorial mental representations, i.e. thinking in pictures. Thinking by generating or manipulating pictures is termed here mental imagery or visualization. Mental imagery is used to make sense of physical experience and interact with the physical environment (Johnson 1987). For example, a child can visualize the trajectory of a moving ball and reach the hands to catch it without using any symbolic formalism. **How imagery relates to physics practice and to physics learning** is the focus of this chapter. We draw on three epistemological resources: Imagery in the history of physics, cognitive science aspects of imagery, and educational research on physics learning with pictorial representations.

The history of physics provides many examples in which physicists' used imagery to achieve scientific breakthroughs: Einstein claimed to achieve his insight into the nature of space and time by means of thought (Gendaken) experiments on mentally visualized systems of light wave and idealized physical bodies (clocks, rulers), in state of relative motion (Holton and Brush 2001; Miller 2000; Shepard 1988; Miller 1987). Another example is Michael Faraday's analysis of electromagnetic fields in terms of field lines (Holton and Brush 2001; Nerssesian 1995 Shepard 1988). The field lines are not only scaffolds, used to construct mathematical formalism, but also an integral part of electromagnetism, which serve as communication tool within the scientific community.

Research in cognitive science suggests that imagery and visual perception are, in many respects, functionally equivalent processes (Richardson 1999; Finke and Shepard 1986). The term 'mental imagery' refers to the ability to generate mental images and to manipulate these images in the mind (Kosslyn 1994). Empirical research shows that mental images could be scanned (Kosslyn 1994) or rotated in a measurable speed (Shepard 1996; Shepard 1988). Imagery is claimed to facilitate performance on a variety of visual tasks (Marks 1990; Finke and Shepard 1986) and in memory tasks (Clark and Paivio 1991; Paivio 1971). Imagery may also participate in cognitive problem solving (Richardson 1999; Kaufmann 1990).

Educational Researchers in physics learning examine the role of pictorial representations from three main perspectives: the first perspective focuses on learning environments, which use static or dynamic pictorial representations (e.g. Clement and Monagham 2000; Mayer et al 1996; Hegarty 1992; Reiner 1998). These studies indicate that pictorial representations and dynamic simulation are effective for conceptualization and problem solving in physics. The second perspective focuses on classification of pictorial and verbal representations, constructed by students for a variety of physical phenomena (Borges and Gilbert 1998; Driver et al 1994). These representations might reflect mental models held and used by students and hence might serve as evaluation tools. The third perspective focuses on cognitive mechanisms, which underlie construction of pictorial representations of physical phenomena. Reiner (1997) has shown that students communicated with each other through pictorial representations in order to construct meaning to electromagnetic phenomena. Clement (1994) analyzed the use of physical intuition and imagistic simulation in expert problem solving and claimed that these processes played an essential role in expert's thought.

The three epistemological resources, mentioned above, are closely related to each other. Students' ideas are, to some extent, parallel to the historical development of those concepts in science (Nerssesian 1995; Gilbert and Zylbersztajn 1985). Hence, we claim that physicists' practices are relevant to understanding processes of naive students' learning. This claim is also supported by current views, which perceive science learning as developing familiarity with the practices of knowledge construction within the scientific community (Lave and Wenger 1991). Cognitive science provides a framework for interpretation of scientists' imagery thought through the history of physics (Miller 2000; Shepard 1996), as well as interpretation of naive students' representations (Reiner 1997; Borges and Gilbert 1998).

Outline of this chapter

The chapter evolves in four parts. The first part provides a *theoretical framework* for analyzing imagery processes, based on the history of physics, cognitive science and educational research. The second part describes *a learning experiment,* designed to explore students' pictorial representations of magnetic phenomena. Results and *analysis of representations of magnetic phenomena* is the focus of the third. We conclude by discussing *the role of pictorial representations in physics learning* and show that naive students build representations, which are to some extent compatible with physicists' representations.

THEORETICAL OVERVIEW

Physicists' practice and imagery

Mental imagery and thought experiments are recognized as central epistemological mechanisms in innovation in physics (Miller 2000; Reiner and Gilbert 2000; Shepard 1988; Holton 1978). Galileo Galilee's theory is based on the extremely counterintuitive assumption that all bodies fall in a vacuum with the same acceleration, regardless of their weight. Galileo reached this conclusion through thought experiments. The term 'thought experiment' refers to scientists' performance of experiments in their "mind's eye". These experiments are run on idealized apparatuses and so require a high degree of abstraction (Miller 2000). Another example is Einstein's railway thought experiment, through which he established the idea of simultaneity in different frame of reference (Einstein 1922). Reiner (2000) identifies a typical structure of thought experiments and that consists of five components: an imaginary world, a problem, an experiment, experimental 'results' and conclusion. The abstraction level of Einstein's thought experiments is much higher then those of Galileo (Miller 2000) but both of them were catalysts to major breakthrough in the history of physics. Reiner (2000) claims that although these experiments are different in their goal, content, context and conceptual framework, they share the same structure. Other physicists such as Newton, Helmholtz, Bohr, Heisenberg and Feynman used the same structure of visual thought (Miller 2000; Shepard 1988; Miller 1987; Holton 1978). In particular many discoveries in electromagnetism are based on visual thinking.

Imagery in electromagnetism

Imagery has an essential role in the development of the electromagnetic theory. For example in the book 'De Magnete' (1600), the scientist William Gilbert concluded that the earth is a huge magnet, by using a visual analogy between a magnetic needle's incline near a spherical magnet and a compass needle's incline (Agasi 1968). Another example is the microscopic model of magnetic substance, suggested by the French physicist Ampere. Ampere (1820) suggested a microscopic model that contains small closed circuits inside a magnetized substance, by using an analogy to macroscopic current in a circular wire (Agasi 1968). A major contribution to the electromagnetic theory was made Faraday and Maxwell.

Michael Faraday's capability of visual imagination lead him to the invention of the term *field lines* and to the discovery of the electromagnetic induction (Holton and Brush 2001; MacDonald 1965). In 1821 Faraday repeated Oersted's experiment and placed a compass around a current-carrying wire. Faraday realized that the force exerted by the current on the magnet was circular in nature. He represented this phenomenon by a set of concentric circular line of force, so that a magnetic pole that is free to move experiences a push in a circular path around a fixed conducting wire. The collection of these lines of force is called the magnetic field (Holton and Brush 2001). Armed with this line-of-force picture for understanding electric and magnetic fields, Faraday joined in the search for a way of producing currents by magnetism. The idea of the line of force suggested to him the possibility that a current in one wire ought to be able to induce a current in a nearby wire, through an action of the magnetic lines of force in the space around the first current. Faraday examined this possibility trough many experiments in which he refined his experimental system (figure 1) and finally came to the conclusion that changing lines of magnetic force cause a current in a wire.

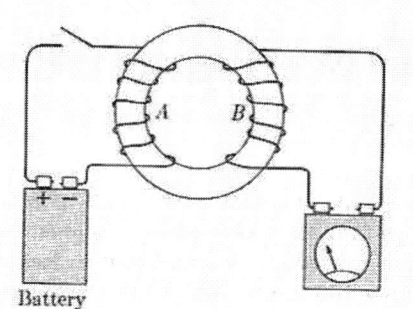

Figure 1: *Faraday electromagnetic induction experiment*

James Clark Maxwell (1860) work on electromagnetic waves was influenced by Faraday's work. In his book "*A Treatise on Electricity and Magnetism*" he describes Faraday's work:

> '...Faraday in his mind's eye saw lines of force traversing all space where the mathematicians centers of force attracting at a distance, faraday saw a medium where they saw nothing but distance: Faraday sought the seat of the phenomena in real actions going in the medium, they were satisfied that they had found it in a power of action at a distance impressed on the electric fluids.' (Maxwell, 1954 p. ix)

Nerssesian (1995) shows that Maxwell used Faraday's visual models and refined them through successive thought experiments. Maxwell developed his electromagnetic equations not by a chain of logical steps but by a series of increasingly abstract hydrodynamic and mechanical models (Shepard 1988). Maxwell considered what happens when an electric current oscillates along a straight piece of wire or circulates in a wire loop. To visualize the interactions between electric currents and magnetic fields, he constructed a mechanical model in which electromagnetic fields where represented by vortices bearing-ball and fluid (figure 2). In this model magnetic fields were represented by rotating vortices in a fluid and charges were represented by tiny spheres-like ball bearing, whose function is to transmit the rotation from one vortex to its' neighbours (Holton and Brush 2001; Miller 2000; Nerssesian 1995).

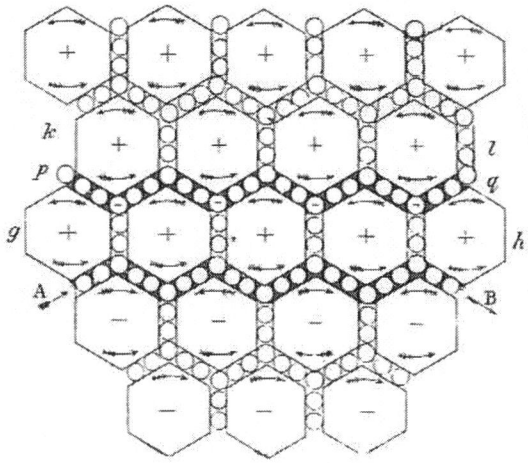

Figure 2: *Maxwell's vortices-ball bearing model of electromagnetic field*

This model allowed Maxwell to generalize the ideas of Oersted, Ampere, and Faraday so that they applied to electromagnetic interactions in a region of space where no current-carrying wire is present. He postulated that such regions contain charges (the ball bearing) that can be moved, or displaced, by changes in the magnetic fields (the vortices). The vortex-ball bearing model suggested that just as a varying magnetic field could generate an electric current in a wire (Faraday's electromagnetic induction), it could also produce a motion of a charge in space. This displacement current then produces a magnetic field (Oersted's effect). That field can then displace other charges, producing more displacement current (Holton and Brush 2001). Maxwell's theoretical conclusion, based on this mode, was that an electric current in a wire must send energy out trough space in a form of magnetic and electric fields. This energy is radiated away from the electric current and spreads out, wavelike in all directions. Maxwell formulated this conclusion in a set of four equations and then abandoned the visual model as if it was a scaffold for a building (Shepard 1988).

The effectiveness of mental imagery

The effectiveness of nonverbal processes of mental imagery is discussed in the context of creative thought in science (Miller 2000; Shepard 1988; Holton 1978) and in the context of problem solving in general (Richardson 1999; Shepard 1996; Clark and Paivio 1991; Kaufmann 1990). Shepard (1988) suggests that the effectiveness of mental imagery relates to four features of these processes: their private nature, their richly concrete and isomorphic structure, their engagement of innate mechanism of spatial intuition and their direct emotional impact. The private nature of imagery process and their departure from traditional verbal thinking explain their contribution to construction of novel ideas. The richness of concrete imagery, together with its isomorphic relation to the external objects, and events that it represents, may permit noticing of significant details that are not adequately preserved in a purely verbal formulation (Miller 2000). The spatial character of visual images makes them accessible to the use of spatial intuition and manipulation that have developed trough sensory interaction with the physical environment (Shepard 1996; Johnson 1987). Finally, vivid mental images provide psychologically more effective substitutes then do verbal encoding for the corresponding external objects and events. Is mental imagery effective for physics learning as well? We claim that it might be. Imagery is claimed to facilitate performance on a variety of visual tasks (Marks 1990; Finke and Shepard 1986) and in memory tasks (Clark and Paivio 1991; Paivio 1971). Imagery may also be part of cognitive problem solving, at list in the early stages of abstraction (Richardson 1999; Kaufmann 1990). Educational researchers suggest that pictorial representations and dynamic simulation are effective in conceptualization and problem solving in physics (e.g. Clement and Monagham 2000; Reiner 1997).

Mental-imagery modes

Miller (Miller 2000;1987) suggests a pattern that relates major breakthrough in physics in the 20th century to transformation of modes of mental imagery. Miller draws a timeline on which he places major discoveries in physics (figure 3). The horizontal axis indicates increasing time and increasing abstraction of intuition.

Galileo, Newton and Einstein's mental images were constructed trough abstraction of sensory experiences. The discovery of the electron in the early years of the 20th century lead to a new mode of imagery which was derived from pure imaginary entities. No one has seen an atom or an electron. Yet Bohr manipulated these entities to ensure the stability of the atom, which was modelled as a mini-scale solar system. He suggested a counterintuitive idea of "allowed orbits" and postulated a lowest allowed orbit below which atomic electrons cannot drop (Miller 2000). Bohr's model could not provide satisfactory explanation for discoveries and disagreements in the 20th and hence visualization was abandoned to be replaced by nonvisualizable mathematical formalism. In 1925 visualization was regain by

Heisenberg who derived the uncertainty principle from quantum equations that describe an electron's motion, by imaging himself measuring quantitative variables (Miller 1987). Following Miller's description of development of imagery modes in the history of physics, we use three modes of mental imagery to analyze visual representations in physics.

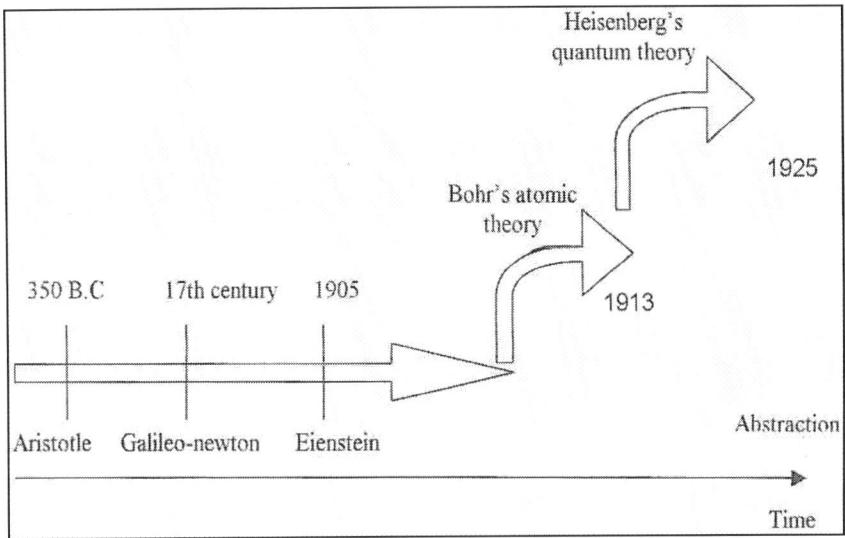

Figure 3: *Development of imagery modes throughout the history of physics*
(Miller A. I. (2000) Insight of Genius Imagery and Creativity in Science and Art, MIT Press)

Sensory-based representation refers to any image, derived from visual sensory experience. Pure imaginary representation– refers to any image which represents a situation which cannot be perceived through the senses. Formalism-based representation- refers to any image, based on mathematical formalism or formal rules.

In order to clarify each of the above imagery modes, we bring physics reasoning examples from the history of physics in table 1.

Table 1: imagery modes and examples from the history of physics

Modes of representations of physical phenomena	Examples of Discoveries in Physics
Sensory based representations	**Galileo** showed that all bodies fall at the same speed using a thought experiment that was based on concrete objects. (Reiner 2000)
Pure imaginary representations	**Boyle** visualized air-particles as tiny springs to explain the compression of the air (Agasi 1968). **Millikan** interpreted the experiment in which he measured the charge of the electron by imagining electrons 'riding' on drops of oil (Holton 1978).
Formalism-based representations	**Dalton** suggested a model, which represented atomic particles as wrapped with a fluid called 'caloric'. Based on Newton's inverse square law, he further assumed that particles should be pushed away from each other. (Agasi 1968)

We suggest that Miller's three modes of imagery might serve as a tool to analyze visual representations in physics learning. For instance, representations of magnetic phenomena can be classified according to the above three modes (see figure 4): a representation derived from the lines of iron-filling round a magnet, is a sensory-based representation; a representation derived from a microscopic model is pure imagery representation (no one really saw a microscopic magnetic dipole); and finally a representation, derived from field lines is a formalism based representation.

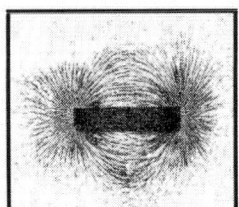 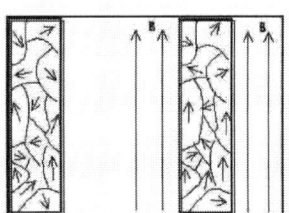 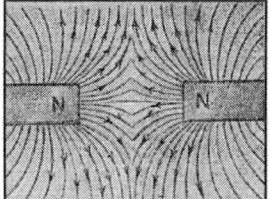

Figure 4: *Representations of magnetic phenomena: Sensory-based, pure-imagery and formalism based*

We used the imagery modes suggested above to analyze naive student's representations in the learning experiment, described as follows.

THE LEARNING EXPERIMENT

Goal

The goal of the learning experiment is to gain an insight into the mental processes, which naive students experience while constructing spontaneously (without teaching intervention) visual representations of magnetic phenomena. The explicit pictorial representations (drawings and gestures) along with the subjects' discourses reflect underlying imagery processes. This study targets the following questions:
What are the:
(1) modes of pictorial representations of magnetic phenomena, constructed by naive students
(2) relations between the pictorial representations and the physics context
(3) developmental patterns of the pictorial representations throughout the learning activity

Procedure

Sixteen ninth grades in Israel, ten girls and six boys, explored magnetic phenomena in the physics laboratory. The students had basic background in mechanic heat and electricity and no background in magnetism. They were placed into eight pairs, according to their achievements in science and math. We refer to the responses of each of these pairs as a case study (eight case studies). The learning experiment took place in three sessions, two hour each, after school hours. Each couple solved, collaboratively, a series of predict-observe-explain (POE) problems, using instructions notes and equipment such as magnets, compasses iron filling and nails. The students were asked to draw and describe verbally magnetic phenomena. They were encouraged to talk freely without worrying about the correctness of their answers.

The learning activities

Subjects were engaged in ten POE activities. In the 'predict' stage subjects were asked to draw an anticipative model (*'What will happened if...'*) In the 'observe' stage they were asked to draw a descriptive model (*'describe what happened when...'*) and in the 'explain' stage they were asked to draw an explanatory model (*Explain according to your understanding the phenomena*). The learning activity included three sessions that differed by the physics context. Physics context is defined as content of the activity and level of abstraction. The content includes the concepts involved in the task and the information provided. The level of abstraction is defined as one of the following three: concrete, microscopic or formal situations. During the first session subjects referred to concrete objects such as magnets, steel nails, compass and iron filling. For example, subjects were asked to respond to the following problem:

> "How can we distinguish between two steel bars that look identical, but one is magnetized and the other is not?"

The second session focused on microscopic situations. For example, subjects were asked to respond to the following problem:

> If we break a magnet, each of the pieces will still have two poles. Suggest a structure that will explain this phenomenon

After suggesting a predictive model, the subjects were introduced to a 'domains model'. A domain is a small naturally magnetized area. Magnetic materials contain large number of such domains, usually arranged randomly. When magnetized, all domains in the material point to one direction, leaving free domains at each end that form the poles of the magnet (Johnston 2001).

During the third session students learnt by manipulating formal representations, mainly constructing field lines. They explored iron-filling patterns to observe the shape of the field lines.

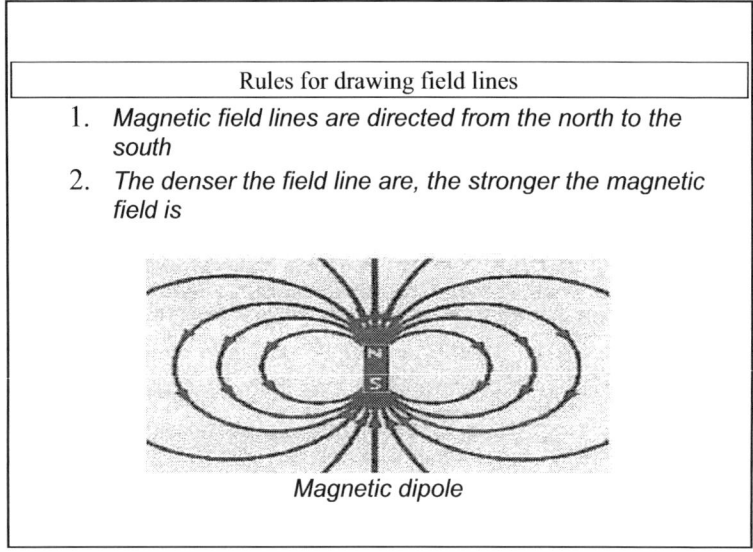

Figure 5: *formal rule for drawing field lines*

Data collection and Analysis

The main research tool was the POE activities, described above. The subjects' physical and verbal interactions were videotaped and field notes were taken. We collected the written responses and the diagrams, including intermediate drafts. In order to validate and complete the collected data, short interviews were taken at the end of the learning activity. The interviews were based on the subjects' responses. Responses were organized according to fine grain content units, episodes ('unit of analysis'). Each episode included a diagram, a written response and a discourse protocol. Analysis was horizontal – across groups, and vertical across sessions. Two independent evaluators validated the analysis.

ANALYSIS OF REPRESENTATIONS OF MAGNETIC PHENOMENA

The results are reported in three parts. The first part describes six modes of pictorial representations. The second part presents a profile of the representations modes in each group of learning experiences: concrete, microscopic and formal. The third part describes developmental patterns throughout the learning activity.

Modes of pictorial representations of magnetic phenomena

Subjects' responses were organized in 112 units of analysis. We analysed representations in two cycles of analysis: the first was based on Miller's modes of imagery, i.e. sensory-based, pure imaginary, and formalism based. The second was a process of refinement of these basic modes into six modes, described in the following section.

Sensory-based representations

We identified three modes of sensory-based representations: photographing sensory experience, projection of a former sensory experience and manipulations of sensory-based image. These modes are describes and exemplified as follows.

(a) Photographing sensory experience: This mode of representations reflects the sensory information as it is as though a photo of the situation was taken. These representations include sensory information, which is relevant to the problem (e.g. a rotation of the compass needle) as well as non-relevant features such as the magnet's colors. Subjects often use *metaphorical* symbols such as straight or circular arrows to describe direction of motion of objects in their pictures (For an overview of metaphorical symbols please see Wise 1988). The following is an example of a photographic representation of two magnets, which are place next to each other, overlaid with straight and rotational arrows.

Both the diagram and words used by the subject, in figure 6 reflect a mere description of the situation of surface features such as the magnets' shape, relative position and colors. The diagram includes non-relevant elements such as the pedestal or the tie.

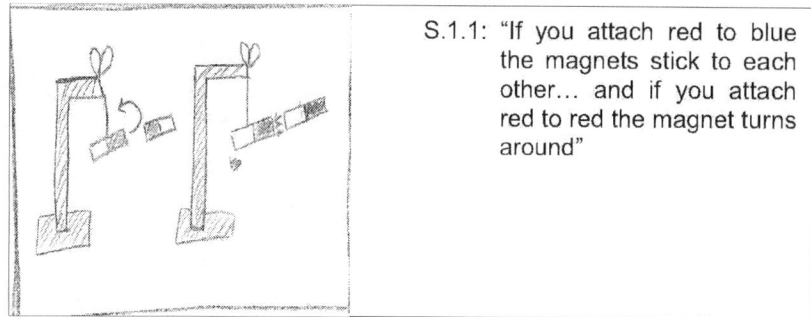

Figure 6: *Photographing sensory experience- interaction between magnets*

(b) Projection of former sensory experiences into a new sensory situation. This mode reflects former sensory experience. These representations include a reflection of sensory information, which was not in front of the subjects. In some cases the subject represented a physical situation, experience in an earlier activity (e.g. imagining a nail attracted to a magnet while predicting the direction of rotation of a compass needle) and in other cases they represented sensory situation, which was perceived outside the learning setup (e.g. imagining a powder sticking to glue, while explaining the iron filling pattern around a magnet). The following example exemplified a projection of glue properties in order to explain magnetic attraction. The subjects interpreted the magnetic force as local phenomena: "here is the attraction force". This miss-interpretation is caused by projection of properties (of glue), which are not relevant to the problem.

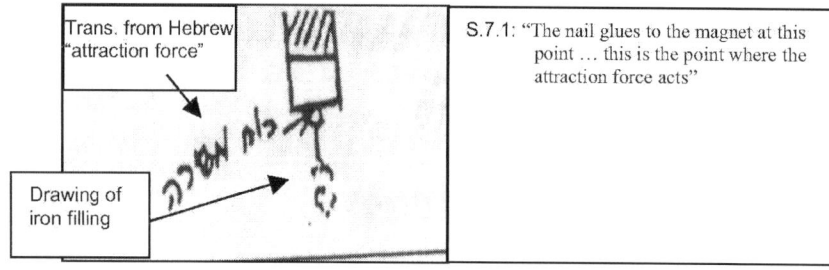

Figure 7: *projection of a former sensory experience (the glue metaphor)*

(c) Transformation of concrete images. This mode of representation reflects visual sensory information, like the former two modes. Yet, in this mode the sensory image in not reflected as is, but transformed, i.e. relocated, rotated or enlarged, so that it fits position of other objects described in the problem. The following example specifies iron-filling pattern around two magnets. The students 'stretched' the pattern of a single magnet to predict the configuration in a new situation.

S.6.1: "It's like a long magnet so it will be the same [as a single magnet] but ah... the 'arch' will be flat"

Figure 8: *Transformation of concrete image-" stretching" of iron-filling pattern*

Pure-imagery based representations

A considerable proportion of the subjects' representations included visual elements that obviously have never seen before as part of the 'real' physical world. The next two paragraphs describe two modes of non-sensory representations: pure products of imagination and physics-formalism-based.

Understanding magnetic phenomena involves construction of mental models of microscopic structures. Visual representations of microscopic structures are not visible, hence are pure-imaginary constructions. We found two sources for such representations: a projection of former sensory (macroscopic) experience into the microscopic world and manipulation of microscopic representations. These modes are describes and exemplified as follows.

(d)   Projection of a former sensory experience into pure –imaginary situation. In the following example (diagram 9) the subject predicted a microscopic structure of magnetic substance, by imposing the two pole macroscopic visible structure of a magnet on the microscopic envisioned structure. The diagram shows un-magnetized object (the right diagram) in which imaginary 'N' and 'S' particles are mixed together and magnetized object in which the particles' arrangement was a projection of the magnet's colors.

Figure 9: *Projection of a former sensory experience into pure-imaginary situation*

(e)   Transformation of microscopic representations.Microscopic representations were previously learnt in mechanics and electricity classes. These representations are transformed, i.e. rotated, stretched or relocated, in order to match the mental image to the drawing presented in the problem. In the following example, subjects were presented with the scientific visual model of magnetic substance and immediately asked to design a compass. The diagram and the discourse reflect a mental rotation of the arrows, which represent pure imaginary entities: microscopic-magnetized domains.

S.5.1. *"If we want the nail's tip to become the north pole we have to arrange all the arrows in the direction of the tip"*

Figure 10: *Transformation of microscopic representation*

Formalism-based representations

We found that the subjects constructed representations, based on verbal rules such as "field lines are directed from the north to the south" or "North pole attracts south pole". This was especially used as a rule for drawing field lines. Thus this category deals integrating formalism in representations.

f) Derivation of representations from formal rules. This mode of representation includes formal symbols such as: plus and minus, 'N' and 'S'. The subjects' justified the representations they constructed by relating to formal roles. (e.g. roles for drawing field lines). In the following example the subjects identified the magnetic poles of a steal nail by using the role "North pole attracts south Pole"

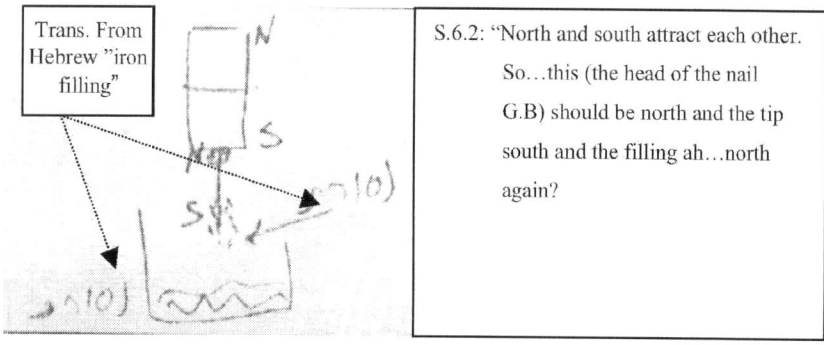

S.6.2: "North and south attract each other. So…this (the head of the nail G.B) should be north and the tip south and the filling ah…north again'?

Figure 11: *Derivation of representations from formal rules*

Representation of magnetic phenomena of the integration of several modes

We found that a pictorial and verbal representation may be associated with more than one mode. Many of the representations reflected sensory information, overlapped by pure imagery entities or formal symbols. For example in figure 10 we can see the shape of the nail, in addition to representation of microscopic entities. In the following example (figure 12) the diagram includes field lines that may be considered as a formal representation, while the discussion reflects two competing mental models of magnetic interaction: concrete (direction of iron filling lines) and formal reasoning based on physics conventions concerning the direction of field lines.

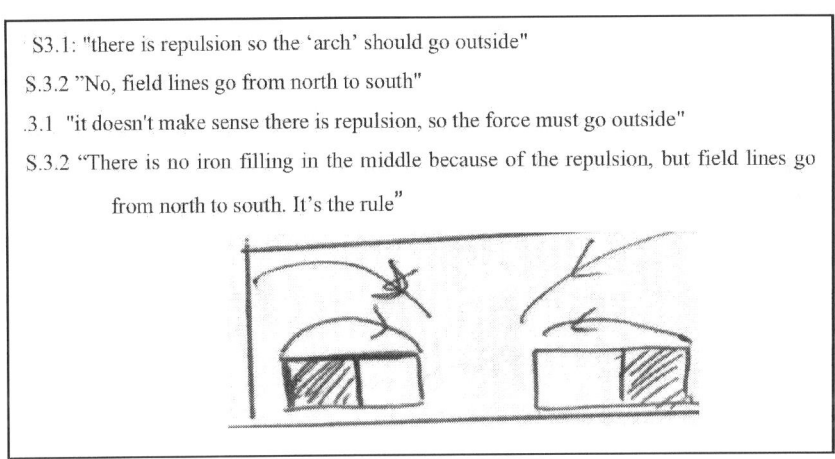

Figure 12: *integration of two modes of representations: sensory and formalism based*

Pictorial representations and context of activity: concrete, microscopic and formal

As mentioned before, the learning sessions differed by progressing level of abstraction – the first was concrete; the second was microscopic, and the third was formal. This section presents profiles of the representations that are typical to each levels of abstraction. We examined the percentage of episodes that included each mode of pictorial representation. Since episodes, often, included more then a single mode of representation, the sum exceeds 100%. Out of the 38 episodes identified during the learning in a concrete context, 70% included sensory-based pictorial representations. Results are described in table 2. The distribution of modes of representation, during the first learning session, while interacting with concrete situations, is described in table 2.

Table 2: *Frequency of modes of pictorial representation according to episodes Session 1:concrete phenomena*

Modes of representations		Frequency [percentage of episodes that included each mode]
Sensory based representations	Photographing sensory experience	70
	Projection of a former sensory experience into a new sensory situation	33
	Transformation of concrete images	28
Pure imagery representations	Projection of a former sensory experience into pure–imaginary situation	10
	Transformation of microscopic representations	5
Formalism based representations	Derivation of representations from formal rules	25

The frequency of the sensory-based representations is the highest. It may be related to the concrete nature of the activity. Most episodes included photographic representations of the situation (70%). This might imply that the subjects did not yet develop a deep structure conceptual model of the physical situation. Some of the concrete representations were overlaid by microscopic and formal symbols such as arrows, + (plus), - (minus), N for north pole, S for south pole. About a quarter of the episodes included applications of the rule that "north pole attracts south pole". The frequency of the microscopic representations is relatively low, since this session was related to macroscopic phenomena. Yet some of the subjects represented, spontaneously, microscopic processes. For example, three groups interpreted magnetization of a nail as "transfer of electrons form the magnet to the nail". This interpretation might be related to former leaning in electricity.

Table 3 specifies the frequencies of each mode in the second learning session, which relates to microscopic magnetic phenomena. The frequency of the pure imaginary mode is the highest. Yet, about three quarters of the episodes reflected projection of a former sensory experience. This implies that subjects' integrated photographic representations with pure imaginary mental models of magnetic substance. 47% of the episodes included transformed microscopic representations. This implies that once a microscopic representation is constructed, it may be transformed (relocated, rotated or enlarged), to fit the situation in the learning activity. Although this session was related to microscopic phenomena a major part of the episodes (about 40%) included sensory-based representations. The concrete equipments (magnets, nails) serve as boundaries in which the microscopic processes were represented.

Table 3: *Frequency of modes of pictorial representation according to episodes Session 2: microscopic phenomena*

Modes of representations		Frequency [percentage of episodes that included each mode]
Sensory based representations	Photographing sensory experience	31
	Projection of a former sensory experience into a new sensory situation	4
	Transformation of concrete images	13
Pure imagery representations	Projection of a former sensory experience into pure–imaginary situation	75
	Transformation of microscopic representations	47
Formalism based representations	Derivation of representations from formal rules	6

The following table (table 4) specifies the frequencies of each mode in the third learning session, which relates to construction of magnetic field-lines.

Table 4: *The frequency of each representation in percents Session 3: construction of field lines*

Modes of representations		Frequency [percentage of episodes that included each mode]
Sensory based representations	Photographing sensory experience	22
	Projection of a former sensory experience into a new sensory situation	28
	Transformation of concrete images	28
Pure imagery representations	Projection of a former sensory experience into pure–imaginary situation	11
	Transformation of microscopic representations	24
Formalism based representations	Derivation of representations from formal rules	77

The frequency of the formalism-based representations was the highest (77%), so was the sensory-based distribution. Subjects overlaid the formal symbols and rules on top of the concrete pattern. The frequency of pure imaginary, microscopic, representations is relatively high (35%) although the task could be performed without microscopic representations.

To summarize, these results show that modes of representation are interrelated with the physics context. Modes of representation profoundly change with the context. The frequency of the sensory based mode is the highest in the concrete session (70%), frequency of pure imaginary representations is the highest in the microscopic session, and the frequency of the formalism-based mode is the highest in the formal session. This implies that learning environment has an impact on types of representations used for learning.

Changes in modes of representations across sessions

The graph in figure 13 is a summary of the frequencies of the modes of representations across sessions. Fig. 13 shows representations vary with context and progress in the learning process. The more one learns, the more formal representations are generated. The usage of photograph-like representations decreases with time while the usage of microscopic and formal representations rises. The variety of the modes of representations increases with time. While in the first session most of the representations are sensory-based, in the third session the representations are a combination of sensory-based, formal symbols and imaginary mental models.

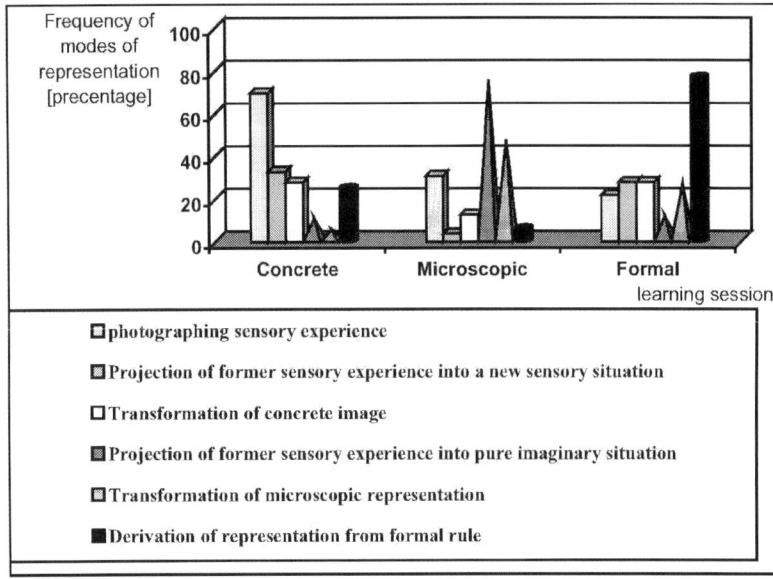

Figure 13: *Frequencies of the modes of representations in three learning sessions*

THE ROLE OF PICTORIAL REPRESENTATIONS IN PHYSICS LEARNING

This paper examined the role of pictorial representations in physics learning in the laboratory on magnetism, in three types of contexts: concrete, microscopic and formal. Results show that naive student use imagery in physics reasoning. These images are externally evident through the generation of pictorial representations. There is a partial overlap between modes of students' imagery in physics and scientific imagery in historical case studies in physics.

Students naive imagery in physics reasoning can be classified into six modes: Photographical sensory experience; Projection of former sensory experience into a new sensory situation; Transformation of concrete images; Projection of former sensory experience into pure imaginary situation; Transformation of microscopic representations; Derivation of representations from formal rules. Although it is common to suppose that knowledge used by naive students is concrete while knowledge used by expert is abstract (Chi et al 1981), in this study we showed a possible pathway for students to develop both.

Students' imagery and corresponding pictorial representations evolve with time and context of interaction: The first is somewhat obvious and expected. The second is interesting. It suggests that the context of learning has an impact on the kinds of imagery used in physics reasoning. Concrete situations call for photographic representations, while formal situations call for a combination of sophisticated overlay of meaningful symbols on top of photo-like pictorial representations. Spontaneously, students developed a sense of conveying messages by using symbolic terminology and scientific conventions. Furthermore, situated mental models, that emerged in macro-hands-on situation were adapted as thinking tools and applied to microscopic relevant new situations. Hence situational imagery tools became general imagery reasoning tools.

The following diagram (figure 14) summarizes some of the results and their relation to the theoretical framework. It displays the physicists' modes of imagery vs. students' modes of imagery, and relates the two to cognitive processes (projection and transformation. It further highlights the two major mental mechanisms involved in construction of mental-visual representations: projection of former sensory experiences and transformation of mental images. These two are widely reported in the cognitive science literature. Projections of former experience into interpretation of a new situation are studied both in cognitive science and in the physics education research community. New experiences are often interpreted by using mental schemas, constructed due to former experience. In particular sensory interaction has essential role in interpretation of physical situation (Smith DiSessa and Rochelle 1993; Johnson 1987). Johnson (1987) suggests that mental reflections of bodily experience and object manipulation have an impact on interpretation of new experiences. The tendency to project former experience into new situations might explain alternative conception of scientific phenomena (Smith DiSessa and Rochelle 1993). Projection of former experience may lead to models that do not match the conventional science. For instance, in this study, students constructed a mental model of gluing to explain magnetic attraction, and magnetic attraction as electric polarization. These models were also reported in earlier educational research.(Borges and Gilbert 1998; Driver et al 1994). Transformation of mental images is widely studied by Sheppard and by Kosslyn: mental images might be scanned, enlarged (Kosslyn 1994) or rotated (Shepard 1988). This transformation enables to fit the representation to the specific feature of the problem.

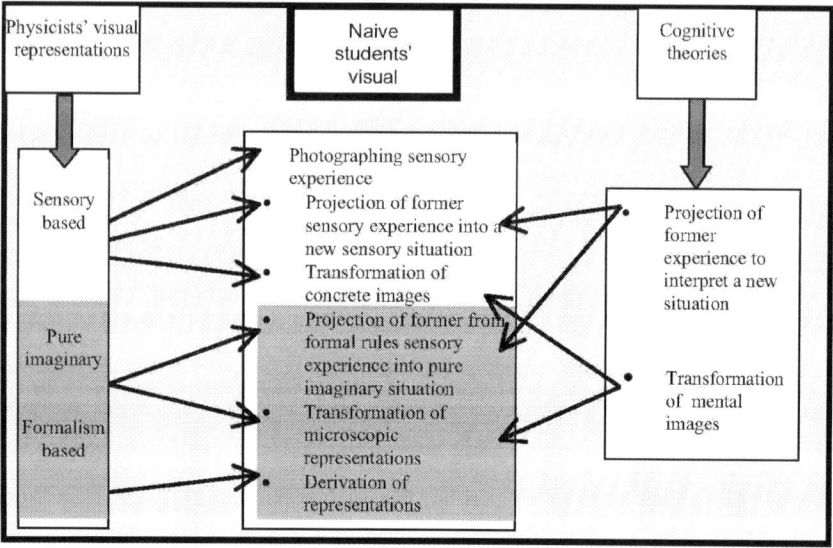

Figure 14: *Framework for interpretation of naive students' representations*

Implication for physics learning

The finding of this study, though restricted to eight case studies, suggest implications for physics learning in general.

Integrating a variety of visual representations in physics learning can facilitate qualitative understanding of physical phenomena (Reiner & Gilbert 2000, Clement & Monagham 1999; Reiner 1997; Clement 1994). In particular, we claim that a gradual progress from concrete representation into microscopic and formal representations might elaborate students' visualizations strategies. Development from concrete to microscopic and formal representations is compatible with the historical progress of visualization strategies in physics, described by Miller (2000).
Using predict-observe-explain problems in phenomenological context can provide students with the opportunity to elaborate their personal mental model into accepted scientific models. It is expected that students' primary predictions will reflect pre-conception which might contradict the scientific models. These preconceptions are common and very persistent (diSessa & Sherin 1998; Smith et al 1993; Gilbert & Zylbersztajn 1985). An appropriate design of the observed situation can provide the student's with sensory information that will support conceptual change.

Collaborative construction of visual representation encourages meaningful discussion over physical phenomena (Clement & Monagham1999; Reiner 1997). Students can share sensory information to construct common representations. These representations serve as a 'window' into student's ideas and might provide teachers with communication and evaluation tool.

REFERENCES

Agassi, J. (1968), The Continuing revolution: A History of physics from Greeks to Einstein McGraw-hill book company

Borges A. T. and Gilbert J. (1998) Models of magnetism International Journal of Science Education Vol. 20 No.5 361-378

Clark, J., M., and Paivio A. (1991), Dual Coding Theory and Education, Educational Psychology Review 3, 149-210

Chi, M., Feltovich, P. J., and Glaser, R. (1981), Categorization and representation of physics problems by experts and novices, Cognitive Science, 5,121-152

Clement J. & Monagham J. M. (1999) Use of a computer simulation, to develop mental simulations for understanding relative motion concepts International Journal of science Education; V21, n9, pp. 921-44

Clement, J. (1994), Use of physical intuition and imagistic simulation in expert problem solving, Tirosh, D. (Ed.), Implicit and Explicit Knowledge, Norwood, NJ: Ablex Publishing Corp

DiSessa, A. A. and Sherin, B. L. (1998). What changes in conceptual change? International Journal of Science Education, 20(10), 1155- 1191.

Driver, R., Squires, A., Rushworth, P., and Wood- Robinson, V. (1994). Making Sense of Secondary Science: Research into Children's Ideas. New York, NY: Routledge.

Einstein, A. (1922) How I created the theory of relativity (A lecture given in Kyoto 14.9.1922), in Weart S. R and Phillips M. (1985) History of Physics: Reading from Physics Today American Institute of physics.

Finke, R. A. and Shepard R. N. (1986). Visual foundation of mental imagery, In Boff K.R., Kaufman L. & Thomas J.P. Handbook of Perception and Human Performance Vol. 2 chapter 37

Gilbert, J.K. and Reiner, M. (2000), Thought experiments in science education: potential and current realization, International Journal of Science Education, 22 (3), 265-283

Gilbert, J. K. and Zylbersztajn, A. (1985). A conceptual framework for science education: The case study of force and movement. European Journal of Science Education, 7(2), 107- 120.

Hegarty, M. (1992). Mental animation: Inferring motion from static diagrams of mechanical systems. Journal of Experimental Psychology: Learning, Memory and Cognition, 18(5) 1084-1102

Holton G. and Brush S. G. (2001) Physics the Human Adventure: From Copernicus to Einstein and Beyond, Rutgers University Press

Holton G. (1978) The Scientific Imagination: Case Studies, Cambridge University Press

Johnson, M. (1987) The Body in The Mind: The Bodily Basis of Meaning, Imagination and Reason. Chicago: The University of Chicago Press.

Johnston, M. (2001), Ed, The Facts on Files, Physics Handbook, The Diagram Group, Checkmark Books.

Kaufmann, G. (1990) Imagery effects on problem solving in Hampson P.J. Marks D. F. and Richardson J.T.E. (Ed), Imagery Current Developments, Routledge New York

Kosslyn, M. S. (1994) Image and brain: The resolution of imagery debate. MIT Press

Macdonald D. K. (1965) Faraday Maxwell and Kelvin. Heinemann Press.

Maxwell, J. K. (1954) A Treatise on Electricity and Magnetism, Dover Ed. Unaltered republication of the third edition of 1891

Mark D. F. (1990) On the relationship between imagery, body and mind, in Hampson P.J. Marks D. F. and Richardson J.T.E. (Eds), Imagery: Current Developments, Routledge New York

Mayer, R., E., Bove, w., Bryman, A., Mars, R. & Tapganco, L.,(1996) . When less is more: meaningful learning from visual and verbal summaries of science textbook lessons. Journal of Educational Psychology, 88, 1, 64-73

Miller A. I. (2000) Insight of Genius Imagery and Creativity in Science and Art, MIT Press

Miller A. I. (1987) Imagery in Scientific Thought Creating 20th Century Physics, MIT Press

Nersessian, N. J. (1995) Should physicists preach what they practice? Constructive modeling in doing and learning physics, Science & Education 4: 203-226

Paivio, A., (1971), Imagery and Verbal Processes, Hillsdale, NJ: Holt Rinehart and Winston

Shepard, R. N. (1996), The science of imagery and the imagery of science, The Annual Meeting of the American psychological Society, San Francisco

Shepard, R., N. (1988), The imagination of the scientist in Egan K. and Nadaner D. (Eds.) Imagination and Education, Teacher College Press, PP. 153-185.

Smith, J. P., diSessa, A. A., & Roschelle, J. (1993). Misconceptions reconceived: A constructivist analysis of knowledge in transition. The Journal of the Learning Sciences, 3(2), 115- 163.

Reiner, M. (2000). Thought experiment and embodied cognition. In Gilbert J. K. and Boulter C.J. Developing Models in Science Education, pp. 157-176, Kluwer Academic Publishers

Reiner, M. (1998) Collaborative thought experiments in physics learning, International Journal of Science Education, 20 (9), 1043-1059

Reiner, M. (1997), A learning environment for visualization in electromagnetism, International Journal of Computers In Mathematics Learning, 2, (2) 125-15

Reiner M. & Gilbert J. (2000) Epistemological resources for thought experimentation in science learning. International Journal of Science Education 22(5) 489-506

Richardson T. E. John (1999), Imagery. Taylor and Francis group

Wise, N. M. (1988), The mutual embrace of electricity and magnetism. In S. B. Brush (ed.) History of Physics: Selected Reprints. (College Park, MD: AAPT).

JOHN CLEMENT, ALETTA ZIETSMAN, JAMES MONAGHAN

CHAPTER 9

IMAGERY IN SCIENCE LEARNING IN STUDENTS AND EXPERTS

University of Massachusetts, USA

Abstract. In this chapter we review three studies done at the University of Massachusetts which have implications for the role of visualizations in learning. We then identify common themes in these studies and their implications for instructional design. We describe two studies of model-based and imagery-based strategies for teaching conceptual understanding in science. To investigate these issues we have conducted learning studies where students are asked to think aloud while learning from innovative science lessons. The goal of these studies is to describe learning processes and teaching strategies that lead to the construction of visualizable models in science. A third study examines similar processes in scientists thinking aloud about an explanation problem. The studies develop observable indicators for the presence of imagery and use them to support a theory of how imagery is being used. A common theme across the studies is that dynamic imagery from one context could be transferred to a new model being constructed for a new context. In some cases new models can be grounded on runnable prior knowledge schemas (e.g. physical intuitions) that can generate dynamic imagery. We suspect that developing students' runnable mental models in this way provides a deeper level of conceptual understanding. Describing expert novice similarities in terms of hidden nonformal reasoning processes and the use of intuitive knowledge structures allows us to build a map of the conceptual terrain in students that should have important implications for instructional design.

EXTREME CASES AS FACILITATORS FOR CONSTRUCTION OF VISUALIZABLE MODELS IN STUDY 1

In a study of students learning about lever concepts, Zietsman and Clement (1997) investigated the role of analogies and extreme cases in fostering the construction of visualizable models. Zietsman conducted an in depth tutoring study with 8th grade students and tracked the strategies that produced conceptual change in the students. In an initial pilot study, she identified a number of nontrivial misconceptions in the area but also discovered positive conceptions that could be used as starting points for instruction. For example for the problem in Figure 1, one student said:

> EE5: Well, in B 1 think. [less effort]..its harder...to control it and stuff [in A]

Many of the students exhibited a common misconception that the closer one's hand is to the load, the easier it is to control - which is true in many everyday situations. This "control" idea predicting that situation B requires less effort conflicts with accepted physical theory for this type of lever and is referred to as the "control conception".

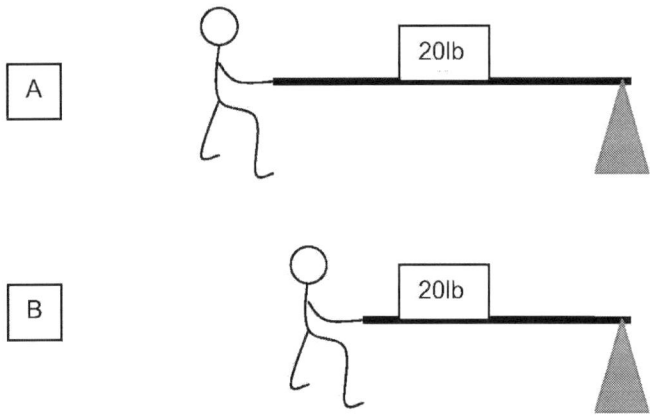

Question: Where would it be easier to keep the board with the 20 lb. load on it level?
(a) In Case A
(b) In Case B
(c) The same force would be needed in Case A and Case B.

Figure 1: *Problem eliciting the "control" misconception*

Extreme case example

One of the strategies used to help students go beyond this "control" conception was the use of an extreme case like the one shown in Figure 2: The teacher asks: In which case would the block be easiest to hold?

> E2: "Oh, in case A, because the 20 lbs. is much farther from him and so... the block is taking more of the weight. And in case B the 20 lbs. is right on his hands basically and I am sure I am right.
> E6: Man A will find it easier. The reason is because... this man A is lifting farther away from it [load] so the 20 lbs. is focusing more on the triangle side.
> I: What do you mean focusing more?
> E6: Well, I would... All the weight, if it was spread out on a 20 lb. bar, then it would be the same, but it's a block so all the weight of 20 lbs. are... Most of the weight is going down on the triangle.
> I: And for Man B?

E6: For man B most of the weight is going down on him... Or his hands whatever and I'm pretty confident on that one. "

Other students were less sure but clearly thinking in the right direction, e.g.:
E3: "And I think that person A [will find it easier]. And it's different from what I said before, but I think that, maybe the, the triangle would help a little. If the 20 lbs. is there [B] nothing could help the person, it's like carrying this thing. And if anything could help keep the weight up then... Then I guess it would be easier closer to the triangle. And I am guessing on that one."

Here the extreme case seems to trigger a correct answer based on the idea that the "fulcrum is helping". Such cases were used to create dissonance with misconceptions and to make conscious and establish intuitions for the idea of "leverage". This reinforced their view that hidden natural reasoning processes and intuitions can be found and tapped in instruction.

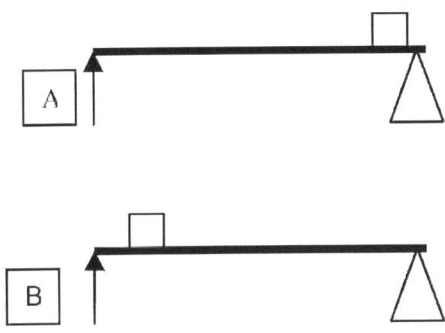

Figure 2: *Example of extreme cases used in instruction on levers*

Zietsman (1990) developed a full set of levers lessons in an authoring cycle that included research interviews for cataloguing students' conceptions, a first round of lesson trials, and a second round of trials with improved lessons. The final set of levers lessons attempted to ground instruction in the students' naive ideas like the "fulcrum helps" conception above that were in agreement with scientific theory. These intuitions were gradually elaborated and extended to more difficult cases, and to other classes of levers. The lessons were effective- even with a small sample of eight students in each group, there were large significant differences between the experimental group and a control group on a post test which included transfer problems that none of the subjects had seen before ($p<.05$). Confidence levels from 1 to 4 for each question were added together with positive values for correct answers and negative values for incorrect answers. Using this method, the total score was 77 for the experimental group and 5 for the control group.

In the first set of lesson trials, extreme cases were originally included only as a supplementary tactic but protocol analysis indicated that the greatest number of significant conceptual shifts occurred during the use of such extreme cases, and so their use was expanded accordingly in the second version. In order to explain the surprising effectiveness of these extreme cases in their protocols, Zietsman and Clement (1997) hypothesized that they were tapping into some powerful intuitions in the students' prior knowledge. They identified evidence for three benefits of using the extreme cases in the transcripts, including one that had been recognized previously in the literature (Weld, 1990):

- Students referred to a new direction of change function (e.g. load closer to the fulcrum means less effort)

In addition there were two newly identified benefits:

- Students referred to a new explanatory model (e.g. the "fulcrum helps support the load")
- Students began referring to a new variable into the system for the subjects that they had not recognized before (e.g. the load-fulcrum distance)

Such changes in the explanations students gave for levers indicated that they had reached a new conceptual understanding of levers or a new way of seeing them, rather than having simply learned a new memorized rule. The investigators were able to account for all three changes above via a hypothesized theory of the students' thinking involving perceptual motor schemas and mental simulations. They found evidence that the drawing of the weight at the extreme, as near to the fulcrum as possible, activates a "support" and/or "helping" schema as applied to the fulcrum. They hypothesized that such a schema allows a primitive form of mental simulation to occur in which the subject can imagine having to exert less force on the other end of the board. In this view the rough, qualitative, functional relationship that "the closer the load is to the turning point, the more it will help to support it." is embodied in the perceptual-motor schema itself and is an emergent feature of its activity. (A more primitive example is that an estimator for "how hard to lift" is incorporated into schema for picking up objects in infants before they have language: the larger the object, the larger the force that is needed (Mounoud & Bower, 1974)). Once the extreme case triggers the "helping" or "supporting" schema, that schema provides the student with a model and a direct way of mentally simulating the situation. Such a simulation is capable of providing implicit information that can eventually be described explicitly by the student. The activation of this schema makes the fulcrum the new reference point for describing the position of the load, allowing the load-fulcrum distance to become a variable. This then allows the construction of a new relationship in terms of this variable.

Imageable Models vs Algorithms

This set of sub-processes provides an initial account for how the three identified benefits of extreme case reasoning are realized together in these transcripts. This view emphasizes the development of the student's model-- via the incorporation of a schema that interprets situations and **projects imagined features into it**-- rather than simply the formulation or confirmation of a statement of the form "A causes an increase in B". This includes the construction of a **new way of seeing** lever situations in the sense that the students (1) are sensitive to new concepts or features of levers that they did not "see" earlier (eg. the load-fulcrum distance); (2) are **able to imagine a hidden visualizable explanatory mechanism and project it into** lever situations.

This result was important, because the authors had previously attributed the power to evoke the construction of a new mechanism only to analogies. It appears that extreme cases can do this as well. Extreme cases are easier to generate than apt analogies, so they may be easier to implement in instruction. However, the students' sensitivity to such small changes in examples in these learning studies is sobering in that it says that considerable research is required to discover what the effect of particular examples is likely to be. Not only do we need a map of common misconceptions to inform instruction, but also a map of anchoring intuitions. And a general description of the intuition is not sufficient- it appears that one needs to know about particular examples that can trigger the intuition--a specialized form of pedagogical content knowledge.

Thus this first study proposed a theory for the effectiveness of instructional strategies like extreme cases, based on the idea that the strategies were able to tap perceptual motor schemas which produced new imagery for the students and a new way of seeing the situations they were studying. We suspect that this is a more meaningful and flexible form of knowledge than simply learning a set of verbal rules, although the instructor was also careful to introduce planned verbal labels for new variables, once they had been identified visually in the examples. In this view the projection of hidden forces and other features into the situations constituted a new imageable model for them and therefore imagery was seen as playing a central role in their learning. This motivated us to look for more explicit evidence of imagery based processes in the next two studies discussed below.

USE OF IMAGERY AND SIMULATIONS BY EXPERT SCIENTISTS IN STUDY 2

More direct evidence for imagery has been found in studies of expert scientists. The goal of our expert studies has been to deepen our theory of learning processes involved in conceptual change so that it can provide new suggestions for student learning. In a continuing strand of research beginning with Clement (1981), he asked scientists and advanced graduate students to give predictions and explanations for unfamiliar situations in which it is to their advantage to construct a model for how a situation works. These studies have shown that nondeductive processes used in instruction such as analogy, model invention, criticism and revision, and extreme case analysis are also important in the thinking of expert scientists, and should therefore not be snubbed by teachers as "too informal" (Clement, 1991). Several learning strategies used by experts have inspired successful teaching strategies for high school lessons, including a strategy for analogy confirmation called bridging analogies (Clement, 1993, 1998).

The data base for this study comes from professors and advanced graduate students in scientific fields who were recorded while thinking aloud about various problems. We have space here to analyze only a brief example from one of the expert protocols. This episode also provides evidence for processes in expert extreme case use that are similar to the processes described for students above. The episode concerns the following spring problem:

> Spring Problem
>
> A weight is hung on a spring. The original spring is replaced with a spring made of the same kind of wire, with the same number of coils, but with coils that are twice as wide in diameter. Will the spring stretch from its natural length more, less, or the same amount under the same weight? (Assume the mass of the spring is negligible.) Why do you think so?

Clement (1989) documented analogies, Aha! insights and cyclical model evaluation and revision processes in these protocols. This section discusses a newer aspect of these expert studies by focussing on the use of imagery and simulations. In a case study of one subject, S2, whether the spring wire is bending or twisting eventually becomes a central issue. Textbooks tell us that it is twisting, whereas many subjects assume bending. This subject examined what the effect of twisting would be in the following thought experiment used to make a prediction for the effort required for short and long rods being twisted:

> S2 (1) If I have a longer (raises hands apart over table) rod and **I put a twist** on it (**moves right hand as if twisting something**), it seems to me-- again, physical intuition--that it will twist more...I think I trust that intuition. I'm **imagining holding something** that has a certain twistyness to it, a-**and twisting it**.
>
> (2) Now I'm confirming (**moves right hand close to left hand,**) that.. As (repeats motion) I bring my hand up closer and closer to the original place where I hold it, I realize very clearly that it will get harder and harder to twist."

(The reader may wish to try this thought experiment with images of short and long lengths of coat hanger wire, bent to have "handles" at each end of the wire for twisting it.) This subject applied this result in constructing a more complex model of the spring as involving twisting elements in the spring wire, predicting correctly that the wider spring would stretch more.

Use of Imagery, Simulation.

Bold type in the above transcript identifies examples of several imagery-related observation categories, in the following order: **personal action projections** (spontaneously redescribing a system action in terms of a human action, consistent with the use of kinesthetic imagery), **depictive hand motions**, and **imagery reports**. The latter occurs when a subject spontaneously uses terms like "imagining" or "picturing" a situation, or "feeling what it's like to manipulate" a situation. In this case it is a **dynamic imagery report** (involving movement or forces).

None of these observations are infallible indicators on their own, but we take them as evidence for imagery. The fact that more than one indicator appears together gives increased support for triangulating on the hypothesis that the subject is using imagery. (There is not space for a review here, but an increasing variety of studies of depictive gestures suggest that they are expressions of core meanings or reasoning strategies and not simply translations of speech. Others indicate that the same brain areas are active during real actions and corresponding imagined actions. Imagery may not always or even usually be accompanied by gestures, but in this study we are primarily interested in depictive hand motions as providers of evidence for internal imagery, when they do appear.)

Taken together with the subject's new predictions, the observations above can be explained theoretically in terms of internal *imagistic simulations* wherein: (1) the subject: has activated a somewhat general and permanent perceptual motor schema that can control the action of twisting real objects; (2) the schema assimilates an image of two rods of different lengths that is more specific and temporary; (3) the action schema "runs through its program" vicariously without touching real objects, generating a simulation of twisting the two rods, and the subject compares the effort required for each. Such a simulation may draw out implicit knowledge in the schema that the subject has not attended to before--e.g. in this case the simulation may draw out knowledge embedded in analog tuning parameters of a motor schema for twisting. In other words, a hypothesis can be made, with initial grounding in data from the above transcript, that the subject is going through a process wherein a general action schema assimilates the image of a particular object and produces expectations about its behavior in a subsequent dynamic image, or simulation.

Imagery Enhancement

But how can considering the extreme case in transcript section (2) above add so much confidence since S2 has already just consulted his knowledge on this issue in section (1) before considering the extreme case? A hypothesis that explains this is the following. Given the above observations it is plausible to interpret the role of the extreme case as providing "imagery enhancement" (or "simulation enhancement")— that is, it enhances the subject's ability to run or compare imagistic simulations with high confidence, and this comes from increasing the difference between the two images being compared and making that difference more detectable under inspection of the images. In this case the main source of conviction in the simulations appears to be the tapping of implicit knowledge embedded in a motor schema and its conversion into explicit knowledge. The extreme case makes differences in implicit expectations more "perceivable" in this case. Thus the imagistic simulation concept can explain a phenomenon difficult to explain in other ways. This theory of imagistic simulation is developed more fully for data that includes additional expert subjects in Clement (1994).

Thought Experiments

The twisting rod episode above is also an example of an **untested thought experiment** in the broad sense, a term we have used to refer to the act of predicting the behavior of an untested, concrete, but absent system (the "experiment") (Clement, 2002). Aspects of the experiment must be new and untested in the sense that the subject is not informed about their behavior from direct observation or from an authority. How can S2 learn anything by focussing on the particular example of the rod and running through the experience of twisting it? How can it give the sensation of "doing an empirical experiment in one's head" (Nersessian, 1991; Gilbert and Reiner, 2000; Reiner and Gilbert, 2000)? This raises what Clement termed the **fundamental paradox of thought experiments**, expressed as: How can findings that carry conviction result from a new experiment conducted entirely within the head? Discussion of this deeper problem is beyond the scope of this chapter, but the imagistic simulation concept has been used as a starting point for dealing with it in Clement (1994a, b), (2003). The theory uses a framework that includes flexible perceptual motor schemas that generate and run imagistic simulations, via the extended application of a schema outside of its normal domain, implicit knowledge, or spatial reasoning. One can point to such sources as potential origins of conviction in thought experiments, to help us begin to explain the fundamental paradox.

Summary of Study 2

In summary, for us this strand of our research program demonstrated the possibility of developing observable indicators of imagery and mental simulation in scientific thinking. Analyses of more complete protocols (Clement, 1989, 2003) found that the simulations for relations in cases like the "twisting a rod" episode above were used as axioms for developing a more complicated mental model of the spring. This emphasizes the importance of runnable intuition schemas as prior knowledge resources for developing runnable, visualizable models in experts. Thus not all expert knowledge is abstract, rather it can include concretely imageable physical intuitions.

The themes of extreme cases, use of imagery and imagistic simulation, and the use of prior knowledge in the form of perceptual motor schemas that act as intuitions for anchoring learning are strongly reminiscent of the explanation given earlier for the effectiveness of extreme cases in teaching students about levers. This perspective also allows one to view the levers instruction in section one of this chapter as learning via thought experiments. While we would be the first to also emphasize the importance of real experiments as well in science classrooms, we believe the role of thought experiments in instruction has been strongly underestimated in areas where students have useable intuitions. To us these parallels underline the importance of expert-novice comparisons in the area of nonformal reasoning.

USE OF A COMPUTER SIMULATION TO DEVELOP MENTAL SIMULATIONS FOR LEARNING ABOUT RELATIVE MOTION IN STUDY 3

Monaghan and Clement (1999) attempted to use some related indicators as evidence for imagistic thought in students as part of a third study. In the area of relative motion, evidence was found for a particularly persistent set of student difficulties which have resisted innovative instructional efforts in the past. They investigated the use of an animated computer simulation of simple relative motion as a method for dealing with these difficulties. The goal was to identify critical factors for designing lessons on relative motion as well as general information on the role of imagery and simulations in science instruction.

In this study high school physics students were videotaped in pairs in a laboratory setting as they performed collaborative activities with computer simulations. In one case study think aloud interview protocols from three of the students were analyzed (Monaghan and Clement, 1999). This study examined students' qualitative and quantitative conceptions and misconceptions of apparently simple one dimensional relative motion problems, and examined the effects on their conceptions of using a computer simulations.

Instruction via Relative Motion Simulations

Subjects were given identical pre and post tests immediately before and after the simulation treatments. During the treatment they were shown several simulations in a predict, observe, explain format. Figure 3 shows a screen snapshot of one of these displaying a black car, a white car, and a plane. The interviewer first ran the simulation from the ground frame of reference. During this time, the black car moved from the far right of the screen to the left, the white car moved from the far left of the screen to the right, and the airplane appeared from the left of the screen and traveled to the right at a higher speed than the white car. A "stopwatch" timer on the screen ran concurrently with the animations. The simulation was then reset and run again until the computer stopwatch read approximately 3 seconds. At this point, the interviewer asked the subject to predict what the direction and comparative speed of each of the cars would be when viewed from the airplane frame of reference. Many of the subjects made incorrect predictions, finding it especially hard to predict that the white car would appear to move from right to left (opposite to its previous motion). However, once they viewed this in the simulation and were asked to account for it, several of the subjects were able to explain to themselves why this would be the case.

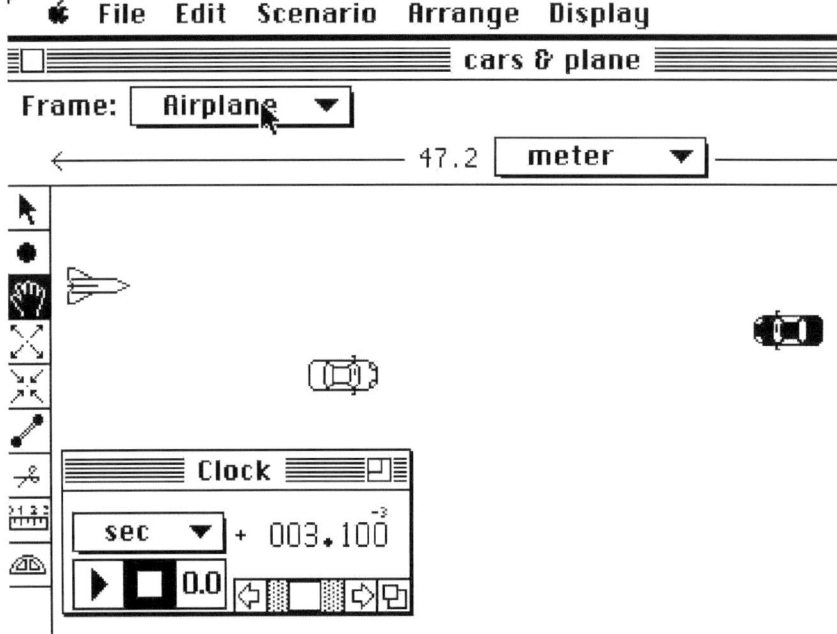

Figure 3: *Screen Shot of Computer Simulation*

Example from the Posttest

Here we will give an example from one subject's response on the post test done off line following the on line simulation. Post test questions 5, 6, 9, and 10 shown in Figure 4 refer to a near-transfer problem that is similar to the plane problem but involves a helicopter travelling in a vertical orientation on the paper rather than horizontal. There is evidence that subject GS10 performs an appropriate mapping of features from the instructional simulation to the post test question shown in Figure 4. Emphasis by underscoring in the following transcript was added by the authors. Numbers 5-10 refer to the picture below:

IMAGERY IN SCIENCE LEARNING 179

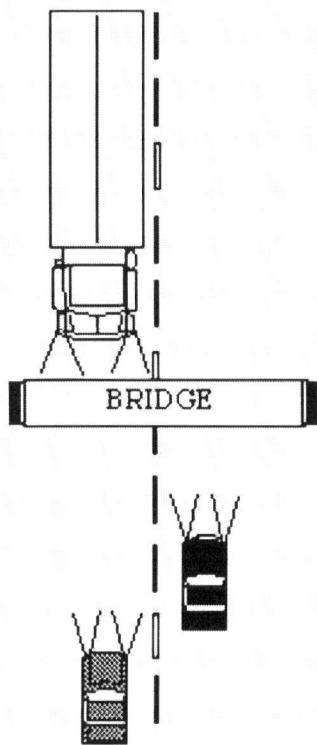

In the picture above, you are in the gray car. Your speedometer reads 40 mph.
5. What is your car's speed relative to a very low flying helicopter going exactly the same direction as your car, at a speed relative to the ground of 200 mph?
6. **What is your confidence in your answer?**
a.) Just a blind guess   c.) Fairly confident
b.) Not very confident   d.) I'm sure I'm right
9. The white truck is traveling toward you. If the truck's speedometer reads 40 mph, what is the truck's speed relative to the helicopter?
10. What is your confidence in your answer?
a.) Just a blind guess   c.) Fairly confident
b.) Not very confident   d.) I'm sure I'm right

Figure 4: *Problem from Pre and Post Test*

GS10   D1 S: OK. Uh, number 5. (Reads question)... Um, OK, now the helicopter is going this way (labels diagram with vertical vector and 200;)at 200 MPH, and you also going this way at 40 MPH (puts 40 mph on the diagram), -- uh (looks at static graphic on computer screen), lets see. From the simulation it showed that like (points right hand holding pencil laterally to the right) when the plane was going the same direction as the

car that was going the same direction, that looked like the car that was going the same direction was going opposite direction (moves right hand holding pencil laterally to the right) [clear memory of simulation]. So, if this car -- the car (draws vertical vector on diagram) probably <u>looks like its going that way</u> at--, uh, -- so you probably have to subtract, so 200 minus 40 equals 160 MPH (writes 200-40=160 mph and circles answer)
D2 I: What are you thinking?
D3 S: I'm thinking that since this is going the opposite direction than .. then uh, .. <u>it [the car] slows down</u> [from the helicopter frame of reference] instead of speeds up, <u>the mph that it looks like from the helicopter</u>, because <u>it is going the opposite direction, that looks like it is from the helicopter</u>, because you're going faster than it, so I subtracted 200 which is the mph of the helicopter from the 40 mph of the car to give me 160 MPH that it looks like it is going [in] the opposite direction
D4 I: How is your confidence?
D5 S: Uh, fairly confident --
D6 I: OK. <u>Why are you more confident?</u>
D7 S: Because from watching the screen <u>I'd seen like a simulation and an actual, like, picture of it, instead of like something like on the paper.</u>

Since this subject's answer is basically correct and she had answered this question incorrectly on the pretest, she apparently learned from the simulation. The subject's statement that the computer simulation provided an " actual ... picture of it" suggests that she found the computer intervention to be a useful tool to introduce a representation of the problem scenario.

3.3 Imagery and Mental Simulation

The detection of mental simulations posed a particularly interesting methodological challenge in this study. There are several indications in the above segment providing evidence that the subject was using imagery. The following observational categories were developed as evidence for dynamic imagery in mental simulations:

a. <u>Depictive hand or pencil motions</u> that are not simply pointing to a word or picture, but that appear to indicate movements of objects. They may also appear near oral references to those movements and appear over a drawing of the problem. Although the drawing provides an external visual representation, it cannot provide a dynamic representation of motion. We interpret these motions as expressions of and evidence for internal motion imagery.

b. <u>References to perceptions</u>. The subject refers explicitly to the sensation of perception while describing visual or other perceptual aspects of the scene during thinking by using phrases such as "the car probably looks like it is going that way" in line D1.

c. <u>Self projection</u>. These occur in phrases like that in line D3 above: "because <u>you're</u> going faster that it," (emphasis added) where the subject uses a personal pronoun indicating that he or she is imagining projecting themselves into a particular object in the problem. Whereas abstract ideas may be "view independent", visual imagery usually occurs from a particular point of view (Kosslyn, 1980). For example, it is extremely difficult to imagine all sides of a house at once. We can see this type of language as originating in her imagining what it is like from the point of view of the helicopter. In fact the wording of problem 5 talks about "your car", and indeed her language early in the solution in line D1 suggests that she accepts the car point of view: "Ok, now the helicopter is going this way at 200 MPH, and you're also going this way at 40 MPH." But later in line D3 her point of view appears to shift to projecting herself into the helicopter.

d. Imagery reports. Other subjects, but not this subject, also exhibited imagery reports where the subject reports imagining or "seeing" a situation, e.g. ... "OK, in my mind I'm seeing the truck [from problem 9] as if it were a black car on the screen, on the computer, and again the helicopter [in problem 9] as if it were the plane [in the computer simulation]"

Again, while none of these indicators are ironclad indicators on their own, they give us reason to believe that the subject is using imagery. And again, there are accompanying indicators that the imagery is dynamic when subjects are describing vehicle movements as they display imagery indicators, and displaying some indicators that are inherently dynamic, such as hand or pencil motions. These observations contrasted with GS10's language below from the pretest on the same problem-- there in giving her incorrect answer she talked about the use of numerical algorithms, and these contained no such personal pronoun indicators of projected points of view:

GS10 Pretest: "I think the velocity equals just 200 um, yeah, relative speed equals the first take away the second I think...because I know it's um one of the speeds is A and the other one is B but I'm not sure if you're supposed to subtract or add when they're going the same direction....so I just decided that that's plus because I know that they are going the same direction and they wouldn't, It would be subtracted if they were going in opposite directions."

The first three imagery indicators listed above are also present in the transcript of this subject's predictions during the treatment. This suggests that the post test solution was reached not just by means of a verbal analogy; based on the above observations, we hypothesize that it involves imagery, and the use of dynamic mental models. Thus there is evidence that she was better able to visualize the problem scenario because of the treatment.

On the basis of the imagery indicators in the above transcript, Monaghan and Clement hypothesized that the students were running imagistic simulations of the moving objects in the problem. It was hypothesized that for successful students, dissonance between their incorrect predictions and simulations displayed by the computer initiated the construction of new ways of thinking about relative motion. In addition, the memory of certain simulations acted as an analogue "framework for visualization" of target problems solved off line after the intervention. In such cases we find that interaction with a *computer simulation on-line* can facilitate a student's appropriate *mental simulations off-line* in related target problems. We believe this is one of the most appropriate and ambitious outcomes one can aim for in an instructional approach that uses computer simulations. This suggests that it is important for teachers to assign and discuss off line application problems that use the imagery from simulations developed on line.

Summary of Study 3

This study found evidence for: students' qualitative and quantitative conceptions and misconceptions of relative motion problems; students' spontaneous visualizations of relative motion problems; and students' memory of the on-line simulation used as a framework for visualization of posttest problems solved off-line. Instances of successful mapping of remembered simulation features onto target problems were presented. Evidence was gathered from hand motions and other indicators suggesting that the subjects were using dynamic imagery in mental simulations during the treatment and post test.

This is one of the first studies where our group was able to point to evidence for the use of imagery in students. Imagery has historically been notoriously difficult to study scientifically because it is hard to acquire data on such a deeply internal process. A difficulty is that, depending on the context and the age group, the aforementioned imagery indicators will not always be very prevalent. John Hayes (19) once wrote that protocols are like a porpoise: you only see them come to the surface 20% of the time, and while they are at the surface you only see 25% of the body at a time. But the whole porpoise is really there all the time, nonetheless. This may be even more true of imagery. So in the classroom we need other less conservative ways of detecting when students are using visualizable models. Teachers may encourage the use of student drawings and/or gestures as one possibility; attending to concrete descriptions of events in a student's language as opposed to formulas or formal laws is another. And further studies of ways to detect imagery use in research are needed as well.

COMMON THEMES

We close with an emphasis on several themes and implications derived from these studies:

- The possibility of developing observable indicators of imagery and mental simulation in science learning was demonstrated. Evidence from video tape for the important role of imagery and mental simulation in science learning in both experts and students is obtainable and useful in analyzing thinking.
- While the value of hand motions and other indicators in the present studies was primarily as a research tool providing a window on student imagery, these indicators may also have enormous value as external representations and communication tools in the classroom.
- The importance of developing dynamic, imageable mental models as a locus of understanding as opposed to learning only verbal and mathematical rules was highlighted.
- The importance of prior knowledge resources for developing such models in both students and experts was highlighted in studies one and two. These appear to include runnable intuition schemas that can generate dynamic imagery. Using these as anchors as sources for developing models may

ground learning in a way that makes it meaningful, comprehensible, and long lasting.
- On the other hand the relative motion instruction may have relied less on finding an established intuition in students. In contrast to the levers instruction, the relative motion exercises may have challenged the students' spatial reasoning and imagery system itself to construct a "mental movie" of relative motion. Although a dynamic animation of the result was shown to the students, we believe that this still involved a form of active learning because the students were required to predict the outcome first, and then to explain the observed result in their own words, which encouraged them to organize and actively regenerate the imagery.
- Implicit in each of the three studies described here was the idea that dynamic imagery from one context could be transferred to a new model being constructed for a new context. Other examples of this theme are discussed in Clement (2003) and Clement and Steinberg (2002). If we can learn to tap these sources of imagery in prior knowledge through exemplars or by analogy or extreme cases we should gain a powerful tool for developing students' runnable mental models.

Learning via imagery is a hidden process, and it requires innovative research strategies to uncover it. Implementation of these processes in classrooms will require that they be integrated into the social dynamics of small and large group processes. However, uncovering a map of the conceptual terrain in individuals in the form of hidden nonformal reasoning processes and hidden intuitive knowledge structures will be invaluable for this task.

REFERENCES

Clement, J, 1981. Analogy generation in scientific problem solving, in: *Proceedings of the Third Annual Meeting of the Cognitive Science Society*, Berkeley, CA. ERIC RIE #SE-048-920.

Clement, J., 1989. Learning via model construction and criticism: Protocol evidence on sources of creativity in science, in: *Handbook of creativity: Assessment, theory and research*. Glover, J., Ronning, R., and Reynolds, C. eds., Plenum: New York, pp. 341-381.

Clement, J., 1991). Non-formal reasoning in science: The use of analogies, extreme cases, and physical intuition. In *Informal reasoning and education*, J. Voss, D., Perkins, & J. Segal, (Eds.), Hillsdale, NJ, Lawrence Erlbaum Associates.

Clement, J. , 1993. Using bridging analogies and anchoring intuitions to deal with students' preconceptions in physics. *Journal of Research in Science Teaching*, 30(10), 1241-1257.

Clement, J., 1994a. Imagistic simulation and physical intuition in expert problem solving, in: *The Sixteenth Annual Meeting of the Cognitive Science Society,* Lawrence Erlbaum, Hillsdale, NJ.

Clement, J., 1994b. Use of physical intuition and imagistic simulation in expert problem solving, in: *Implicit and explicit knowledge*. D. Tirosh, ed., Ablex Publishing Corporation: Hillsdale, New Jersey, pp. 204-244.

Clement, J., 1998, Expert novice similarities and instruction using analogies, *International Journal of Science Education*, 20(10), 1271-1286.

Clement, J., 2000. Analysis of clinical interviews: Foundations and model viability, in: *Research methods in mathematics and science education*. Kelly, A. E., & Lesh, R. eds., Erlbaum: Englewood Cliffs, NJ, pp. 341-385.

Clement, J., 2002. Protocol evidence on thought experiments used by experts, in: *Proceedings of the Twenty-Fourth Annual Conference of the Cognitive Science Society,* Wayne Gray and Christian Schunn, eds., Erlbaum: Mahwah, NJ.

Clement J., 2003. Imagistic simulation in scientific model construction. In *Proceedings of the Twenty-Fifth Annual Conference of the Cognitive Science Society*,. R. Alterman and D. Kirsh, Editors, Mahwah, NJ, Erlbaum, 258-263.

Clement, J. and Steinberg, M., 2002, Step-wise evolution of models of electric circuits: A "learning-aloud" case study. *Journal of the Learning Sciences* 11(4), 389-452.

de Jong, T., 1991. Learning and instruction with computer simulations. *Education and Computing*. 6: 217-229.

Gilbert, J. and Reiner, M., 2000. Thought experiments in science education: Potential and current realization. *International Journal of Science Education* 22, 3: 265-283.

Hayes, J. 1978, *Cognitive psychology: Thinking and creating*, Homewood, Ill, Dorsey.

Horwitz, P., Feurzeig, W., Shetline, K., Barowy, W., & Taylor, E.F., 1992. *RelLab* [Computer software]. Cambridge, MA: Bolt Beranek and Newman Inc.

Kosslyn, S., 1980, *Image and mind*, Cambridge, MA, Harvard University Press..

Monaghan, J. M., 1996, Use of collaborative computer simulation activities by high school science students learning relative motion. Unpublished doctoral dissertation, University of Massachusetts.

Monaghan, J. M. & Clement, J., 1999. Use of a computer simulation to develop mental simulations for learning relative motion concepts. *International Journal of Science Education*. 21(9): 921-944.

Monaghan, J. M. & Clement, J., 2000. Algorithms, visualization, and mental models: High school students' interactions with a relative motion simulation. *Journal of Science Education and Technology*, 9 (4): 311-325.

Mounoud, P. & Bower, T. G. R., 1974. Conservation of weight in infants. *Cognition*, 3(1): 29-40.

Reiner, M. and Gilbert, J., 2000. Epistemological resources for thought experimentation in science learning. *International Journal of Science Education*. 22(5): 489-506.

Weld, D., 1990. Exaggeration, in: Qualitative reasoning about physical systems, D. Weld and J. de Kleer (eds.), San Mateo, CA: Morgan Kaufmann Publishers, Inc.

White, B. Y., 1993a. Intermediate Causal Models: A Missing Link for Successful Science Education. *Advances in Instructional Psychology*. R. Glaser. Hillsdale, NJ, Lawrence Erlbaum Associates, Publishers.

Zietsman, A. and Clement, J., 1997. Extreme cases as mechanisms of learning: Teaching seventh graders about levers. *The Journal of the Learning Sciences*, 6(1): 61-89.

Zietsman, A., 1990, Case studies of cycles in developing a physics lesson. Unpublished doctoral dissertation, University of Massachusetts.

SECTION C

INTEGRATING VISUALIZATION INTO CURRICULA IN THE SCIENCES

YEHUDIT JUDY DORI[1,2], JOHN BELCHER[2,3]

CHAPTER 10

LEARNING ELECTROMAGNETISM WITH VISUALIZATIONS AND ACTIVE LEARNING

[1]Department of Education in Technology & Science, Technion, Israel Institute of technology, Haifa 32000, Israel; [2]Center for Educational Computing Initiatives, Massachusetts Institute of Technology, Cambridge, MA 02139, USA; [3]Department of Physics, Massachusetts Institute of Technology Cambridge, MA 02139, USA

Abstract. This chapter describes learning electromagnetism with visualizations and focuses on the value of concrete and visual representations in teaching abstract concepts. We start with a theoretical background consisting of three subsections: visualization in science, simulations and microcomputer-based laboratory, and studies that investigated the effectiveness of simulations and real-time graphing in physics. We then present the TEAL (Technology Enabled Active Learning) project for MIT's introductory electromagnetism course as a case in point. We demonstrate the various types of visualizations and how they are used in the TEAL classroom. A description of a large-scale study at MIT follows. In this study, we investigated the effects of introducing 2D and 3D visualizations into an active learning setting on learning outcomes in both the cognitive and affective domains. We conclude by describing an example of TEAL classroom discourse, which demonstrates the effects and benefits of the TEAL project in general, and the active learning and visualizations in particular.

INTRODUCTION

The issues related to problematic learning in large passive undergraduate physics classes were first identified and researched over a decade ago (McDermott, 1991). This problem is still under investigation (Hake, 1998; Maloney, O'Kuma, Hieggelke & Van Heuvelen, 2001). Some researchers cite the lack of a common language between mathematicians and physicists as the root of the learning difficulties of students who study physics (Dunn & Barbanel, 2000). Other researchers place the blame on traditional teaching methods that reward memorization over conceptual thinking (Mazur, 1997). Yet others claim that instructors do not adequately address the individual needs of students (Novak, Patterson, Gavin & Christian, 1999). Hestenes (2003) emphasized the necessity of developing teaching models which stress conceptual understanding in physics classrooms. Woolnough (2000) noted that active learning which integrates experimental work plays an important role in helping students create cognitive links between the world of mathematics and that of physics. We agree with these later researchers. We also strongly believe that teachers and students need to incorporate visualizations in the teaching and learning of scientific phenomena and processes, especially when dealing with abstract concepts such as electromagnetism (Dori & Belcher, 2004).

In this chapter we argue and demonstrate that visualization technology can support meaningful learning by enabling the presentation of spatial and dynamic images. Such images can portray and illuminate relationships among complex concepts. The Technology-Enabled Active Learning (TEAL) Project at MIT (Belcher, 2001) involves the use of visualization in a collaborative learning environment of a freshman electricity and magnetism (E&M) course. The objective of the TEAL Project was to transform the way physics is taught in large enrollment physics classes at MIT in order to decrease failure rates and increase students' conceptual understanding. The approach is designed to help students visualize, develop better intuition about, and develop better conceptual models of electromagnetic phenomena. Patterned in some ways after the Studio Physics project of Rensselaer Polytechnic Institute (Cummings et al., 1999) and the Scale-Up project of North Carolina State University (Beichner, et al., 2002; Handelsman et al., 2004), TEAL extends these efforts by incorporating advanced 2D and 3D visualizations to enable students to construct a deeper view of the nature of various electromagnetism phenomena. Such visualizations allow students to gain insight into the way in which fields transmit forces by watching how the motions of objects evolve in time in response to those forces. They also allow students to intuitively relate the forces transmitted by electromagnetic fields to those transmitted by more familiar agents, for example, rubber bands and strings. This makes electromagnetic phenomena more concrete and more comprehensible, because it allows the students to appreciate the correspondence between electromagnetic stresses and phenomena that they already understand

Following the theoretical background section, this chapter focuses on the value of concrete and visual representations in teaching abstract concepts. We will use the TEAL project in the E&M course at MIT as a case in point to demonstrate the various types of visualizations and how they are used in the classroom. We conclude by briefly presenting the effects of introducing 2D and 3D visualizations in an active learning setting on undergraduate students' learning outcomes in both the cognitive and affective domains.

THEORETICAL BACKGROUND

The AAAS (1989) report "Science for all Americans" recommended that reform in science education be founded on "scientific teaching." Scientific teaching approaches the teaching of science with the same rigor as science is approached at its best. However, changes in this direction have neither progressed rapidly, nor have they been driven by research universities as a collective force (Handelsman, Ebert-May, Beichner, Bruns, Chang, DeHaan, Gentile, Lauffer, Strwart, Tilghman, & Wood, 2004). While there is agreement that active participation in lectures and discovery-based laboratories help students develop science-oriented thinking, most introductory courses still rely on "transmission-of-information" lectures and "cook-book" laboratory exercises.

Visualization in Science

Science and technology develop through the exchange of information, much of which is presented as still and moving images, diagrams, illustrations, maps, plots, and models that summarize information, and help others to understand scientific data and phenomena (Mathewson, 1999). To fully understand the nature and implications of scientific phenomena and their models, students should be exposed to a broad array of types of models, such as concrete, verbal, symbolic, mathematical, and visual models (Boulter & Gilbert, 2000; Gilbert & Boulter, 2000; Treagust, 1996).

Teaching science requires that the learning materials be presented not only through words—in spoken or printed text—but also through pictures—still and moving pictures, visual simulations, 2D and 3D visualizations, graphs, and illustrations. Pictures are superior to words for remembering concrete concepts (Rieber, 2002). The dual coding theory (Paivio, 1990; Sadoski & Paivio, 2001) suggests a model of human cognition divided into two dominant processing systems—verbal and nonverbal. The verbal system specializes in "language-like" processing, while the nonverbal system includes vision and emotions.

The cognitive theory presented by Mayer (2002) offers three theory-based assumptions about how people learn from words and pictures. The first assumption is the *dual channel* assumption, which is essentially the same as Paivio's dual coding theory (1990). Mayer suggested that knowledge is represented and manipulated through both through the visual-pictorial channel and the auditory-verbal channel. Mayer's second assumption, called *limited capacity* emphasizes that the channels can become overloaded when a large volume of spoken words or pictures are presented. The third assumption, *active processing*, is that meaningful learning occurs when students engage in active learning. They select relevant words and pictures, organize them into pictorial and verbal models, and integrate them with prior knowledge. These active learning processes are most likely to occur when corresponding verbal and pictorial representations are in working memory at the same time.

Visualization technologies were initially pioneered for the representation of complex scientific data and models. Eventually, these visualizations have found their way into science education, focusing on enhancing students' scientific inquiry while they are involved in data analysis and modeling (Jacobson, 2004). Scientists and science educators have therefore argued that visualization of phenomena and laboratory experiences are important components for understanding scientific concepts (Cadmus, 1990; Escalada, Grabhorn, & Zollman, 1996). Demonstrations, simulations, models, real-time graphs, video, and visualizations can contribute to students' understanding of scientific concepts by providing mental images to go with abstract concepts. They enable the capture and perseveration of the essence of such phenomena more effectively than verbal or textual descriptions. The National Science Education Standards describes the learning of science as an active process (National Research Council, 1996) that implies a physical ("hands-on") and mental ("minds-on") activity.

According to Mathewson (1999), visual-spatial thinking includes both vision and imagery. Vision refers to the process of identifying, locating and thinking about objects in the world. Imagery is the more abstract process, carried out in the absence of a visual stimulus, in which we engage in formation, inspection, and transformation of images in the "mind's eye". Suwa and Tversky (2002) claimed that representations such as diagrams, sketches, charts and graphs serve not only as memory aids but also as aides for inference, problem solving, and to facilitate thinking about entities or elements and their spatial array. Thinking about functional aspects of the situation from these representations requires associations and inferences from the visio-spatial display to more abstract concepts. Their study showed that experts are more adept than novices at perceiving functional thoughts from representations.

Discussing dynamic mental models in learning science, Frederiksen, White, and Gutwill (1999) characterized two levels of models. Lower-level models focus on concrete phenomena, such as the electrostatic interaction of charged particles, and employ causal reasoning about object interactions. Higher-level models focus on more abstract representations of variables and the relationships among them. Developing conceptual links from lower- to higher-level models helps students explain interactions and behaviors and improves their performance in solving problems.

Focusing on the role of problem representation in physics, Larkin (1983) attributed many differences in the problem-solving performance of experts and novices to the use of different representations. A naïve representation contains direct, familiar, visible entities, found directly in the problem. More advanced physical representations include qualitative, time-independent, and redundant inference rules and intangible artifacts (forces, moments, fields, etc.) that invoke fundamental principles of physics. When solving physics problems, novices base their representations and approaches on the problem's literal features, while experts apply physical principles and integrate the representation with prior knowledge (Chi, Feltovich & Glaser, 1981). Experts also spend time constructing abstract representations for such variables as force and energy, to reduce the amount of information that must be attended to at any one time (Larkin, McDermon, Simon, & Simon, 1980).

Simulations and Microcomputer-Based Laboratory in Physics

During the past decade, a number of physics curricula for undergraduate students have been developed that utilize outcomes from educational research. These include *Physics by Inquiry* and *Tutorials in Introductory Physics* (McDermott, 1991; McDermott & Shaffer, 2002), *Tools for Scientific Thinking* (Thornton & Sokoloff, 1990), *Matter & Interaction* (Chabay & Sherwood, 2002), *Peer Instruction*, (Mazur,1997) and the calculus-based physics without lectures course *Workshop Physics* (Laws, 1991). Projects that incorporated Information & Communication Technologies (ICT) include *Real-Time Physics* (Sokoloff, Thornton & Laws, 1999), *Studio Physics* at Rensselaer Polytechnic Institute (Cummings, Marx, Thornton & Kuhi, 1999), and the SCALE-UP project at North Carolina State University (Beichner, Bernold, Burnsiton, Dali, Gastineau, Gjertsen & Risley, 1999), as well as the TEAL Project at MIT (Dori, Belcher, Bessette, Danziger, McKinney & Hult, 2003). Common characteristics of these innovative courses included activity-based pedagogies, collaborative learning, the integration of curricula, context-rich problems, and the use of technology.

ICTs, which are in daily use in physics research, are slowly finding their way into physics teaching. Although many textbooks and software devoted to computational physics exist, many physics courses in college and universities are still taught in the traditional manner. In the best cases, special courses on simulation and computational physics are taught, but the introductory courses usually make little use of the possibilities opened up by ICT.

In what follows, we present several simulations and applications of Microcomputer-Based Laboratory (MBL) in physics. In the following section, we describe studies on the effectiveness of simulations and MBL in physics education.

Martinez-Jimenez and Casado (2004) developed a computer-aided education package, *Electros*, especially designed for introductory physics courses. The software evolved from the experience with physics and engineering undergraduate students. *Electros*, which focuses on electrostatic fields and potentials produced in the presence of dielectrics, has, in addition to the simulation, a tutorial, an initial test, and a final test.

Cox, Belloni, Dancy and Christian (2003) developed interactive Physlet-based curricular material to help students learn concepts of thermodynamics, with a particular focus on kinetic theory models. The simulations help students to visualize ideal gas particle dynamics and engine cycles, make concrete connections between mechanics and thermodynamics, and to develop a conceptual framework for problem solving. Students can determine the efficiency of a simplified version of real-world combustion engine and this helps them connect the abstract *PV* diagrams of physics and chemistry to a real-life car engine.

Escalada, Rebello and Zollman (2004) developed another package, *Visual Quantum Mechanics–The Next Generation*, that includes simulation, computerized tools, and instructional strategies for undergraduate physics majors. This package focuses on enabling students to make observations, develop mental models consistent with quantum principles, and apply these models to related situations. For example, while students carried out visualization activities, they developed a model of waves of matter, explained discrete energy states, and examined applications to the electron microscope.

Chabay (2002) supplemented an electromagnetism textbook (Chabay & Sherwood, 2002) with short 3-D QuickTime movies of electric and magnetic fields in space. The movies, which depict spatial configurations of electric and magnetic fields that are frequently discussed in introductory physics courses, are designed to address beginning students' difficulties in visualizing fields throughout space. Two-dimensional diagrams or static images are not sufficient for three-dimensional reasoning about fields. This is so because magnetic field and force vectors are computed using vector cross products, and thus these fields and vectors inherently involve three dimensions. A common misconception, which is exacerbated with time-varying fields, involves thinking in general terms of "the field" rather than precisely of the field vector at a particular location in space and time.

The MBL, also called *desktop experiments*, is yet another means to foster students' understanding of physical phenomena through visualization. Designed to collect data via various probes such as motion, light, temperature, force, pressure, sound, or current, typical MBL applications store or feed the data into a computer. This data can then be analyzed and displayed using many different representations, enabling the student to gather the data in real-time and then graph it either at a later time or simultaneously (Kown, 2002; Lapp, 1999).

The Studio Physics at Rensselaer Polytechnic Institute (RPI) (Cummings, Marx, Thornton & Kuhi, 1999) combined MBL with the studio approach, where students are actively involved in class experiments and small group discussions. At RPI, the Introductory Studio Physics was a calculus-based, two-semester sequence equivalent to the standard physics for engineers and scientists, in which large classes were broken into small sections.

Studies on Simulations, MBL, and Real-time Graphing

Having reviewed several examples of physics simulation, we now turn to a more theoretical and empirical-based discussion on this subject. Steinberg (2000) investigated the impact of using a computer simulation in a classroom studying the behavior of objects moving under the influence of air resistance by comparing two classes which both had interactive learning environment. One class used the simulation, while the other used only paper and pencil activities. Although there were differences in how students approached learning, students' performance on a common exam question on air resistance was not significantly different. The author claimed that a computer simulation that quickly and transparently delivers exact answers can result in students learning science passively, and there is a danger that students will see no need to take responsibility for their own understanding, verification, or inquiry. On the other hand, many students used the air resistance simulation actively and productively, considering more complex problems than were possible without the computer.

Several studies compared the effects of simulation-based-learning to various types of expository teaching. Some of these studies which dealt with physics focused on Newtonian mechanics (Rieber, Boyce, & Assad, 1990; Rieber, Noah & Nolan, 1998; Rieber & Parmley, 1995; White, 1993) and on electrical circuits (Carlsen, & Andre, 1992). These studies showed that learning with computer simulation is as effective as expository instruction; however, they also suggest that effects of simulation-based-learning are stronger when the simulation is embedded in an environment that intends to support specific aspects of discovery learning.

Learners using simulations can be given assignments or be exposed to model progression, i.e., viewing the simulation gradually, avoiding exposure to its full complexity from the beginning. De Jong, Martin, Zamarro, Esquembre, Swaak, and van Joolingen (1999) evaluated a simulation of collisions in a discovery environment. They concentrated on the effect of providing support for the planning process by means of presenting freshmen with assignments and model progression. One group of students worked with assignments-enhanced simulation. Students in the second group were free to select their assignments while using a model progression in addition to the assignments. The control group used an unsupported simulation. Research results contributed to the understanding of how to support learners in their discovery learning process. Assignments helped learners gain intuitive knowledge, but the role of model progression was unclear.

Several studies on the effect of MBL on students' learning outcomes focused around real-time graping. The immediacy of graph production can help students develop the cognitive linkage between the occurrence of an event and the graphic representation, while focusing more on what was happening in an activity. However, misconceptions in using and interpreting graphs may include referring to the graph as a picture and confusing slope with height and shape with path of motion (Dunham & Osborne, 1991; Mokros & Tinker, 1987; McDermott, Rosenquist & van Zee, 1987; Goldberg & Anderson, 1989). Brasell (1987) showed that the standard MBL approach was significantly better than delayed MBL or pencil-and-paper approaches due to the immediacy of the graph production experienced in the standard MBL groups. Beichner (1989) suggested that the simultaneity of the physical event and its graphical representation is not the only feature that makes MBL effective; the ability of the student to control the environment may also play a vital role in understanding of physical events.

In high school physical science, Stein (1987) investigated features of an MBL implementation and observed a high degree of student autonomy in troubleshooting and sharing of data when students formed consulting groups for cooperative remediation of problems. In chemistry, Nakhleh and Krajcik (1994) found that high school students constructed more concepts using MBL. In addition, careful analysis of the task, directed teaching, and class discussion were needed to prevent the formation of misconceptions. In physics, Trumper and Gelbman (2000) described an MBL experiment designed for high-school students that analyses very accurately Faraday's law of electromagnetic induction. The authors stated that the MBL tools provided a clear and concrete illustration of this law and that the learning provided a link between a concrete experience and the symbolic representation.

The objective of the research conducted by Cummings et al. at RPI (1999) was to determine if incorporation of research-based activities into Studio Physics would have a significant effect on conceptual learning gains. The researchers used the Force Concept Inventory and the Force and Motion Conceptual Evaluation to verify the effectiveness of interactive lecture, demonstrations, and cooperative group problem solving. The study implied that the standard Studio format used for introductory physics instruction at Rensselaer is no more successful at teaching fundamental concepts of Newtonian physics than traditional instruction. This result is in line with an earlier study (Redish, Saul, & Steinberg, 1997).

The SCALE UP project (Beichner, Bernold, Burnsiton, Dali, Gastineau, Gjertsen & Risley, 1999) was accompanied by research aimed at investigating the effect of the approach on students' achievements and attitudes. During class activities, students had to make predictions, develop models of physical phenomena, collect and analyze data from MBL-type probes, and work on design projects. They were responsible for reading material from the textbook and asking questions. The assessment tools included the Force Concept Inventory, the Test of Understanding Graphs in Kinematics, problem-solving tasks and a final examination. The type of instruction used in the experimental courses, a highly collaborative, technology-rich, activity-based learning environment, had a substantial positive effect on the students' conceptual understanding and confidence levels. This project demonstrated that when implemented properly, the Studio Physics approach is viable and effective. This result encouraged us to propose the TEAL Project at MIT.

THE TEAL PROJECT

In this chapter, we describe the Technology-Enabled Active Learning (TEAL) Project at MIT (Belcher, 2001) as an illustration of the use of visualizations in teaching science. Founded on theoretical underpinnings of cognition (Paivio, 1991; Mayer, 2002), the TEAL project uses both passive and interactive visualizations of phenomena that cannot ordinarily be seen, integrated with textual explanations that immerse the students in an active learning environment. The TEAL visualizations of electromagnetism phenomena (Belcher & Bessette, 2001; Belcher & Olbert, 2003; Dori et al., 2003) are created using such technologies as Java 3D simulation of desktop experiments, ShockWave visualizations, and passive movies made using 3ds max, a commercial 3D modeling and animation software program.

Motivation and Setting

The TEAL Project represents an innovative style of teaching introductory physics at MIT. An important objective of the TEAL project was to transform the way physics is taught in large enrollment physics classes at MIT in order to decrease failure rates and to increase students' conceptual understanding.

The motivation for the transition from the traditional lecture and recitation mode to this different mode was threefold. First, the traditional format for teaching the mechanics and electromagnetism courses at MIT have had a 40-50% attendance rate, even with spectacularly good lecturers, and a 10% or higher failure rate. Second, a range of educational innovations in teaching freshman physics over the last two decades have demonstrated that any pedagogy using "interactive-engagement" methods results in higher learning gains than the traditional lecture format (e.g., Halloun and Hestenes 1985, Hake 1998, Crouch and Mazur 2001). Third, unlike many educational institutions in the United States and around the world, the mainline introductory physics courses at MIT have not had a laboratory component for over three decades, and TEAL was designed to re-introduce a laboratory component to the mainline courses. Visualization technologies can support meaningful conceptual learning by enabling the presentation of spatial and dynamic images that portray relationships among complex concepts. This is particularly true in electromagnetism, in that the topics discussed in this subject are highly abstract in nature and visualization can potentially alleviate difficulties in students' understanding of electromagnetic field concepts.

The TEAL project is centered on an "active learning" approach, aimed at helping students visualize, develop better intuition about, and better conceptual models of, electromagnetic phenomena (Dori & Belcher, 2004). Taught in a specially designed classroom with extensive use of networked laptops, this collaborative, hands-on approach merges lecture, recitations, and desktop laboratory experience in a media-rich environment. In the TEAL classroom, nine students sit together at round tables (see Figure 1), with a total of thirteen tables.

Figure 1. Undergraduate physics students in the d'Arbeloff Studio Classroom

In five hours of class per week (two two-hour sessions and one one-hour problem-solving session led by graduate student TAs), the students are exposed to a mixture of presentations, visualizations, desktop experiments, problem solving and collaborative exercises. The desktop experiments and computer-aided analysis of experimental data provide the students with direct experience of various electromagnetic phenomena.

Electromagnetism Visualizations in TEAL

The visualizations of electromagnetic phenomena embedded in the TEAL project allow students to gain insight into the nature of electromagnetic fields in several ways.

First, the students can appreciate the manner in which fields transmit forces between material objects by watching how the motions of those objects evolve in time in response to the forces. Such visualizations allow students to intuitively relate the forces transmitted by electromagnetic fields to more tangible forces that are directly part of their daily experiences. Examples of such experiences include the forces exerted by strings and rubber bands. Moreover the TEAL visualizations are in many cases *virtual* models of *real* desktop experiments that the students interact with in the class. This makes the physical phenomena associated with fields more concrete by enhancing the feel for reality through adding such artifacts as magnetic field lines embedded in a virtual version of a real experimental setting.

Second the TEAL visualizations employ the display of time-varying graphs of physical parameters such as electric current, which are obtained in real time by sensors hooked to laptops. The data are captured and recorded while students perform the experiments, and are amenable to graphic display modes and further analysis.

The TEAL visualizations are both passive and interactive. The passive visualizations are in the form of mpeg movies generated with 3ds max. They have the advantage of being highly detailed and representing the interaction of material objects with fields at high resolution. These visualizations achieve graphical quality that cannot be reached with a real-time simulation.

The interactive visualizations display lower resolutions than the passive mpeg movies and are of lower graphical quality. However, unlike a passive visualization, interactive visualizations enable the viewer to manipulate elements on the screen and/or control variables so he or she can get a more direct feeling of the phenomenon being demonstrated and experience it with higher involvement than in a passive visualization. Interactive visualizations in the TEAL project further split into simulations and graphs resulting from hands-on desktop experiments.

Following are several black and white examples of visualizations developed especially for and used in the TEAL Electricity and Magnetism (E&M) course. For a wider selection of color and animated passive and active visualizations see http://web.mit.edu/8.02t/www/802TEAL3D or the accompanying CD. Each visualization example is supplemented with an explanation of the underlying physical principles and equations that explain the phenomenon being visualized.

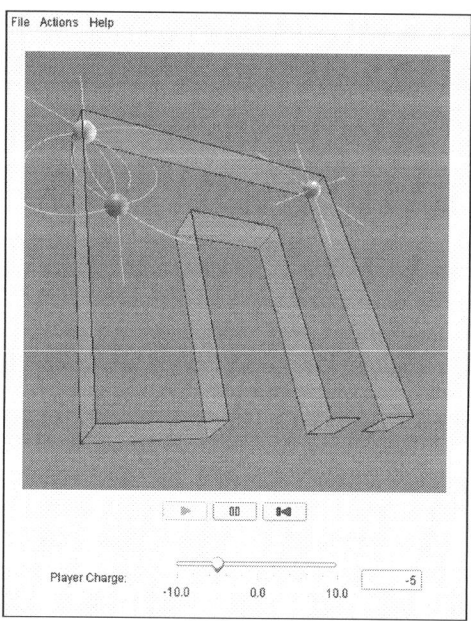

Figure 2. *An interactive electrostatic video game illustrating how the interaction of charges is mediated by their electric fields*

As an example of an interactive visualization in electrostatics, Figure 2 shows an "electrostatic video game" Java 3D applet that helps the student understand how electrostatic fields mediate the attraction of unlike charges and the repulsion of like charges. The student maneuvers a ball around an enclosure by changing the charge on the ball from positive to negative and back again, using the slider shown in the figure. As the charge on the ball is changed, the ball can be moved around the enclosure because of its repulsion or attraction from two stationary and fixed positive charges whose charge does not change. The object of the game is to maneuver the ball through the "gate" at the lower right of the enclosure in the shortest possible time. The educational purpose of this game is to illustrate in a visceral way the electrostatic field "tension" or "pressure" that leads either to attraction or repulsion between particles of unlike charge or like charge, respectively. The student maneuvering the ball by changing its charge sees the constant interplay of tension versus pressure as he or she moves the charge around the enclosure.

Turning to the topic of magnetostatics, we first present the "Floating Coil" passive visualization. This visualization simulates the magnetic fields generated by a permanent magnet and a current-carrying coil of wire suspended by a spring above the magnet. In the configuration shown, the north pole of the magnet is up and the direction of the current in the coil is clockwise as viewed from above. The coil of wire is repelled by the magnet and tends to "float" on the magnetic field of the magnet.

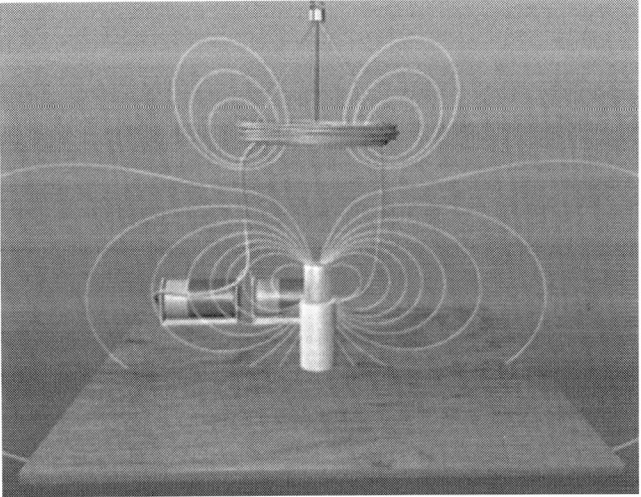

Figure 3. *A passive visualization showing magnetic fields generated by a floating coil*
http://web.mit.edu/8.02T/www/802TEAL3D/visualizations/magnetostatics/floatingcoil/floating
coil.htm

The second example in magnetostatics is an interactive one (see Figure 4). This Java 3D visualization also illustrates the forces on a current carrying coil sitting on the axis of a permanent magnet. For current flowing in one direction in the coil, the force on the coil will be upward. If the current is strong enough, the coil will levitate, floating on the magnetic fields of the coil plus the magnet. This is the way one form of the Magnetic Levitation (MagLev) train works. For the other direction of current, the coil is attracted to the magnet.

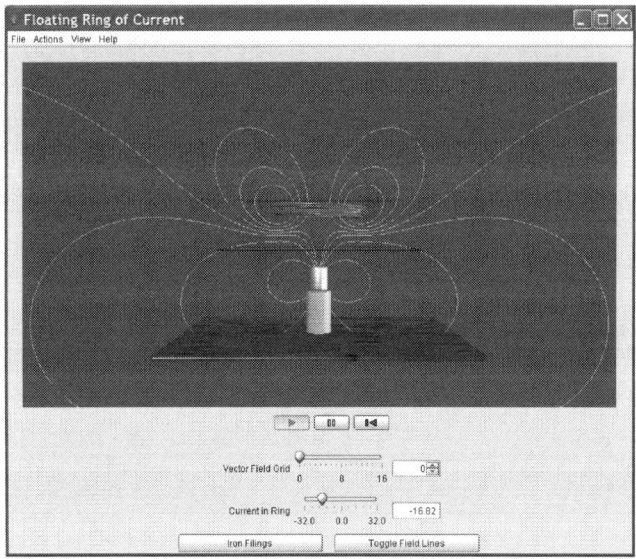

Figure 4. *An interactive 3D visualization showing magnetic fields generated by a floating coil*
http://web.mit.edu/8.02T/www/802TEAL3D/visualizations/magnetostatics/floatingcoilapp/floatingcoilapp.htm

The user interacting with the Java applet 3D visualization shown in Figure 4 can vary the amount of current flowing in the coil. A related homework problem asks the student to calculate the amount and direction of the current necessary to get the coil to levitate above the magnet at a given distance. This calculation mimics a real experiment that the students do in class, where they directly observe the levitation of the coil in this circumstance.

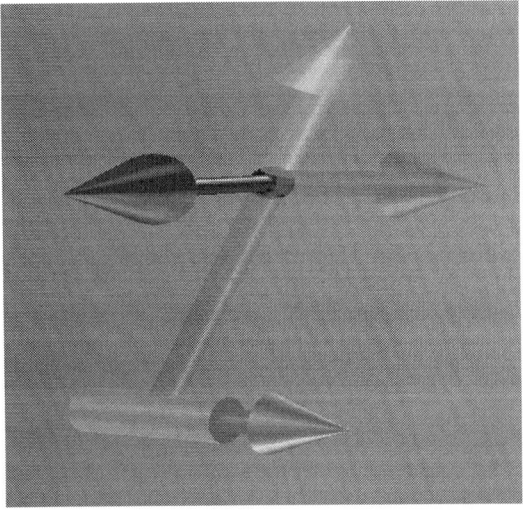

Figure 5: *An interactive visualization of a magnetic field generated by a single current element*
http://web.mit.edu/8.02T/www/802TEAL3D/visualizations/magnetostatics/CurrentElement3d/CurrentElement.htm

A Shockwave interactive visualization of the Biot-Savart Law in Figure 5 shows a magnetic field that a single current element generates. The top arrow pointing to the left represents the magnetic field generated by the current element (the bottom cylinder) at the observation point (the black sphere at the intersection of the three arrows). The observation point can be moved using keyboard arrow keys to sample the field at different positions relative to the current element. The bottom arrow at the current element and the top, right-pointing arrow at the observation point indicate the direction of current flow. The upward-pointing arrow at the observation point indicates the direction from the current element to the observation point. By left-clicking and dragging in the window, the student can change the perspective of the view to get a sense of the 3D properties of this construction.

Figure 6 is another Shockwave interactive visualization, this time of the magnetic field generated by a ring of current. This visualization demonstrates the Biot-Savart Law applied to many current elements. By the principle of superposition, a continuous current distribution can be thought of as the sum of many discrete current elements (30 in this case). The large arrow in Figure 6(a) whose tail is at the observation point (black sphere) shows the total magnetic field at that point. The total field is the sum of the fields due to each element of the ring of current, represented here by the small vectors attached to the observation point. As Figure 6(b) shows, by shift clicking on any current element, the student can change the shading of both that current element and the field that it produces at the observation point. This helps identifying the contribution of that current element to the total field at the observation point. By left-clicking and dragging in the window, the student can change the perspective of the view to get a feel for the 3D properties of this assembly.

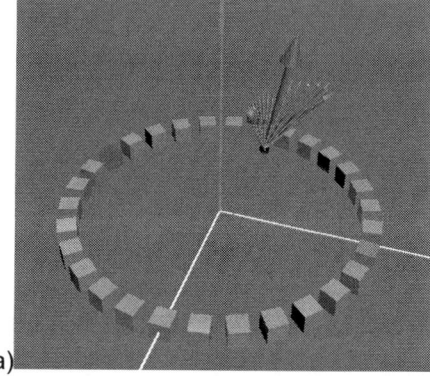

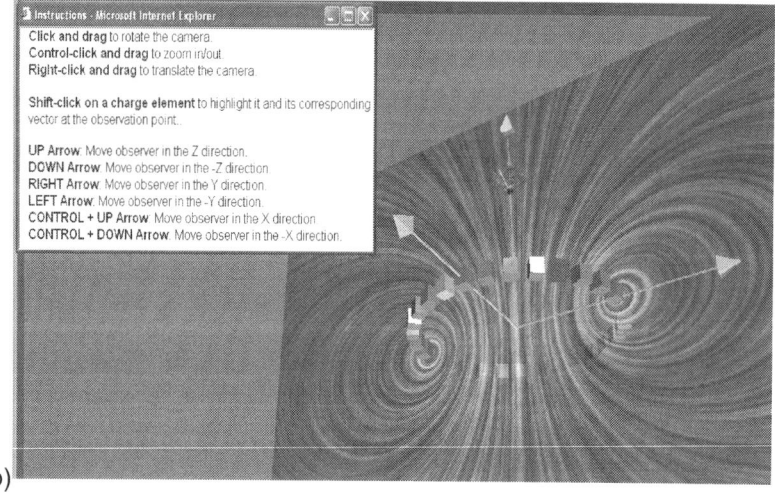

Figure 6. *An interactive visualization showing the magnetic field generated by a ring of current*
http://web.mit.edu/8.02T/www/802TEAL3D/visualizations/magnetostatics/RingMagField/Ring MagField.htm

Figure 7 is another interactive visualization, this time of Faraday's Law. The magnetic field in this JAVA applet is generated by a permanent magnet and by the current in an isolated coil of wire made of conducting non-magnetic wire (e.g., copper) that falls along the axis of the magnet. The only current in the coil is due to eddy currents arising in the coil as a result of the time-changing magnetic flux through the coil as it falls past the permanent magnet.

The student can vary both the total resistance of the coil of wire and the magnetic dipole moment of the magnet. The frame above is taken after a coil of fairly large resistance has fallen past the magnet from above. The curve in the graph box is the current in the wire as a function of time. The time profile here matches closely the time profile of an experiment that the students carry out in class to illustrate Faraday's Law.

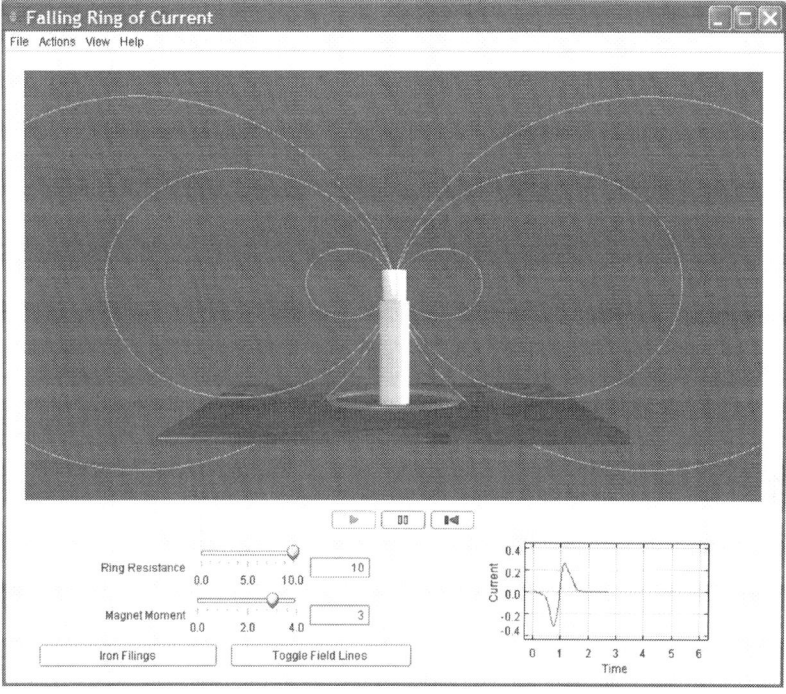

Figure 7. *Farady's Law interactive visualization – the Falling Coil applet*

http://web.mit.edu/8.02T/www/802TEAL3D/visualizations/faraday/fallingcoilapp/fallingcoilapp.htm

Figure 8. *Faraday Ice Pail desktop experiment*

Finally, returning to the subject of electrostatics, Figure 8 is an example of a hands-on desktop experiment dealing with electrostatic induction. The visualization here is of the data collected in the experiment (Figure 8(b)) when the student moves charged paddles in and out of the "Faraday" pail (Figure 8(a)). The two paddles are first rubbed against each other to produce electrostatic charges and are then put into the Faraday pail. The attached charge sensors measure the outer charge distribution that shows up on the laptop screen (Figure 8(b)).

FINDINGS

Quantitative Learning Outcomes of the TEAL Project

We carried out a study that focused, among other factors, on students' learning outcomes and their perceptions about the visualizations and how they help to comprehend abstract concepts. The educational impact of the TEAL project was assessed using conceptual tests and a perception questionnaire. Results from students who took the TEAL Spring 2003 course (the experimental group) and Spring 2002 course (the traditional E&M course, the control group) are reported in detail by Dori and Belcher (2004). To underscore the contribution of the visualizations to the TEAL project students, we next briefly present the main findings of the assessment.

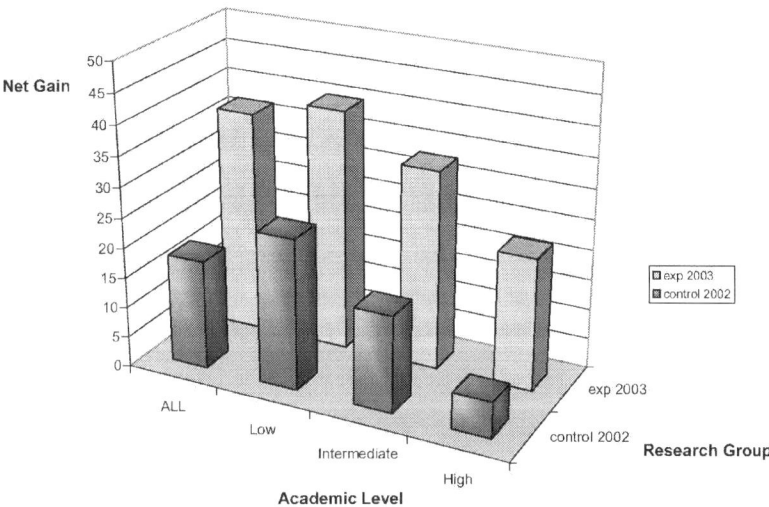

Figure 9. *Students' net gain scores in the conceptual tests by academic levels for Spring 2003 experimental group – TEAL course (N= 514) andSpring 2002 control group – traditional course (N= 121)*

The TEAL Spring 2003 course included about 600 students and six new instructors, who were not part of the development team (Belcher, 2001; Dori & Belcher, 2004; Dori et al., 2003). The students consisted of 90% freshmen and 10% upper classmen, so most of them had never been exposed to the E&M learning material before. The control group of Spring 2002 consisted of 121 students who took the traditional on-term E&M course, which was based on traditional lectures with demonstrations in a large lecture hall and smaller recitation sessions. The control group students were volunteers who were asked to respond to the pre- and posttests for monetary compensation.

To assess the effect of the visualizations and the pedagogical methods implemented in the TEAL project, we examined the scores in conceptual pre- and posttests for the experimental and the control groups. Both the pre- and posttests consisted of 20 multiple-choice conceptual questions from standardized tests (Maloney et al., 2001; Mazur, 1997) augmented by questions of our own devising (Dori & Belcher, 2004). Figure 9 presents the net gain results of the conceptual tests of both experimental and control groups. Based on their pretest scores, students in both research groups were divided into three academic levels: high, intermediate, and low.

The difference in the net gain between the experimental and control groups was significant (p<0.0001) for each academic level separately as well as for the entire population. These results strongly suggest that the learning gains in TEAL are significantly greater than those obtained by the traditional lecture and recitation setting. The results are consistent with several studies of introductory physics education over the last two decades (Beichner et al., 1999; Hake, 2002). It is also in line with the much lower failure rates for the TEAL course of Spring 2003 (a few percent) compared to traditional failure rates in recent years (from 7% to 13%).

In order to investigate Spring 2003 students' perceptions, we asked them to list the most important elements which contributed to their understanding of the subject matter taught in the TEAL project and explain their selection. We divided their responses into four categories: Oral explanations in class, technology, written problems, and the textbook (Serway & Beichner, 2000). The technology category included desktop experiments performed in groups, 2D and 3D visualizations, individual Web-based home assignments turned in electronically, and individual real-time class responses to conceptual questions using a personal response system (PRS) accompanied by peer discussion. Written problems included both individual problem sets given as home assignments and analytic problems solved in class workshops.

The Spring 2003 questionnaire was completed by 308 students. The results showed that about 40% favored the problem solving method, about 22% selected the technology, 22% selected the textbook, and 16% favored the professor's oral explanations. Typical explanations students gave to their selection of technology-based teaching methods (see Table 1) included elements of visualization, desktop experiments, PRS-based conceptual questions, and Web-based assignments. Still, the teacher turned out to be indispensable for both the oral explanations and the problem solving workshops.

Table 1. *Sample of 2003 experimental students' explanations to their selection of the various teaching methods*

Teaching Method		Student's Explanation
Oral explanations by the professor		"Having teachers at our disposal for when we have questions with specific problems is possibly the best aspect of TEAL"
Technology	Desktop Experiments	"Desktop experiments help to really grasp the conceptual background of various problems while integrating calculations and quantitative analysis."
	2D & 3D visualizations	"The visuals and simulations were great for conceptualizing and visualizing how electric and magnetic fields interact with charged particles/wires/etc and what affects, creates and changes them." "The 3D visualizations are the one thing that I can't get from a book or learning on my own." "Visualizations and conceptual questions help explain what is really happening behind all the numbers, especially in an abstract topic as magnetostatics."
	Web-based home assignments	"I think the readings for the Web assignments were really important. They forced me to actually do the readings before class"
	Conceptual questions with PRS	"PRS was the best part of class because it took general concepts and shrunk them down into concise multiple choice questions that both reviewed old stuff and taught new things." "We get to test our knowledge without fear of failure."
Written Problems	Home problem sets	"The problem sets offered the main opportunity to connect material presented in class and figure out how it related to actual material covered on exams."
	Class workshops	"The workshops help me most because I seem to be learning a great deal from working with other students and discussing questions with them."
Textbook		"I learn the most from the textbook because I can learn at my own pace and go back over concepts that I don't understand as many times as I want."

Class Discourse with Various Types of Visualizations

Having presented several examples of visualizations using various techniques and their objectives, as well as quantitative results of TEAL students' learning outcomes, in this section we turn to a description of class discussions. The discourse, which took place while students were engaged in viewing and manipulating various visualizations, gives the flavor of diverse types of students' engagements with visualizations. It also provides some explanations as to how the learning outcomes were achieved.

The first example is related to visualization of a single current element based on Biot-Savart Law. This visualization is designed to encourage students to apply the right hand rule to determine relations between the directions of current and the magnetic field it produces. The second example is a magnetostatics experiment, in which a magnetic field is created by a permanent magnet. The students are asked to predict and draw a graph of the field of that magnet. They then use a Hall Probe to measure and verify or refute their prediction.

Next, we describe a brief "mini-lecture" demonstration of magnet levitation. Finally, an audio demo is described, in which students can hear the difference in signal strength when the orientation of two coils is changed so that their magnetic flux linkage goes from large to zero.

Demonstration of the Biot-Savart Law for a single current element

The professor began the session by providing an explanation of the electric field of a point charge and then drawing an analogy to magnetic field of a current element using PowerPoint slides. Students are quiet, some look at printouts of slides.

Professor: *OK, we're going to look at a visualization of the magnetic field generated by a current element.*

[Some students navigate on their laptops to the visualization, a Shockwave application that demonstrates the Biot-Savart Law for a single current element (see Figure 5).]

Professor: *This is the current element. Out in space, it will create a magnetic field.* [Draws vector out to a point] *Then it takes the cross product at the observation point. If I move this vector that goes out to the observation point, then it gives me the field. Watch what happens when I cross this point.* [Demonstrates on screen]. *It changes direction. This is not very intuitive so people have a hard time with it.* [Navigates back to lecture slides]. *If you look at the Biot-Savart visualization, we can look at what was going on.* [Students act out right hand rule with hands, similar to what is shown in Figure 10]. *Now let's talk about the force on a current-carrying wire in a magnetic field. This is our desktop experiment for today. First let's look at a simulation of what we expect when we put a coil of wire carrying current in the field of a strong permanent magnet, like what you have in the experiment.* [Demonstrates on screen first the passive visualization of Figure 3, then the interactive visualization of Figure 4. Some students watch these visualizations on their laptop screens while others look at one the many large screens surrounding the classroom.]

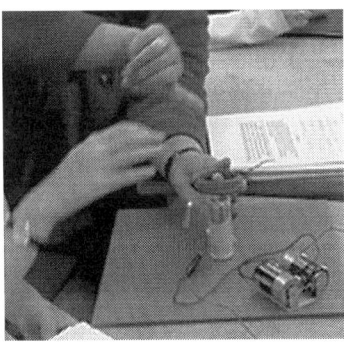

Figure 10. *Students acting out the right hand rule*

[Students watch the visualization, use right hand rule to determine forces, some play with the applet on the laptops. Visualization gets more attention than the slides.]

Professor: *Now let's do the real experiment with a real loop of current. Use your batteries* [Students carry out experiment illustrated in Figure 9 and get the coil to be either attracted or repelled by the magnet, depending on the direction of the current.]

Magnetostatics experiment: Magnetic field due to a permanent magnet

After exploring the force between the current-carrying coil and the magnet in the above experiment, the class turns to a more detailed look at the magnetic field of the current-carrying coil of wire. They first use the Shockwave visualization in Figure 6 to explore virtually how this field is generated. They then use a Hall Probe and make measurements near the actual coil in predetermined ways. In addition to the Hall Probe, they also have a current sensor that measures the current through the coil. Before making the measurements, they are asked to predict what signal they would see. One of the predictions they are asked to make is shown in Figure 10.

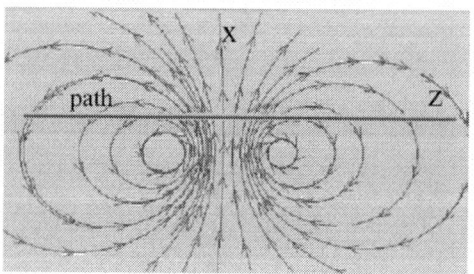

Figure 6 The magnetic field lines of a single coil of wire.

Prediction 7 (*answer on your tear-sheet at the end*): Suppose you move from left to right along the horizontal path indicated in Figure 6 above. Predict the behavior of the *x-component* (i.e. the vertical component) of the magnetic field as you move along that path, and draw it in the panel of Figure 7 and on your tear-sheet at the end.

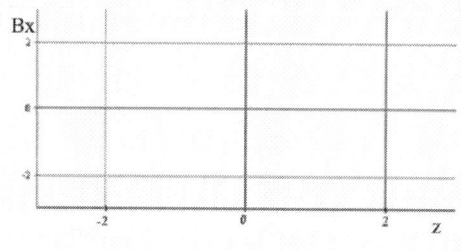

Figure 7 Your prediction of the behavior of the *x -component* (the vertical component) of the magnetic field as you move along the path shown in Figure 6.

Figure 10. The worksheet used by students to predict the magnetic field of a current-carrying coil of wire

Student B: [reading] *Predict magnetic field as you move along the path using the drawing.* [Refers to Figure 10]. *Well it goes negative to zero to positive. See, it's going to get stronger and then weaker.*
Student A: [thinks about this]
Student B: [draws out his prediction]
Student A: *That's convenient.*
Student B: *Well, it looks really weird but… Well, I definitely know it is zero here.*
[Student A looks skeptical, Student B explains his drawing more].
[Students start measuring].
Student A: *How do we measure this?*
Student B: *With these graphs.*
Student A: [reads] *with the current sensor.*
Student B: *Do we need the current sensor?*
Student A: *yeah.*
Student B: *So do we need this thing?*
Student A: *We need both.* [reading] *Usually it tells you. Put that in series with the sensor.*

Student B: [plugs in wires] *So this goes in here?*
Student A: *yup.*
Student B: *OK.*
Student A: *Wait* [pointing]—*this is the current sensor, right?*
Student B: *Yes.*
Student A: *Then it goes in here.*
Student B: *X-direction or Radial on the current sensor?*
Student A: [Looking at handout]. *It says radial.*
Student B: *OK. Good decision.*
Student B: *OK* [reading] *"Measure the vertical component of the magnetic field. Make sure the white dot is up."*
Student A: *Let's just see how it goes.* [They begin testing].
Student B: *Shouldn't the current be the same?*
Student A: *The current is changing.*
Student B: *Maybe we have our things switched.*
Student A [turning to the laptop]: *Stop making me mad!* [to Student B] *I love how they tell us one thing and it is really another.*
Student B: *I know. Do we need to calibrate it?*
[Student A moves sensor across coil, they watch the graph being drawn on the laptop screen. A & B are puzzled by resulting graph].
Student B: *Is that what we just did?*
Professor [to class]: *If you haven't made a measurement, raise your hand.*
Student B: *Don't worry about the numbers, just the slope matters* [then consults with a neighboring group about their results].
Student A: *I think it was right.*
Student B: *If we keep going out, it would get small again.*

Magnet levitation demonstration

This is a brief transcript of a magnet levitation demonstration. Students can watch a movie of this demo on the Web as well.

Professor: *We start with a magnet levitation demonstration: A small magnet levitates above a superconducting disk in liquid nitrogen.* [This is the rare earth magnet levitating above the high temperature super-conductor, see Figure 11]. *There are applications in magnetic levitation trains in Japan and France.*

Figure 11. *A rare earth magnet levitating above a high-temperature superconductor in a bath of liquid nitrogen*

> Student F: [upon seeing magnet levitate] *Wow—that's cool!*
> Professor: *So we're using liquid nitrogen with a high temperature superconductor. At a low temperature, the resistance goes to zero. So I am taking the magnet and putting it here above the conductor with absolutely zero resistance.*

Coil-radio mutual inductance audio demonstration

A radio amplified is hooked up to a large coil of wire. A second coil, placed near the first one, is connected to loudspeakers, with the two coils facing one another. The professor turns on the radio, and a commercial is suddenly audible in the room. When the professor rotates one of the coils by 90 degrees with respect to the other coil, the voice is silenced. The abrupt end of the cookie commercial gets a big laugh from almost all of the students.

> Professor: *So you see the effect of mutual inductance.*
> Student H: *Maybe they should have explained that the radio was related to the demonstration.*
> [Professor does the demonstration again, and students laugh at the sudden end to the voice again.]
> Professor: *By rotating the coil, the signal is not received.*

Summarizing the TEAL class discourse, it was apparent that the social aspect was an important factor in the construction of knowledge and contributed to establishing new insights and sharing knowledge with peers. The class discourse showed that the TEAL approach encouraged multiple-aspect interactions, including technical, affective, and cognitive. The transcript presented in this section illustrates an incremental shift in students' cognition from ambiguous, partially constructed knowledge to repaired, shared knowledge (Roschelle, 1992), much of which is gained through visualazation, engagement with an MBL setting, and social interaction.

DISCUSSION AND SUMMARY

Information and communication technology, as applied for instruction, has struggled with designing appropriate and effective learning environments for teaching science, especially at a higher education level. This chapter has focused on visualizations and simulations in undergraduate physics education and the value they may bring to elevating the level of deep understanding of the laws and phenomena in electricity and magnetism in particular. The visualizations included visual simulations (still and moving, 2D and 3D, active and passive computer-rendered images) and MBL (desktop experiments and the graphs produced) in physics education. In addition to describing the development and usage of these ICT-based means, we also investigated the value they may bring to elevating the level of deep understanding of laws and phenomena in physics in general and electricity and magnetism in particular.

Rieber (2002) discussed learning within computer-based simulations and brought up a key question of when to defer to instruction as an intervention. Presenting too much instruction with a simulation too soon may strengthen the perception of science as accumulation of facts and principles and inhibit a student from discovery learning—using the scientific method to think like a scientist. However, most students are ill-equipped to handle the full discovery learning process without some level of support. Therefore, instructional intervention is a viable and desirable approach, along with suggestions during the use of simulation, as long as the teacher or the professor carefully chooses models and multiple representations and the order in which they are presented.

It is interesting to compare the level of visualizations in physics and chemistry. In chemistry (Russel & Kozma, 2004, in this book), visualizations have been extensively developed, applied and studied in the context of computerized molecular modeling. However, the MBL elements of visualizations in chemistry have not been investigated thoroughly. In physics, on the other hand, the situation is almost reversed. While MBL has been developed, applied and researched extensively (Kown, 2002; Lapp, 1999; Trumper & Gelbman, 2000), the other facet of visualizations, namely visual simulations, is much less prevalent. Most of the literature describes simulations in mechanics and the learning effects of some of these simulations were investigated with inconclusive results (e.g., de Jong et. al., 1999; Steinberg, 2000; Hake, 1998). In electricity, fewer simulations exist and these are even less investigated in terms of their learning gains. In contrast to the TEAL outcomes, Escalanda and Zollman (1997) did not find significant differences in final exam scores between the participants in interactive digital video activities in introductory college physics course and the non-participants. However, as in the TEAL project, the researchers found that the majority of the participants felt that the activities and the visualizations were effective in helping them learn the physics concepts studied in that course.

The TEAL project has been described in detail since it focuses on electromagnetism and incorporates both visual simulations and MBL. The research outcomes of the TEAL project indicate that students significantly improved their conceptual understanding of the subject matter since they were engaged in high levels of interaction, as can be seen from the class discourse above. The results strongly suggest that the learning gains in TEAL are significantly greater than those obtained by the traditional lecture and recitation setting. The positive results are consistent with several studies of introductory physics education over the last two decades (Beichner et al., 1999; Hake, 2002). It is also in line with the much lower failure rates for the TEAL course of Spring 2003 (a few percent) compared to traditional failure rates in recent years (from 7% to 13%).

While Brasell (1987) emphasized the importance of real-time analysis with microcomputer-based laboratory tools, Brungardt and Zollman (1995) demonstrated through qualitative and quantitative data that the simultaneous-time effect in videodisc activities is not a critical factor in improving student learning of kinematics while using their graphing skills. However, they found that students were aware of the simultaneous-time effect and seemed motivated by it: students decreased eye movement between computer screen and video screen as subsequent graphs were produced, and demonstrated more discussion during graphing than delayed-time students. Russell, Lucas & McRobbie (2004) claimed that positive research findings in support of MBL activities facilitating science learning can be explained in terms of the increased opportunities for student-student interactions and peer group discussions about familiar events.

Some of the inconclusive results in the literature regarding the effect of rich visualization environments on cognitive skills can be explained by Bonham, Deardorff and Beichner (2003) who claimed that the change in the medium itself has limited effect on student learning. Their research findings indicate that although web-based homework affects the nature of feedback students received, there was no significant differences between students who used web-based homework and those who used paper-based homework - when comparing students performance on regular and conceptual exams, quizzes, laboratory and homework.

When educational technology is intertwined with social interaction and proper mentoring, improvement in both the cognitive and affective domains can be achieved not only for middle and high school science education, as shown in previous studies (Blumenfeld Fishman, Krajcik, Marx & Soloway, 2000; Barnea & Dori, 2000; Dori, Sasson, Kaberman & Herscovitz, 2004; Elyon, Ronen & Ganiel, 1996; Krajcik, 2002; Linn, 1998), but also in undergraduate chemistry (Adamson, Nakhleh & Zimmerman, 1997; Dori, Barak & Adir, 2003; Kozma, Chin, E., Russell, J., & Marx, 2000) and physics education (Dori & Belcher, 2004).

As Lee, Nicoll, and Brooks (2004) indicated, students tend to approach physics mathematically, not conceptually, when they work on traditional textbook problems. However, they think about phenomena conceptually before making calculations when they interact with problems in a visualizations setting. More studies are needed to further establish the educational value of the various types of visualizations for teaching physics in general and electromagnetism in particular.

ACKNOWLEDGEMENT

The TEAL project is supported by the d'Arbeloff Fund, the MIT/Microsoft iCampus Alliance, NSF Grant #9950380 and the MIT School of Science and Department of Physics. We thank Steve Lerman, Director, Center for Educational Computing Initiatives (CECI), MIT for hosting and supporting this research. We also thank the faculty, staff and students of the CECI who contributed to the TEAL project. Special thanks to Erin Hult who helped collecting and processing the assessment data and to Ayala Cohen of the Faculty of Industrial Engineering and Management, Technion, for her valuable statistical analyses.

REFERENCES

AAAS (1989). Science for All American. Washington, DC: American Association for the Advancement of Science.

Adamson, G., Nakhleh, M.B. & Zimmerman, J. A (1997). Computer-interfaced O_2 Probe: Instrumentation for undergraduate chemistry laboratories. Journal of Computers in Mathematics and Science Teaching, 16(4), 513-525.

Barnea, N. & Dori, Y.J. (2000). Computerized molecular modeling the new technology for enhancing model perception among chemistry educators and learners. *Chemistry Education: Research and Practice in Europe,* 1(1), 109-120. http://www.uoi.gr/conf_sem/cerapie/2000_January/pdf/16barneaf.pdf

Beichner, R.J. (1989).The effect of simultaneous motion presentation and graph generation in a kinematics lab. Dissertation Abstracts International, 50, 06A.

Beichner, R.J. et al. (2002). Scale-Up Project. www.ncsu.edu/per/scaleup.html

Beichner, R., Bernold, L., Burnsiton, E., Dali, P., Gastineau, J., Gjertsen, M. & Risley, J. (1999). Case study of the physics component of an integrated curriculum. Phys. Educ. Res., Am. J. Phys. Suppl., 67, 16-24.

Belcher, J.W. (2001). Studio Physics at MIT. MIT Physics Annual. http://evangelion.mit.edu/802teal3d/visualizations/resources/PhysicsNewsLetter.pdf

Belcher, J W. & Bessette, R. M. (2001). Using 3D animation in teaching introductory electromagnetism. Computer Graphics 35, 18-21.

Belcher, J. W. & Olbert, S. (2003). Field line motion in classical electromagnetism. The American Journal of Physics 71, 220-228.

Blumenfeld, P., Fishman, B. J., Krajcik, J., Marx, R. W., & Soloway, E. (2000). Creating usable innovations in systemic reform: Scaling up technology-embedded project-based science in urban schools. Educational Psychologist, 35(3), 149-164.

Bonham, S.W., Deardorff D.L. & Beichner, R.J. (2003). Comparison of student performance using web and paper-based homework in college-level physics. Journal of Research in Science Teaching, 40, 1050-1071.

Boulter, C. J. & Gilbert, J. K. (2000). Challenges and opportunities of developing models in science education. In J. K. Gilbert & C. J. Boulter (Eds.), Developing Models in Science Education (pp. 343-362). Dordrecht: Kluwer.

Brasell, H. (1987). The effect of real time laboratory graphing on learning graphic representation of distance and velocity. Journal of Research in Science Teaching, 24, 385-395.

Brungardt, J.B. & Zollman, D. (1995). Influence of interactive videodisc instruction using simultaneous-time analysis on Kinematics graphing skills of high school physics students. Journal of Research in Science Teaching, 32, 855-869.

Cadmus, R.R. Jr. (1990). A video technique to facilitate the visualization of physical phenomena. American Journal of Physics, 58(4), 397-399.

Carlsen, D.D., & Andre, T. (1992). Use of microcomputer simulation and conceptual change text to overcome students' preconceptions about electric circuits. Journal of Computer-based Instruction, 19, 105-109.

Chabay, R. (2002). Electric & Magnetic Interactions: The Movies. Available: http://www4.ncsu.edu/%7Erwchabay/emimovies/

Chabay, R. & Sherwood, B. (2002). Matter & Interactions, Vol. II: Electric & Magnetic Interactions, New York: John Wiley & Sons.

Chi, M.T.H., Feltovich, P.J. & Glaser, R. (1981). Categorization and representation of physics problems by experts and novices. Cognitive Science, 5, 121-152.

Cox, A.J, Belloni, M., Dancy, M. & Christian, W. (2003). Physlets in introductory physics. Physics Education, 38(5), 433-440.

Cummings, K., Marx, J., Thornton, R. & Kuhi, D. (1999). Evaluating innovation in studio physics. Phys. Educ. Res., Am. J. Phys. Suppl., 67, 38-44.

de Jong, T., Martin, E., Zamarro, J.M., Esquembre, F., Swaak, J. & van Joolingen, W. R. (1999). The integrating of computer simulation and learning support: An example from the physics domain of collisions. Journal of Research in Science Teaching, 36, 597-615.

Dori, Y.J., Barak, M. & Adir, N. (2003). A Web-based chemistry course as a means to foster freshmen learning. Journal of Chemical Education, 80, 1084-1092.

Dori, Y.J. & Belcher, J.W. (2004). How does technology-enabled active learning affect students' understanding of scientific concepts? Accepted to The Journal of the Learning Sciences.

Dori, Y.J., Belcher, J.W., Bessette, M., Danziger, M., McKinney, A. & Hult, E. (2003). Technology for active learning. *Materials Today*, **6**(12), 44-49.

Dori, Y.J., Sasson, I. Kaberman, Z. & Herscovitz, O. (2004). Integrating case-based computerized laboratories into high school chemistry. The Chemical Educator, 9, 1-5.

Dunn, J.W. & Barbanel, J. (2000). One model for an integrated math / physics course focusing on electricity and magnetism and related calculus topics" American Journal of Physics, 68(8), 749-757.

Dunham, P.H. & Osborne, A. (1991).Learning how to see: Students' graphing difficulties. Focus on Learning Problems in Mathematics, 13(4), 35–49.

Escalada, L.T., Grabhorn, R. & Zollman, D.A. (1996). Applications of interactive digital video in a physics classroom, Journal of Educational Multimedia and Hypermedia, 5(1), 73-97.

Escalada, L.T., Rebello, N.S. & Zollman, D.A. (2004). Students' explorations of quantum effects in LEDs and luminescent devices, The Physics Teacher, 42, 173-179.

Escalanda, L.T. & Zollman, D.A. (1997). An investigation on the effects of using interactive digital video in a physics classroom on student learning and attitudes. Journal of Research in Science Teaching, 34, 467-489.

Eylon B., Ronen M. and Ganiel U. (1996). Computer simulations as a tool for teaching and learning: Using a simulation environment in optics. Journal of Science Education and Technology, 5(2), 93-110.

Frederiksen, J.R., White, B.Y. & Gutwill, J. (1999). Dynamic mental models in learning science: The importance of constructing derivational linkages among models. Journal of Research in Science Teaching, 36, 806-836.

Gilbert, J. K., & Boulter, C. J. (Eds.). (2000). Developing Models in Science Education. Dordrecht: Kluwer.

Goldberg, F.M. & Anderson, J.H. (1989). Student difficulties with graphical representations of negative values of velocity. The Physics Teacher, 4, 254-260.

Hake, R.R. (1998). Interactive-engagement versus traditional methods: A six-thousand-students-survey of mechanics test data for introductory physics courses. American Journal of Physics, 66, 67-74.

Handelsman, J., Ebert-May, D., Beichner, R., Bruns, P., Chang, A., DeHaan, R., Gentile, J., Lauffer, S., Strwart, J., Tilghman, S. & Wood, W. (2004). "Scientific Teaching", Science Magazine, 304, 521-522.

Hestenes, D. (2003). Oersted Medal Lecture 2002: Reforming the mathematical language of physics. American Journal of Physics, 71(2), 104-121.

Jacobson, M. J. (2004). Cognitive visualizations and the design of learning technologies. International Journal of Learning Technology, 1, 40-62.

Kown, O.N. (2002). The effect of calculator-based ranger activities on students' graphing ability. School Science and Mathematics, 102. 57-67.

Kozma, R.B., Chin, E., Russell, J., & Marx, N. (2000). The role of representations and tools in the chemistry laboratory and their implications for chemistry learning. Journal of the Learning Sciences, 9(3), 105-144.

Krajcik, J.S. (2002). The value and challenges of using learning technologies to support students in learning science. Research in Science Education, 32(4), 411-415.

Lapp, D.A. (1999). Using calculator-based laboratory technology: Insights from Research. Proc. ICTMT4 - The Fourth International Conference on Technology in Mathematics Teaching, Plymouth, England. http://www.tech.plym.ac.uk/maths/CTMHOME/ictmt4/P40_Lapp.pdf

Larkin, J.H. (1983). The role of problem representation in physics. In: Gentner, D. & Stevens, A.L. (Eds). Mental models. Lawrence Erlbaum Associates, London. Pg. 75-98.

Larkin, J., McDermon, J., Simon, D.P. & Simon, H.A. (1980). Expert and novice performance in solving physics problems. Science, 208, 1334-1342.

Laws, P.W. (1991). Calculus-based physics without lectures. Physics Today, 44, 24-31.

Lee, K.M., Nicoll, G. & Brooks, D.W. (2004). A comparison of inquiry and worked example Web-based instruction using Physlets. Journal of Science Education and Technology, 13(1), 81-88.

Linn, M.C. (1998). The impact of technology on science instruction: Historical trends and current opportunities. In: B Fraser & K. Tobin (eds.) International handbook of science education. Dordrecht: Kluwer Academic Publishers, pp. 265-293.

Maloney, D., O'Kuma, T., Hieggelke C. & Van Heuvelen, A. (2001). Surveying students' conceptual knowledge of electricity and magnetism. Physics Education Research, American Journal of Physics Suppl. 69(7), S12-S23.

Martinez- Jimenez, P. & Casado, E. (2004). Electros: Development of an educational software for simulation in electrostatic. Computer Application in Engineering Education, 12, 65-73.

Mathewson, J.H. (1999). Visual-spatial thinking: An aspects of science overlooked by educators. Science Education, 83, 33-54.

Mayer, R. E. (2002). Cognitive theory and the design of multimedia instruction: An example of the two-way street between cognition and instruction. In D. F. Halpern & M. D. Hakel (Eds.), Applying the science of learning to university teaching and beyond (pp. 55-72). San Francisco: Jossey-Bass.

Mazur, A. (1997). Peer Instruction. Prentice Hall, Upper Saddle River, NJ.

McDermott, L.C. (1991). Millikan Lecture 1990: What we teach and what is learned – closing the gap. American Journal of Physics, 59, 301-315.

McDermott, L.C., Rosenquist, M.L. & van Zee, E.H.(1987). Student difficulties in connecting graphs and physics: Examples from kinematics. American Journal of Physics, 55(6), 503-513.

McDermott, L.C. & Shaffer, P.S. and the Physics Education Group (2002). Tutorials in Introductory Physics. Upper Saddle River, NJ: Prentice Hall.

Mokros, J.R. & Tinker,R.F. (1987).The impact of microcomputer-based labs on children's ability to interpret graphs. Journal of Research in ScienceTeaching,24(4),369–383.

Nakhleh, M. B. & Krajcik, J. S. (1994). Influence of levels of information as presented by different technologies on students' understanding of acid, base, and pH concepts. Journal of Research in Science Teaching, 31(10), 1077-1096.

Novak, G.M., Patterson, E.T., Gavrin, A.D. & Christian, W. (1999). Just-In-Time Teaching: Blending Active Learning with Web Technology Prentice Hall, New Jersey.

NRC - National Research Council. (1996). National Science Education Standards. Washington, D.C.: National Academic Press.

Paivio, A. (1990). Mental representations: A dual coding approach (2nd ed.). New York: Oxford University Press.

Redish, E. F., Saul, J. M. & Steinberg, R. N. (1997). On the effectiveness of active-engagement microcomputer-based laboratories. American Journal of Physics, 65, 45–54.

Rieber, L.P. (2002). Supporting discovery-based learning with simulations. Invited presentation at the International Workshop on Dynamic Visualizations and Learning, Knowledge Media Research Center, Tubingen, Germany, July 18-19. Available: http://www.iwm-kmrc.de/workshops/visualization/rieber.pdf

Rieber, L.P., Boyce, M., & Assad, C. (1990). The effects of computer animation on adult learning and retrieval tasks. Journal of Computer-based Instruction, 17, 46-52.

Rieber, L.P., Noah, D. & Nolan, M. (1998, April). Metaphors as graphical representations within open-ended computer-based simulations. Paper presented at the annual meeting of the American Educational Research Association, San Diego.

Rieber, L.P., & Parmley, M.W. (1995). To teach or not to teach? Comparing the use of computer-based simulation in deductive versus inductive approaches to learning with adults in science. Journal of Educational Computing Research, 14, 359-374.

Roschelle, J. (1992). Learning by collaborating: Convergent conceptual change. Journal of the Learning Sciences, 2(3), 235-276.

Russell, D.W., Lucas, K.B. & McRobbie, C.J. (2004). Role of the Microcomputer-based Laboratory display in supporting the construction of new understanding in thermal physics. Journal of Research in Science Education, 41, 165-185.

Stein, J.S. (1987). The computer as lab partner: Classroom experience gleaned from one year use of microcomputer-based laboratory. Journal of Educational Technology Systems, 15, 225-235.

Steinberg, R.N. (2000). Computers in teaching science: To simulate or not to simulate? American Association of Physics Teachers, 68, S37-S41.

Suwa, M. & Tversky, B. (2002). How do designers shift their focus of attention in their own sketches? In: Anderson, M., Meyer, B. & Olivier, P. (eds.) Diagrammatic Representation and Reasoning, 241-254, Springer Verlag, London.

Sadoski, M. & Paivio, A. (2001). Imagery and text: A dual coding theory of reading and writing. Mahwah, NJ: Lawrence Erlbaum Associates.

Sokoloff, D.R., Thornton, R.K. & Laws, P.W. (1999). Real Time Physics Active Learning Laboratories. New York, NY: Wiley.

Thornton, R.K. & Sokoloff, D.R. (1990). Learning motion concepts using real-time microcomputer-based laboratory tools. American Journal of Physics, 58, 858-866.

Treagust, D. F., Harrison, A. G., Venville, G. J., & Dagher, Z. (1996). Using an analogical teaching approach to engender conceptual change. International Journal of Science Education, 18(2), 213-229.

Trumper, R. & Gelbman, M. (2000). Investigating electromagnetic induction through a microcomputer based laboratory. Physics Education, 35, 90-95.

White, B.Y. (1993). ThinkeTools: Casual models, conceptual change, and science education. Cognition and Instruction, 10, 1-100.

KATHY TAKAYAMA

CHAPTER 11

VISUALIZING THE SCIENCE OF GENOMICS

The University of New South Wales, Australia

Abstract: The term 'genomics' broadly refers to the study of the genome, or the complete genetic inheritance of an organism. The genome sequence of an organism provides the equivalent of a complete genetic map; yet, knowledge of the sequence itself does not reveal how this map manifests itself into the physical characteristics or phenotypes observed for an organism. Genomics research is dependent upon comparative analyses of extraordinary volumes of data. Whilst visualizations may facilitate the significance and understanding of such comparisons, the complexity and scope of the information provides a challenge for the classroom learner. This chapter examines the roles of representations in genomics 'visual literacy', and addresses the challenges associated with distilling a rapidly progressing research area into pedagogical frameworks that can accommodate the dynamic nature of the field. The chapter also presents an application of visualizations in genomics education within the context of a tertiary level international collaborative research project. The student-centred project, 'Visualizing the Science of Genomics', presents a novel example of inquiry-based teaching in genomics in an online environment.

INTRODUCTION

The 'biological sciences' have traditionally been a visually-rich discipline, from the sub-microscopic to the biome level. Educators have depended upon the combination of theory with laboratory or field-based teaching to provide contextual frameworks for student learning. The era of genomics has introduced novel challenges and opportunities for the incorporation of visualizations in science education. The 'exemplar phenomena' as described by Gilbert in the first chapter of this book take on a different context in genomics. For, in genomics, we have not exemplar phenomena, but the field itself lays the foundation for the discovery of exemplars. It is an approach whereby systematic discovery, integration and analysis of genetic information are transformed into multidisciplinary contexts. The models are cognitively stratified as the 'interpretation of the genomic script' continues to reveal new layers of complexity. Whilst scientists readily embrace rapid improvements in sophisticated visualization and analytical technologies in genomics research, they are not practicable in the teaching laboratory. The research-teaching nexus risks disconnection if fidelity of context cannot be attained through an authentic learning experience. Visualizations can facilitate the understanding of genomics if they are strategically presented, learning with the visualization is appropriately scaffolded, and educators consider the requirements for visual literacy of the learner in specific contexts.

GENOMICS: WHAT, WHY, HOW

The term 'genomics' broadly refers to the study of the genome, which encompasses the complete genetic inheritance (DNA) of an individual organism. Ever since the elucidation of the double helical structure of DNA by Watson and Crick in 1953 (Watson and Crick, 1953), scientists have been analysing individual genes from various species long before the term 'genomics' entered the public vernacular. Yet it was not until more than 40 years after Watson and Crick's discovery that the first genome, from the bacterium *Haemophilus influenzae*, was deciphered, revealing the complete sequence of A, G, C and T bases comprising its DNA code (Fleischmann et al., 1995). The era of genomics was hence launched by this achievement, and has contextualised research and teaching in the life sciences into an entirely new framework. The availability of the complete DNA blueprint for an organism paves the way for global analysis of how the genetic, metabolic, and physiological processes in the organism are dictated, and how these networks interact. Subsequently, the much-anticipated announcement of the completion of the human genome sequence in 2001 (Lander et al., 2001) marked an historical milestone. As of September 2004 a total of 215 completed genome sequences had been published, and 963 genome sequencing projects were in progress (Bernal et al., 2001; Krypides, 1999, 2004) and the number of sequenced genomes continues to increase at a rate of one per week.

The genome sequence of an organism provides the equivalent of a complete genetic map. However, knowledge of the genome sequence itself does not reveal how this map manifests itself into the physical characteristics or behaviours (phenotypes) observed for an organism. Some would argue that we are now well ensconced in a *post-genomic* era, as we focus on the analyses of the enormous datasets that comprise these genetic maps. Such analytic approaches include functional genomics (study of the expression patterns for all of the genes in a genome), structural genomics (study of physical characteristics of the genome), comparative genomics (comparisons of genomes from different organisms), proteomics (study of expression patterns for all proteins of a cell/organism and pharmacogenomics (application of genomics toward the identification of drug targets). Taxonomic jargon notwithstanding, the enormity and scope of these approaches render them crucially dependent upon bringing together communities of diverse scientists contributing their distinctive talents. Consequently, the genomics/post-genomics eras have led to a major international shift in research and educational strategic initiatives, resulting in the development of interdisciplinary centres and degree programmes.

The biology student studying genomics is presented with a contextually rich perspective on how living systems function at a global molecular and cellular level. That is, whilst the development of techniques in molecular biology over several decades facilitated our understanding of how specific genes or pathways functioned individually, genomics provides us with a holistic view of how all of these pathways function simultaneously. In so doing, genomics enables the elucidation of entire networks, and for the educator, genomics presents a forum for the exploration of a broad range of contexts. Hence, biology education in the 21st century has evolved towards what has come to be known as 'systems biology', embracing multi- and interdisciplinary frameworks.

SIGNIFICANCE OF VISUAL LITERACY IN GENOMICS EDUCATION

In research practice, the scope of the field of genomics has enhanced interactions: i) between the sciences, including biology, mathematics, computer science, chemistry, and statistics; ii) between the basic and clinical sciences; and iii) between the life sciences, social sciences and humanities. In practicality, however, an authentic learning experience that reflects this multidisciplinarity is not as readily created in a 'traditional' classroom or course structure. The challenge for the biology educator to teach genomics in a contextually relevant framework which simultaneously is reflective of research practice and societal value becomes dependent on alternate approaches incorporating visualization and application.

Genomics research is dependent upon visual comparative analysis of extraordinary amounts of complex information. Most of this primary visual analysis, however, is performed by high-precision detection equipment and computers. The complexity and scope of such data, whilst daunting enough for the researcher, is prohibitively impractical for classroom students. The 'visual literacy' of the learner is significantly distinct from that practiced by the genomics researcher. Such visual literacy skills are acquired through extensive experience and contextual practice. According to Christopherson, visual literacy includes the ability to interpret and comprehend the meaning and significance of visual representations, effective communication through the application of the basic principles and concepts of visual design, the production of visual messages through appropriate technologies, and the application of visual thinking toward problem-solving (Christopherson, 1997).

The effectiveness of visual representations in genomics education pivots on distillation of the essential and useful 'visual messages' from the breadth and depth of options available. Metavisual competence as described in Chapter 1 by Gilbert, may be described in the area of genomics as specifically encompassing the following practices:

- Recognition and comprehension of modes of visual representation of genomic data.
- Communication of genomic information through visual representation.
- Comparative application of visual representations between sets of genomic data.
- Transfer of genomic information from one visual mode of representation into another visual mode of representation.
- Development of models representing relationships based on quantitative and/or qualitative comparisons of visual genomic data analysis.
- Predictions of behaviour of new genomic data based on previous visual models.

These skills are gradually acquired through active experience in genomics research, whereby the researcher's understanding evolves as knowledge is synthesised within appropriate contexts. However, the classroom student who does not have the opportunity to access this experiential learning must develop these metavisual skills through alternate frameworks that are often subject to limitations of time and resources.

The development of genomics visual literacy for the learner is facilitated through structured contextual exemplars. For example, given a new dataset of genome sequences from a novel bacterial species, the student may first utilise bioinformatics tools to analyse the data. S/he then learns to recognise specific visual motifs in the output that indicate functional, taxonomic, or structural information about the genomic sequence. This process might be conducted within the context of investigating evolutionary relationships between genes or organisms, or within the context of following the epidemiology of a disease-causing pathogen. Such frameworks provide relevance and application, key criteria for authentic learning experiences (Chinn and Malhotra, 2002; Herrington and Herrington, 1998; Kolb, 1984; Meyer, 1992). Oftentimes the visual exemplars employed under the umbrella of genomics education do not represent first order concepts that stand alone, but rely inherently on the integration of conceptual understanding, identification of relationships between genomic concepts, and applications of several concepts toward the discovery of new information or the formulation of new hypotheses. The cognitive skills associated with this visual literacy are indicative of a higher order level. The student engaged in these visual representations would not only have to comprehensively integrate several genomics concepts associated with the visual representation, but would also proceed to apply this relational interpretation toward the synthesis of a new visual schematic. The complex nature of visual literacy in science and its dependence on the synthesis of knowledge from different contexts has indeed been highlighted by others (Ferk et al., 2003; Vrtacnik et al., 2000a, 2000b; Wu and Shah, 2004).

To ultimately facilitate visual literacy at this level, it is imperative to ensure that the process is appropriately scaffolded by assessing competence in 'prerequisite' visual literacy in the genomic context. That is, in order to perform cognitive operations in a spatial domain, the learner must be competent in the visuospatial skills that are required for each of the conceptual steps that comprise genomics visual literacy. If one were to deconstruct the metavisual processes described above into specific elements of 'basic' visual cognition, they could be described as follows:

- Gene structure: the visual conceptualisation of the structural and regulatory components of a gene, the basic unit of a genome. A corollary to this is the visual cognition of the structural components of a protein, which is a product encoded by a gene.
- Gene orientation and organisation: the ability to recognise or identify the orientation and organisation of genes relative to each other. There are three possibilities for genetic arrangement (orientation), as depicted below, whereby the arrows indicate the direction in which a gene is 'read':

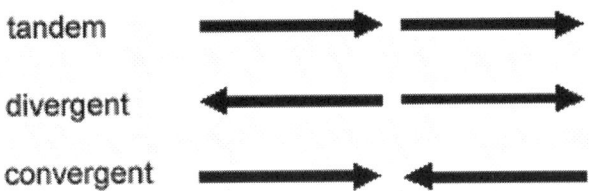

- Gene relationships: integration of spatial and temporal conceptualisations of the regulation of one or more genes by another gene(s). For example, in the diagram below, Gene A can positively influence the expression of Gene B. Gene B in turn negatively regulates Gene C. However, Gene A cannot be said to directly negatively regulate Gene C, because if a mutation results in the loss of Gene B, Gene A has no direct control over Gene C:

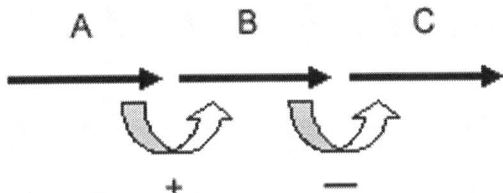

- Sound understanding of which visual representation/application/model (or a combination thereof) is most appropriate for conveying specific genomic concepts, observations, or relationships.
- Genome relationships: visual recognition of patterns of similarity or differences between: i) genome sequences; ii) patterns of the expression of genome sequences; or iii) functional consequences of alterations in genome sequences.

The development of visual literacy in genomics education builds upon these foundations. Layers of complexity are added to contribute contextual depth and breadth to one's metavisual cognition through the integration of experience and practice. The significance of this learning process becomes apparent if we reflect on the scientific method itself.

The practice of scientific inquiry is traditionally represented as a progression of distinct steps which are connected in a mainly sequential order:

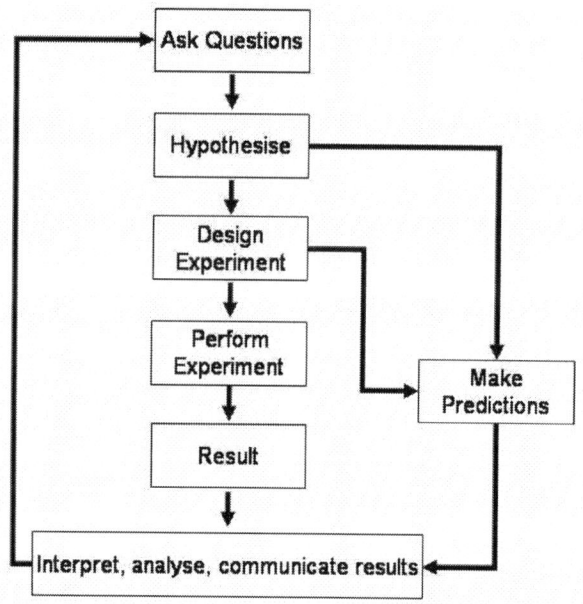

Figure 1. *Traditional representation of the process of scientific inquiry.*

The same process, when considered from a metavisual perspective, becomes highly interactive through the inherently dynamic role that visual cognition plays throughout every phase of inquiry (Figure 2). The visualizations created and employed in the process are themselves fluid in that they both inform, and are informed by each stage of the cycle. Furthermore, the formulation of predictions becomes more central in a metavisual context, because visual predictions can be conceptual (mental models; as discussed by Gilbert in Chapter 1, and reviewed in Gilbert, 2002) and/or factual (visual representation of data).

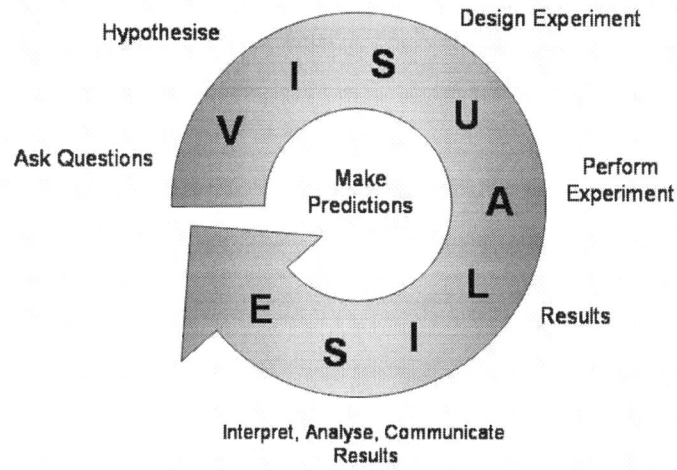

Figure 2. *The process of scientific inquiry represented from a metavisual perspective. The model is a modification of that proposed by Keefe et al. (2004), in which visualization is one of the sequential steps of the scientific process.*

Kirby and co-authors emphasise the importance of the definition of explicit visualization goals in order to create effective visualizations (Keefe et al., 2004). In the model above, the visual goals would evolve with each cycle of the scientific process. Indeed, the definition of visualization goals has been described as an iterative process dictated by the underlying scientific applications (Brooks Jr., 1996), and as progress is made in the elucidation of the scientific problem, the goals for the visualizations that address or communicate the problem are modified accordingly. The field of genomics presents a prime example of how such a formative metavisual approach continues to inform both research practice and pedagogy.

THE ROLE OF VISUALIZATION IN TEACHING GENOMICS

The rapidly increasing number of sequenced genomes continues to add to the enormity and diversity of genomic information in global databases. The researcher or classroom learner is challenged with interpreting and communicating a study focused on a gene(s) or genome of interest with peers who may approach the analysis of a gene/genome from a different contextual perspective. The Gene Ontology Consortium has developed a taxonomic system to facilitate a universal dialogue based on ontologies describing the products of genes in terms of their: i) molecular functions; ii) associated biological processes; and iii) cellular components (Consortium, 2000). According to the Consortium, the *molecular function* describes *what* the gene product does; that is, its molecular activity. The *biological process* refers to the *significance (how)* of this activity within "one or more ordered assemblies of molecular functions" (Consortium, 2000). Whilst a biological process is not equivalent to a pathway, it is dependent upon several or large networks of pathways to achieve a broader aim. Examples range from specific processes such as pyrimidine metabolism to broader processes such as cell growth and maintenance. The *cellular component* simply describes *where* the activity of the gene product takes place in the cell (Consortium, 2000).

These groupings are comprehensively inclusive for any gene regardless of species. I would propose that all genomic visualizations can also be categorised under the same contextual groupings. In identifying the category(ies) to which a given visualization belongs, the educator or learner develops a metacognitive awareness of the goals and interpretive practices invoked by the visualization. This in turn facilitates the processes associated with conceptual understanding of the content embedded in the visualization.

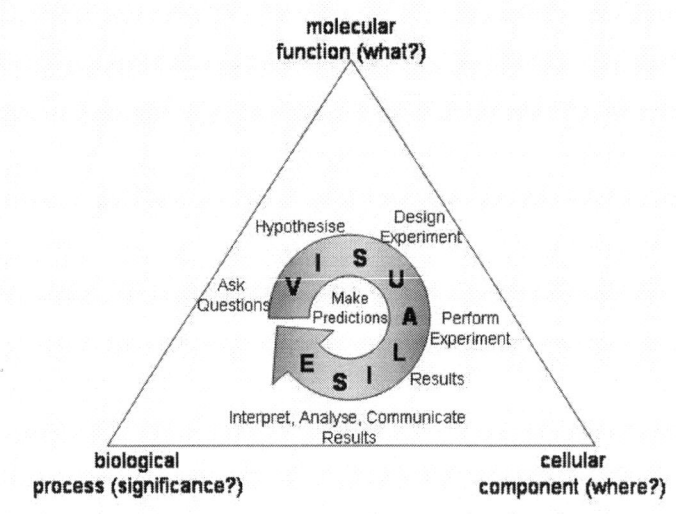

Figure 3. *Integration of the metavisual process within the context of the pedagogical goals for genomics visualizations.*

In the following sections, commonly used examples of visualizations in genomics research are discussed from a pedagogical perspective. The examples selected are by no means exhaustive; especially, as the field of genomics continues to rapidly advance, modelling environments will also continue to evolve to accommodate the vast datasets that are generated.

VISUAL REPRESENTATIONS OF GENOMIC INFORMATION: CONTEXTUAL, VISUAL AND PEDAGOGICAL GOALS

Sequence analysis: comparative genomics

A genomic sequence can be compared to other genomic sequences for various investigative purposes. Comparisons can also be performed for protein (gene product) sequences. Examples of such visualizations are shown in Figure 4A and B (a colour version of this figure is available in the accompanying CDROM).

The contextual goal of the visualization representing the comparison in Figure 4A would be to demonstrate i) similarities/differences in the gene/protein sequences between species or within a single species; or ii) identification of conserved sequence motifs, which can in turn be extrapolated and analysed to assign function. In the specific case presented here, which represents alignments of protein sequences from the Human Immunodeficiency Virus (HIV-1), the first goal is realised. The visual goals are to illustrate patterns of similarity or differences amongst a set of sequences. For the student to navigate the figure the metavisual knowledge required includes:

- knowing the convention for reading sequences: left to right;
- knowing that each line of the sequence comparisons, or alignments, represents an individual dataset (whether it represents a gene(s) or protein sequence) in the comparison;
- knowing that each letter represents a 'basic unit' of the sequence (whether it is a nucleotide (for DNA) or an amino acid (for proteins))

The pedagogical goal of the figure would be a merger of these two goals, and would be summarised: to identify patterns of similarity/differences in a sequence alignment that would represent changes/conservation of genome sequences.

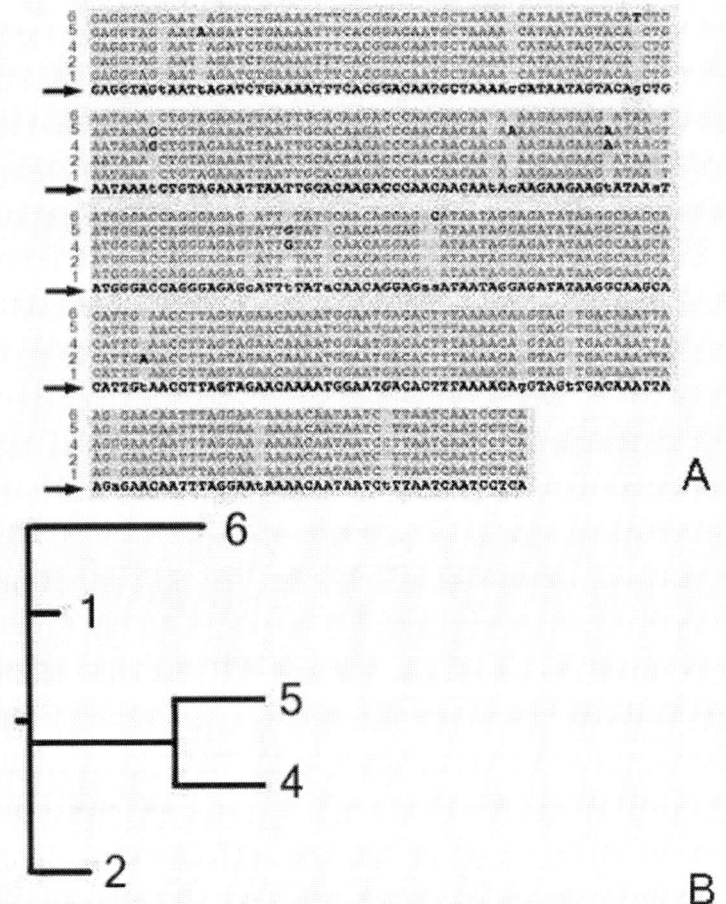

Figure 4. A: Alignment of nucleic acid sequences from a hypervariable (highly prone to mutation) region of a specific Human Immunodeficiency Virus (HIV-1) clone (subspecies) found in a patient. The alignment continues from the top set of grouped rows through to the bottom set of grouped rows, and is read from left to right. Each row represents the nucleic acid sequence of the virus present in a blood sample at a specific visit. The visits began at seroconversion[1] and continued at 6-month intervals; the numbers refer to visits in sequential order. The arrow indicates a consensus sequence, which represents the most common nucleotides found at each position. B: Phylogenetic tree diagram representing the evolutionary relationships amongst the HIV-1 clones from Figure 4A. Branch lengths and proximity are indicative of how related specific clones are to one another. Refer to the text for further explanation.

---

[1] Seroconversion: After an infection with HIV-1, antibodies to the virus can be detected in the blood. This is called seroconversion (converting from HIV-negative to HIV-positive).

Interestingly, Figure 4B is a computer-generated visualization that is derived from the sequence alignment data of Figure 4A. The contextual goal of such a figure may be: i) to reveal relative sequence similarities amongst genes or proteins; or ii) to reveal relative gene/protein sequence similarities, and in so doing, demonstrate the evolutionary (phylogenetic) relationships amongst species. The same visual representation is assigned distinct terminologies depending upon the contextual goal of the diagram. If the biological context is limited to the first goal, the diagram is called a 'dendrogram'; if the contextual goal extends to the latter, the visualization is a 'phylogenetic tree'. The specific example presented here is a phylogenetic tree showing evolutionary relationships amongst mutational variants of HIV-1. Because a phylogenetic tree inherently results from comparisons of sequence similarities, it is a specific type of dendrogram. Dendrograms are an exemplar of the integration of the metavisual process within the scientific and pedagogical contexts (see Figure 3). They organise, using a visual process, the relationships between species/genes/proteins attained through sequence comparisons. The specific visual literacy skills necessary for a student to interpret this organisational map are:

- to know that each node represents sequences from an individual species/gene/protein;
- to know that species/genes/proteins with the most similar sequences are closest together on the branched diagram;
- to know that the branch distances are a measure of how similar or divergent two sequences are.

The pedagogical aim of a visualization such as one depicted in Figure 4B is to map and identify the relationships amongst a group of species/genes/proteins based on comparisons of sequence similarities. The visual organisation of these comparisons facilitates the learner's achievement of this goal. The dendrogram distills the computational process involved in determining relational data amongst a set of sequences, and presents the learner with the opportunity to interact with this complex analysis. The student can readily engage in comprehending, analysing, applying, and synthesising new models from data presented in this visual format.

The third visualization in this series is presented in Figure 5 (a colour version of Figure 5, and a 3D movie file of a similar coloured model is available in the Colour Section.

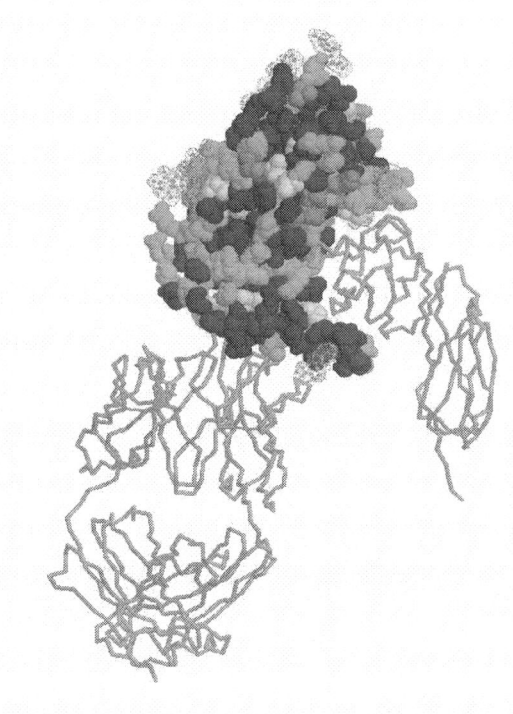

Figure 5. *Screen capture of 3D Protein Explorer representation of an HIV protein sequence alignment using the ConSurf program. For a coloured version of Figure 5, please refer to the Colour Section.*

The image is a screen capture of a three-dimensional (3D) computer-derived structure representing the HIV-1 protein molecule encoded by the sequences from Figure 4A. The coloured 3D image can be viewed in the accompanying Colour Section. Each amino acid residue of the protein structure is colour-coded to indicate whether that specific residue is identical amongst all HIV-1 sequences analysed, or whether it is highly variable. The colour gradations that fall between the two extremes are indicative of varying degrees of sequence variability. Collectively referred to as sequence homology, this type of qualitative and quantitative data can be obtained from the first visualization in Figure 4A. The 3D figure, however, maps this information onto a biological structure. From a metacognitive perspective, this representation integrates the scientific process of analysing abstract data (in which letters represent protein sequences) into a more relevant visual context. For the learner, the outcomes of sequence variability or similarity are modeled onto a tangible representation of the HIV-1 protein. The 3D visualization draws on an additional component of visual literacy, that of spatial cognition and application.

This chapter will not engage in a discussion of spatial literacy, as it has been comprehensively covered in other sections of this book and by numerous other studies (chapters x, y, z, etc- to be cross-referenced later; (Ben-Zvi et al., 1988; Keig and Rubba, 1993; Kozma and Russell, 1997; Seddon et al., 1985; Tuckey et al., 1991; Wu and Shah, 2004), except to succinctly recap in general terms, metacognitive spatial literacy requires the learner to:
- be able to translate fluently between modes of representation;
- be able to mentally and conceptually visualize changes in perspective for, as well as perform mental operations on, a 3D representation.

Engagement with the 3D visualization of the HIV-1 protein involves cognitive function at several levels. The student must be able to conceptually translate between the sequence comparison (Figure 4A) and the 3D model. The student must visually and conceptually recognise that each molecule in the 3D structure represents a 'variable position' whereby variability is indicated by colour. Furthermore, the student should be able to appreciate the conceptual implications when studying the 3D visualization from several perspectives: i) focusing on variability of specific amino acid residues; ii) focusing collectively on variability vs conservation for the entire 3D structure; iii) focusing on relationship between variability of amino acid residues and topology of the molecule; iv) mentally considering biological function and interaction with other molecules in the context of amino acid variability.

Edelson and Gordin (1998) have highlighted the significance of scientific visualizations as a means for students to participate in the practice of science using an inquiry-based approach. The representation of genomic analysis in the above examples can be integrated into an inquiry-based learning process through the creation of appropriate contextual frameworks. The project discussed below, 'Visualizing the Science of Genomics', revolves around this approach.

Global expression analysis using microarrays: functional genomics

From a biological perspective, the examples above may be considered visualizations of 'static' genomics analysis. That is, the visualizations capture what can be inferred from sequence data without taking into consideration the actual expression (or absence thereof) of these sequences. The student has gained an appreciation for the inherent characteristics of a genome(s) but has yet to 'see' it in action. The molecular functions (what?), cellular components (where?), and/or biological processes (significance; how?) associated with the genomic information are predicted but not assessed experimentally or visually. Expression analysis introduces a new perspective on genomics for the student. In this case, the *what*, *where* and *how* become primarily relevant as the focus is directed to the products of gene sequences; i.e., whether the genes are expressed, when they are expressed, and to what extent they are expressed. These parameters comprise the goals of functional genomics. The student is now challenged with visualizing the application of a genome sequence with added quantitative and qualitative cognitive layers.

DNA microarrays (also referred to as DNA chips) are utilised by researchers to simultaneously detect the level of expression (in the form of messenger RNA, or mRNA) of every gene in a genome. Each spot on a microarray represents a specific gene of the genome. The reader unfamiliar with microarray technology who wishes to gain further insight into the scientific methodology is referred to several excellent microarray teaching resources (Campbell, 2004; Campbell and Heyer, 2003; Shi, 1998). In brief, microarrays are hybridised with a target mRNA population (or a cDNA 'copy' of the population) that has been labelled with a fluorescent dye. If a gene is 'expressed', its corresponding mRNA would be present in the population, and this would be indicated on the microarray by hybridisation of the fluorescent mRNA molecule to a specific spot on the array corresponding to the gene sequence. Two different target mRNA populations would represent two different conditions to be tested, and the populations would be labelled with two distinct fluorescent dyes in order to distinguish between the two.

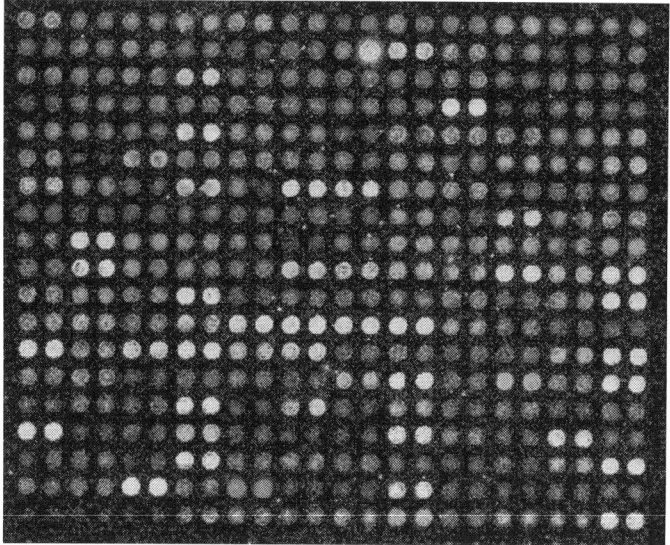

Figure 6. *Subsection of a microarray. Image courtesy of The Ramaciotti Centre for Gene Function Analysis. A coloured version of this image is available in the Colour Section.*

Figure 6 is a subsection of a laser-scanned microarray (A coloured version of this figure is available in the Colour Section). The experimental goal for this visualization is to determine differences in the levels of expression of every single gene of a cell or organism when it is subjected to an experimental condition and a control (or reference) condition. For example, the microarray in Figure 6 was utilised to measure changes in global gene expression profiles for rat cells grown under a specific experimental condition A vs. condition B. Each spot represents an individual gene and the genes have been spotted in duplicate on the microarray slide to demonstrate reproducibility within the experiment. Redder spots indicate genes that are expressed at higher levels under condition A whilst greener spots represent genes expressed at higher levels under condition B. When expression levels between the two conditions are equivalent, the red and green are 'merged' and the spot is yellow.

The visual goal of this representation is not inherently apparent, because it is, in effect, 'raw data'. The experimental and visual goals are only realised when the specific red *vs.* green intensities of each spot are converted into numerical ratios that are subsequently analysed by log transformations. Biological significance is portended by experts through practiced comparisons of significant trends in patterns of expression amongst networks of genetic pathways. Furthermore, the expert also may have prior knowledge of the significance of specific relationships between these networks. Yet, educators will often present similar microarray visualizations in the classroom as a demonstration of functional genomics without providing the rest of the pedagogical framework. This framework requires the following components:

- conceptual and visual link between a genomic sequence and a microarray image.
- visual and temporal understanding of the procedures and steps involved in a functional genomics analysis study.
- conceptual understanding of the scientific inquiry process in microarray analysis.
- visual literacy with regard to identification of visual goals for microarray analysis representations and the ability to select appropriate visualizations to convey contextual goals.

Such an example of the categorical delineation of contextual teaching goals associated with functional genomics can also provide a framework to guide the development of mental practice tasks that can further enhance learning (Cooper et al., 2001). The metacognitive process becomes crucial for the student to benefit from the multitude and variety of visualizations that are directly taken from research environment to the classroom setting. The cognitive apprenticeship model described by Collins and colleagues (1991) is particularly apt in describing how this can be achieved at both cognitive and metavisual levels:

- identify the biological question and make it visible to the student
- provide an authentic context for an abstract task (like functional analysis of a genome) so students understand the relevance
- vary the diversity of situations, as well as the perspectives and contexts of visual presentations, and articulate common aspects so that students can transfer what they learn (Collins et al., 1991).

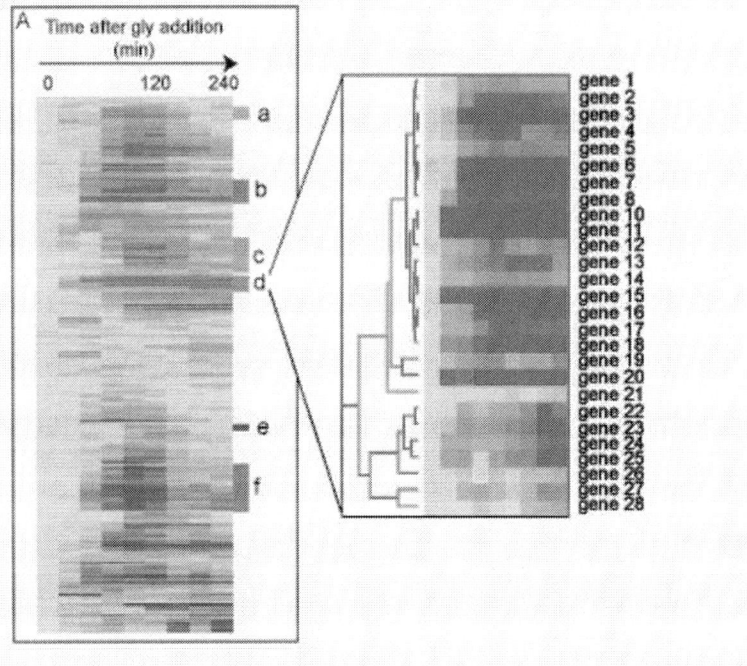

Figure 7. *Representation of the changes in the expression patterns for all genes of a yeast microarray over time. Please refer to the text for further details; a coloured version of this figure is available in the Colour Section. Figure adapted from image courtesy of I. Dawes, The University of New South Wales and The Ramaciotti Centre for Gene Function Analysis*

Microarray raw data as exemplified in Figure 6 is quantified through computer analysis, and collectively presented as in Figure 7. A coloured version of Figure 7 is available in the Colour Section. This image represents the changes in the expression patterns for all genes on a microarray over time. Visual cognitive load is significantly increased as there are several conceptual representations that need to be linked and integrated into the student's thought process. Each column in the left panel represents the expression status for all genes at a given timepoint. Reading from left to right, these columns represent microarray data analysed for samples from 0 minutes to 240 minutes after the induction of the experimental condition (addition of glycine). As in the scanned microarray, the expression status is depicted by coded colour. The panel on the right is a detailed 'zoom-in' of a specific section of the left panel. Here, the individual genes are indicated and their expression patterns over time (reading left to right) become more obvious. An additional level of visual representation is included toward the left of the righthand panel; this dendrogram indicates functional relationships amongst these genes. Hence, the student must visually interpret and integrate into a framework, the categorical concepts of time, qualitative and quantitative expression, and functional relationships.

2D vs 3D visualizations in genomics and contextual authenticity

Functional relevance of genomics analysis can also be represented through the creation of 3D visualizations of molecular structures. In order for a 3D image to be pedagogically successful, Allen (1991) stresses the importance of the inclusion of spatial information (spatial cues). Hence in order to achieve visual literacy, not only must learners be able to navigate 3D space and construct mental models of their environment, but they must master complex rules for inferring 3D relationships from 2D cues (Allen, 1991).

Wu and Shah (2004) have suggested five principles for the design of visualization tools that support students' visuospatial thinking and in so doing, enable them to understand chemistry concepts and develop representational skills:
- providing multiple representations and descriptions;
- making linked referential connections visible;
- presenting the dynamic and interactive nature of chemistry;
- promoting the transformation between 2D and 3D; and
- reducing cognitive load by making information explicit and integrating information for students.

Interestingly, Shepard (1978) notes that the discovery of the three-dimensional double helical structure of DNA was in large part dependent upon the integration of spatial visualization into the deductive process. Watson's own account of the journey toward this discovery highlights the role that visualizations and the connections inferred from various incarnations of visual representations played in his cognitive processes and those of his collaborator, Francis Crick (Watson, 1968). Whilst this seminal discovery is an exemplar of visuospatial reasoning at the expert level, the processes invoked by Watson and Crick would have included those described above as the visual interpretation of new information also necessitated the reinterpretation of previous work and revision of hypotheses (see also Figure 2).

For students, the successful transition through the developmental stages of metacognitive capability described earlier in this book by Gilbert (Chapter); i.e. *acquisition, retention, retrieval* and *amendment*; would necessitate strategic scaffolding of the cognitive processes reflective of these principles. In the context of genomics, the amendment stage is particularly central to interpretation and knowledge construction because the field continues to evolve at a rapid pace with advances in the research technology. Traditional representations utilised by learners in genomic analysis are information-rich but comprehensively distant from the 3D world of molecular interactions and relationships they are meant to convey. Contextual relevance, authenticity, and experience in traversing these domains are therefore paramount in facilitating the transition from learner to expert. Whilst experts are able to easily navigate and make connections between multiple representations of concepts and processes, the learner is still in the process of mapping their visual and conceptual knowledge. Kozma has observed that when chemistry students are presented with a 3D representation of a molecular structure, they tend to focus on surface features without thinking about the underlying chemical principles (Kozma, 2003; Kozma and Russell, 1997).

In genomics, 2D and 3D representations are not necessarily alternate representations of the identical molecule or chemical structure. For example, a Chime-rendered 3D structure (Figure 5) can represent a coded compilation of a series of 2D alignments of sequences from multiple genomes (Figure 4A). In other words, the 3D and 2D images contain the same information, but their respective interpretation can vary significantly depending upon the context in which they are presented and the experience of the learner. When undergraduate biology students are presented with informationally equivalent 2D and 3D representations of a viral surface protein, we have observed marked differences in their descriptive explanations of concepts relevant to virus-receptor interaction (See and Takayama, unpublished). For a 3D representation to be effective in supporting student understanding, the prior 3D literacy can significantly influence successful student engagement (Richardson and Richardson, 2002; Wu and Shah, 2004). Correct perception of the 3D structure is crucial for further cognitive operations, and problem solving tasks of increasing complexity should be scaffolded strategically to achieve optimal learning outcomes (Ferk et al., 2003).

The metavisual process in interpreting the 2D alignment (Figure 4A) is distinct from that utilised in understanding the 3D protein structure (Figure 5). Pattern recognition is often the initial simplest process the researcher engages in when first examining a 2D alignment. It is important to note, however, that what may appear to the novice as a simple scanning process belies the underlying reasoning and 'visual cues' (known as 'sequence motifs') that the researcher integrates into a working model based on interpretation of functional significance. That is, certain sequence motifs, which are indicative of evolutionary and functional conservation, are identically preserved amongst and sometimes between species. The student must learn how to link these visual cues into a conceptual framework. Boyle and Boyle (2003) have explored alternate visual representations of genomic alignment data to facilitate student engagement and inquiry. In their case, alignments are not analysed by pattern recognition but by graphical representation of alignment frequencies.

In summary, metavisual ability in genomics is inherently embedded in the learning process as the student becomes increasingly familiar with the interpretation of comparative genomic analysis. Visual literacy also necessitates the ability to not only utilise the appropriate representation to convey specific functional information, but to be cognisant of scale and relationship.

COLLABORATIVE VISUALIZATION IN GENOMICS: INQUIRY-BASED TEACHING

Edelson and Gordin (1998) have suggested the key criteria that must be fulfilled for visualizations to succeed in the construction of a successful framework for inquiry-based learning are: motivating context, learner-appropriate activities, data selection, scaffolding interfaces and support for learning. The complexity of the cognitive processes that would be required for the student to actively process multiple representations of data-rich genomic information presents a challenge to the educator. Our goal was to develop an inquiry-based project that reflected authenticity and emphasised the process of scientific inquiry rather than content and outcomes. A key strategy in this project was to facilitate collaborative learning whereby the students were instrumental in creating the visual representations of genomic analyses and in so doing, directed their knowledge construction.

Visualizing the Science of Genomics: an international online research project

Visualizing the Science of Genomics (VSG) was developed as an international research project to engage students in the active process of collaborative scientific inquiry. This unique project was conducted entirely online amongst geographically distanced participants who worked in 'research teams' of five students, each from a different country. Participants represented a diversity of scientific backgrounds including: microbiology, bioinformatics, medicine, chemical engineering, biotechnology, pharmaceutical sciences, molecular biology, medical chemistry, genetics, biochemistry, mathematics and computer science. The international and multidisciplinary composition of each research team provided the context for scientific research as a concerted global effort dependent upon contributions by scientists with specific areas of expertise.

Students worked in teams to analyse, hypothesise, reflect, predict, visualize and formulate models based on genomic sequence data from the Human Immunodeficiency Virus-1 (HIV-1), the causative agent of AIDS. Contextual relevance was provided through the creation of case studies based on actual data. The goal of VSG was to allow students to assess and interpret available information, and to develop their own research questions and methodology. This approach contrasts significantly from the traditional university laboratory practical, in which the student learning experience is dictated by the 'aim of the experiment' and the prescribed methodology in the lab manual. The biological sciences have traditionally placed emphasis on laboratory classes to promote inquiry-based learning. In principle, the laboratory introduces the student to the *practice* of biology, whereby the learner is provided with the opportunity to apply his/her theoretical knowledge. One of the goals of the biology educator is to teach students how to 'think like scientists'; we aim to engage the student in a cognitive apprenticeship as a researcher. Paradoxically, in most laboratory courses the lab manual specifies the 'aims' or 'hypotheses', and the student follows an established protocol to conduct the experiment. A true cognitive apprenticeship, however, must include development of the thought processes that facilitate the formulation of a hypothesis, as well as the reasoning processes invoked in the development or application of appropriate methodology to test the hypothesis. Hence whilst technically the laboratory provides a tangible context, focus on content and outcome may override learning *how*. Kozma (2003) has highlighted the argument by Dunbar (1997) that naturalistic studies of scientific practice (that is, cognitive studies of how scientists think and solve problems) require the process to be examined in an authentic setting whereby scientists solve complex, extended scientific problems as they interact with colleagues and with resources in their research environment. In other words, a true cognitive apprenticeship must engage the student in what biologists would qualify as an *in vivo* (natural) setting as opposed to an *in vitro* setting (in an artificial environment). The VSG project endeavoured to foster authentic inquiry through the creation of a collaborative research community.

In view of the diverse backgrounds of the participants, preliminary information was send to all participants prior to the start of the project. The information included background reading on HIV-1 as well as a CD-ROM tour and necessary technical information for the web-based work. The students were also encouraged to post brief introductions about themselves in their team sites to initiate students into the social framework of their learning community. Indeed, the merits of a constructivist approach to learning in a networked community, whereby the social construction of knowledge is engendered through collaboration and opportunities for the transfer of authority to the student, have been demonstrated by others (Brown et al., 1989; Greeno, 1998; Greeno et al., 1996; Resnick, 1988).

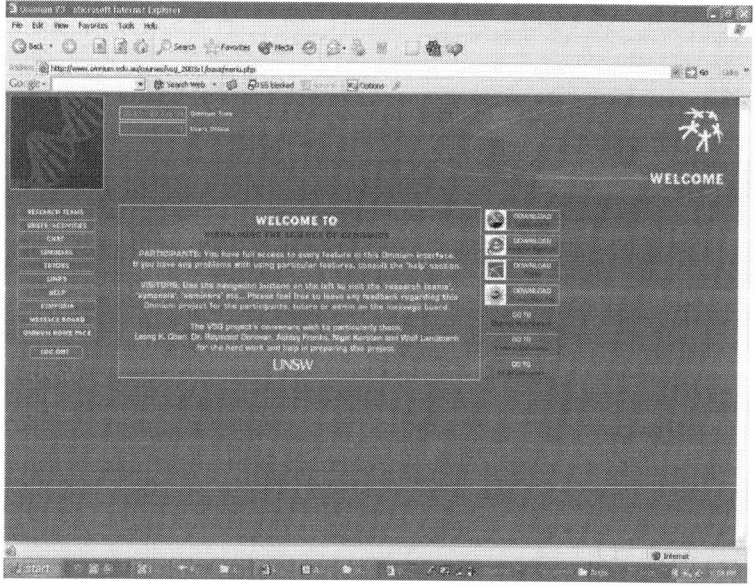

Figure 8. *Homepage for VSG interface. A coloured version of this figure is available in the Colour Section. The navigation buttons on the left enable students to access their individual research team areas for collaborative work, as well as the areas common to all teams. The VSG site can be visited at: http://www.omnium.edu.au/courses/vsg_2003s1/base/index.php*

HIV-1 genomics analysis as an authentic inquiry-based approach

A two-week pilot project was conducted with 42 participants from 24 universities, based in eleven different countries. Each team was provided with a case study modelled on HIV-1 genomic sequence datasets obtained from the GenBank international public database (Benson, 2000). The sequences represented the clones (types) of HIV-1 viruses present in blood samples drawn from HIV-1-positive subjects. Students analysed sequences derived from a region of the HIV-1 genome encoding gp120, a key viral surface protein involved in virus-host interaction. The interaction between gp120 and the host immune cell is a core concept that serves as a 'launching point' for student discussions on their ideas about significance of the data and strategies for further exploration. Online tutorials and 'seminars' were readily available, allowing students to refer to them as needed at their own pace. After assessing the available information, the teams faced the challenge of developing their own research question(s) and appropriate methodology for investigation. The greatest initial challenge was 'What is our question?' or 'What is our hypothesis?' before teams could proceed to analyse, interpret, and create visual representations of their analyses. Student discussions sought to determine appropriate strategies for comparative genomic sequence alignments (as in Figure 4A) to test their hypothesis; this became an iterative process for some teams as they re-visited their initial hypothesis after performing preliminary alignments to seek patterns or trends in HIV-1 genomic mutations (which could be depicted as in Figure 4B).

VSG was modelled as a unique online collaborative research model for teaching and learning the scientific process. By shifting the emphasis from a content- and outcome-focused approach to a process-oriented approach, reflection and analysis can be facilitated. One of the greatest challenges for the educator is to teach students to apply what they have learned to a new situation; the process of transfer. The VSG project aims to identify and clarify the scientific process and situate abstract concepts in **relevant** and **authentic** contexts to aid students' understanding of the 'big picture' as well as the functional details.

The pedagogical approach focused on open-ended inquiry, so that students were not pressured to 'produce results', allowing them to explore, research and pursue approaches that were of interest to them. Open exploration can encourage intrinsically motivated learning; Vollmeyer and colleagues (1996) have demonstrated that learning from computer-based biology problems was retarded when tertiary students were directed to solve problems with a specific goal in comparison to students who were instructed to 'explore the system'. In teaching biology students about the relationship between genomic sequence analysis and an organism's evolutionary history, Parker has also argued that an open-ended question-driven approach motivates students and is a more realistic way to teach science (Parker et al., 2004).

Modes of visual and textual interaction online

Collaborative learning was facilitated through a highly interactive interface originally created to teach students in graphic design (Bennett, 1999). The Web-based platform was ideally suited to facilitate visual communication as well as textual communication amongst participants. Figure 9 shows the different modes of interaction amongst research team members (A) and between research teams (B). A coloured version of this figure is available in the Colour Section.

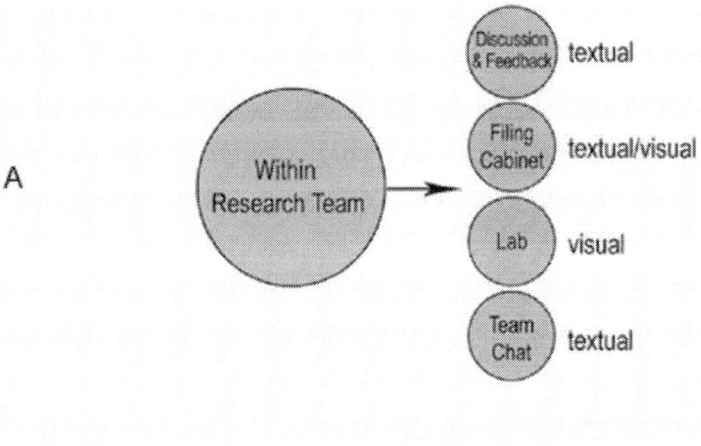

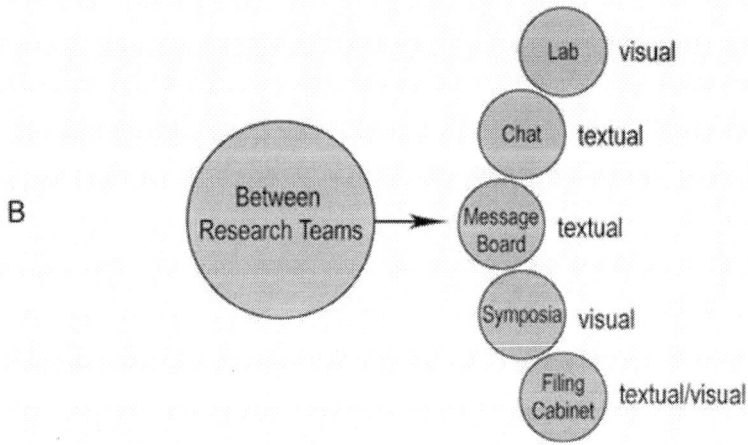

Figure 9. *Modes of interaction amongst the students in each VSG research team (A) and between research teams (B). A coloured version of this figure is available in the Colour Section.*

The availability of both visual and textual areas for communication amongst these multidisciplinary teams allows students to determine appropriate modes of interaction. The learner-centred approach also sought to promote facility in engagement of students from various disciplines with different cognitive learning styles. Studies on multiple representations have demonstrated that text information is better remembered when it is illustrated by images than without illustrations

(Levie and Lentz, 1982; Levin et al., 1987). The basis for such findings can be ascribed to Paivio's dual coding theory which proposes that cognitive processing of verbal information occurs in a separate system from cognitive processing of pictorial information (Paivio, 1986). Hence, the engagement of both systems for illustration + text enhances the memory. Furthermore, as verbal and pictorial explanations are processed in distinct cognitive systems, they are organised into mental models in different parts of the working memory (Chandler and Sweller, 1991; Mayer, 1997). Successful integration of the two processes results when components of the textual model and the corresponding components of the visual model are simultaneously activated (Chandler and Sweller, 1991; Sweller, 1999). How the text is integrated with the image is important; Mayer (1997) proposes that visual representations optimally support comprehension when text and images are explanatory, when textual and visual information are presented closely together spatially and temporally, their respective content is related to each other, and when learners have low prior knowledge about the subject but posses high spatial cognitive ability. Yet, others caution that the design of the integration of textual and visual information should be considered carefully to avoid learning interference due to task-inappropriate graphics (Kirby, 1993; Schnotz and Bannert, 2003). As the structure of graphics affects the structure of the mental model, the form of visualization employed should support the construction of a task-appropriate mental model, and well-designed visualizations can indeed facilitate this process in learners with high prior knowledge (Schnotz and Bannert, 2003).

The VSG approach takes into consideration all of the above arguments by focusing on the *student's primary role in the construction and iterative adaptation of integrated visual and textual models.* An examination of the team discussions reveals that the students have utilised visualizations in several ways:

- To represent the process of their genomic analysis.
- To facilitate the creation and interpretation of conceptual links between HIV genomic sequences, HIV protein structure and function, HIV evolution, and clinical implications for patients.
- To provide appropriate representations (2D alignment *vs* 3D model) to convey specific concepts.

The following excerpt from a Team Capella discussion, in which a team member proposes a strategy for their case study, exemplifies the integration of visual cognition into the scientific process:

"Summary of the team's reunion on Sunday 25:

Choosing the subjects and how to use the data obtained:

1. Select 2 patients to focus on: one slow progressor, one fast progressor (as defined by their CD4 profiles)

2. Either: a) select one to follow over time, and align the sequences from this clone for all visits; b) do this for several or all clones

You will be performing (2) for both patients.

3. Visualize the divergence over time by phylogenetic tree-building of your alignments. This function is easily available in Bio Workbench. This allows you to measure evolutionary distances and to see whether there are marked new 'branches' that form as the viral clones are mutating.

4. Model your alignments onto a 3D structure of gp120 in Protein Explorer to examine where the mutating regions are located, and where the conserved regions are located in relation to the actual shape of gp120, and their functional interactions with CD4, neutralising antibodies, etc."

Another student from Team Avior represented his hypothesis of the presence of a converging selective pressure by linking the patterns of variability in genomic sequence at two distinct onsets of disease, and statistical analysis of evolutionary distances. His integration of visual and statistical information demonstrates his ability to integrate his cognitive processes to create his conceptual model, as shown in Figure 10 (a coloured version of this figure is available in the Colour Section):

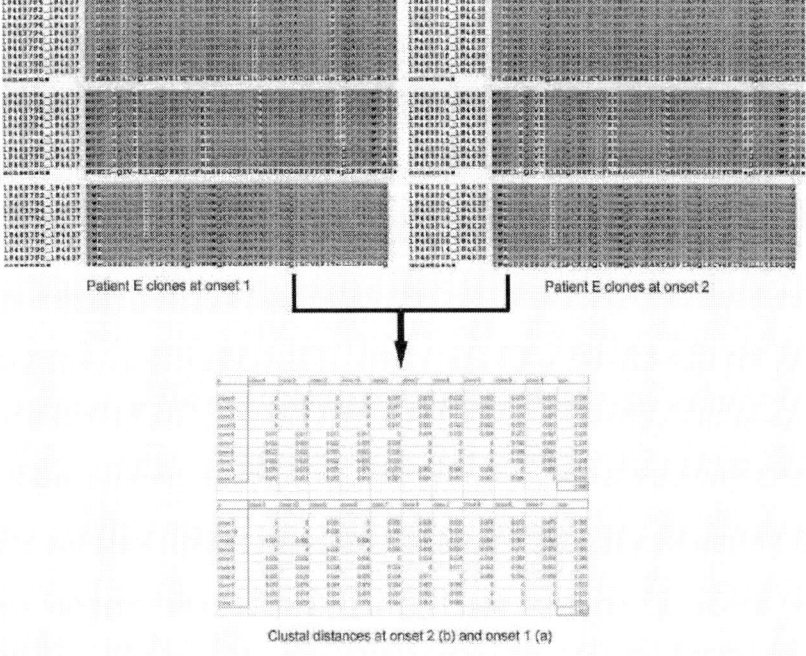

Figure 10. *Conceptual model by a Team Avior member (Samiran) integrating patterns of variability in genomic sequence at two distinct onsets of disease, and statistical analysis of evolutionary distances based on HIV-1 gp120 sequence information derived from clinical blood samples. A coloured version of this figure is available in the Colour Section.*

The collaborative outcomes of the student teams also extended to 3D representations of mutational patterns of HIV-1 viruses. HIV-1 protein sequence alignments were mapped onto three-dimensional (3D) structural models of gp120. The 3D visualization facilitates the understanding of how mutations in the HIV-1 gp120 genome may ultimately affect protein structure and functional interaction with the $CD4^+$ T cell receptor. The open source web-based program, Protein Explorer (Martz, 2002), was utilised to model protein sequence alignments and test predictions. Protein Explorer is a versatile program that enables 3D modelling, visualization, and manipulation of protein structures. Multiple sequence alignments were mapped using Protein Explorer and ConSurf (Glaser et al., 2003) to identify amino acid residue positions that remained highly conserved in comparison to those that mutated significantly.

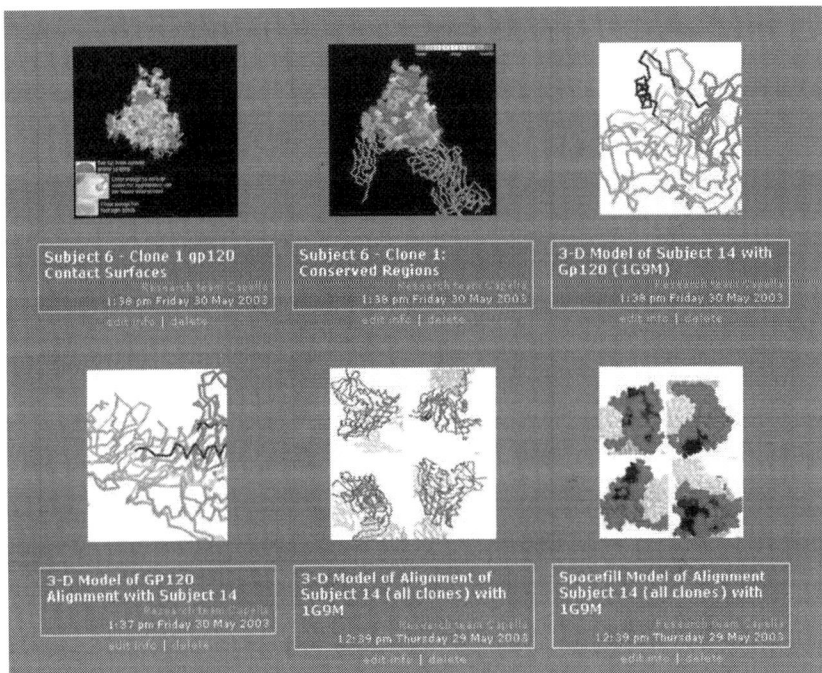

Figure 11. *Screen capture from the Symposium entitled 'Week 2: 3D Models of HIV-1 GP120'. Each panel represents a team submission of a 3D model representing comparative analysis of protein sequences derived from HIV-1 gp120 genomic information. A coloured version of this figure is available in the Colour Section.*

The images in Figure 11 represent screen captures of the students' 3D models of the HIV-1 gp120 protein structure (a coloured version of this figure is available in the Colour Section). The models represent variations in scale, depth, perspective, and context; all of these aspects are outcomes of conceptual frameworks created through team collaboration. Kozma's observations of students using a chemical modelling package have suggested that materials features of representations can support student learning if they correspond to certain characteristics of abstract, scientific entities that do not otherwise have a concrete, visible character (Kozma, 2000). Similarly, these computer-generated 3D models are visual representations of genomic data analyses which provide structural insight into the consequences of the high frequency of mutations in the HIV-1 genome. The case study further provides students with a biological and clinical context for HIV-1 genomic mutations, and the colour-coded comparisons demonstrate how the consequent changes in viral protein structure makes the development of vaccines and therapeutics extremely difficult.

The VSG learning community

According to Shaffer and Anundsen (1993), 'community' is defined as a dynamic entity that emerges when a group of individuals share common practices, are interdependent, make decisions jointly, identify themselves with something larger than the sum of their individual relationships, and make a long-term commitment to well-being (their own, one another's, and the group's). The online community is dependent upon these same attributes in the absence of face-to-face contact or a voice. Indeed, the initial challenge of a project like VSG is the development of the community itself, as learning goals are concurrently being frameworked.

The social interactions in the VSG community were integral to the collaborative learning efforts of its members. Palloff and Pratt (1999) stress the importance of the development of shared goals that are related to the learning process in an online community. The integration of these goals into the social dialogue amongst the students was indeed reflective of this effort. VSG reflects a situative approach to learning (Brown et al., 1989; Greeno, 1998; Resnick, 1988), whereby participation in processes shaped by interactions in the VSG community construct students' conceptual understanding of genomics. Such situative approaches focus on the construction of knowledge within the context of student interactions with others, with the learning material, and with their learning environment (Brown et al., 1989; Greeno, 1998; Greeno et al., 1996; Resnick, 1988).

Social interactions in the VSG community were assessed using a modified rubric based on Sringam and Geer's Cognitive Development Interactive Analysis Model (Sringam and Geer, 2000), with the inclusion of an additional category, 'socialisation'. The seven categories of interaction in the rubric were as follows:
- Socialisation
- Planning
- Sharing/comparing/contributing information
- Identifying or clarifying inconsistency of ideas, concepts or statements
- Negotiation of meaning/co-construction of knowledge
- Testing and modification of proposed synthesis or co-construction of knowledge
- Agreement statement(s) and application of newly constructed knowledge

Figure 12 summarises the distributions of the categories of social interactions that occurred amongst students in the Discussion and Feedback interface (a coloured version of this figure is available in the Colour Section).

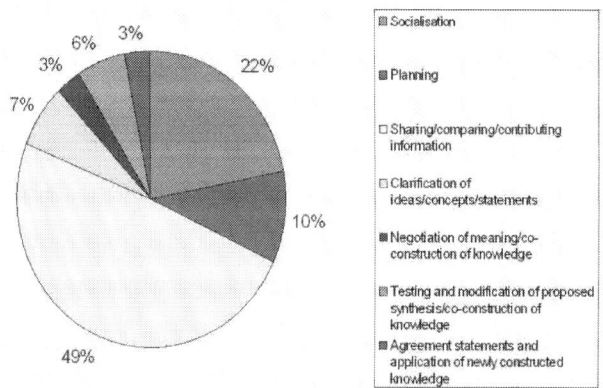

Figure 12. *Categories of social interaction amongst all students in the VSG Discussion and Feedback area. Interactions were scored using a modified rubric based on Sringam and Geer's Cognitive Development Interactive Analysis Model (Sringam and Geer, 2000). Data analysis was performed by J. Ang (2003). A coloured version of this figure is available in the Colour Section.*

The two main categories of social interaction were i) socialisation and ii) sharing, comparing and contribution of information. Whilst socialisation (22% of interactions) plays a significant role in facilitating collaborative learning, task-related interactions (78%) were predominant in the VSG community. This is indicative of student commitment and strong involvement in the investigation. Furthermore, a common observation amongst the groups was the negotiation and development of a shared understanding re: 'the problem' (task). While this sometimes took the majority of the students' time during the 2-week project, this process was in itself a conduit toward cognitive development. This is evidenced by the dialogic progression of group discussions which fostered continual reflection and process-oriented critical analysis integrated with visual cognition.

The diversity of experience and background of the students were instrumental in creating the rich tapestry of this international online research community. The VSG community was further enhanced through the participation of five world-renowned HIV researchers, who appeared as 'guest tutors'. The guest tutors interacted in synchronous chat sessions to provide students with the opportunity to interact with 'real scientists' in the field, each of whom represented a distinct specialisation. This experience underscored the students' appreciation for the multi-disciplinary nature of research.

Analysis of cognitive interactions in the VSG community

One of the key criteria for an authentic learning experience is that of fidelity of context (Herrington and Herrington, 1998; Meyer, 1992; Reeves and Okey, 1996; Wiggins, 1993). For many of the students, it was the first time in their academic careers that they found themselves immersed in collaborative authentic inquiry, whereupon they were driven by intrinsic motivation.

A rubric for assessing cognitive interactions was developed, based on Biggs' and Collis' Structure of Observed Learning Outcomes (SOLO taxonomy) (Biggs and Collis, 1982). The majority of the interactions assessed via SOLO taxonomy were indicative of higher levels of cognitive ability (relational and extended abstract). The cognitive interactions are summarised in Figure 13 (a coloured version of this figure is available in the Colour Section).

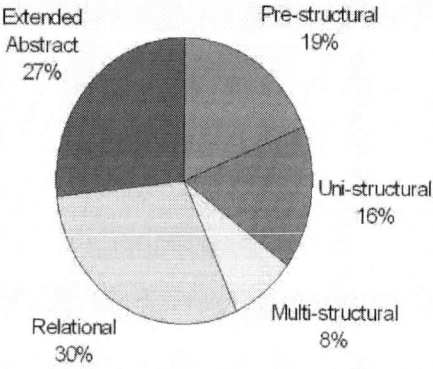

Figure 13. *An example of cognitive assessment of student discussions. Transcripts from the VSG Discussion and Feedback area were analysed by SOLO taxonomy. Data analysis was performed by J. Ang (2003). A coloured version of this figure is available in the Colour Section.*

Student discussions were characterised by analytical, contextual, and social dialogue. The instructor provided feedback and facilitation when appropriate. Whilst most teams initially performed similar preliminary 'experiments', the defined goals, specific datasets chosen, and strategies developed by the teams varied. This diversity of approaches exhibited by teams to develop their 'question' and 'process' may be reflective of the different perspectives, analyses, and expertise provided by the members of each research team. It is possible also that students may exhibit variations in metavisual competence, but this was not rigorously assessed.

The cognitive levels of interaction in VSG were higher than that observed in the instructor's classroom teaching. Whilst this may have been due in part to the calibre of students that had volunteered to participate in the project, the approach utilised was characterised by several qualities that may also have strongly contributed to the learning outcomes:

- Student-centred collaborative approach
- Open-ended scientific inquiry process
- The creation of a strong online community of students and instructor
- Contextual visualizations

Charlin and colleagues (1998) emphasise a learner-centred approach towards problem-solving as being of key importance, and define four principles related to their effect on learning:

1. Learners are active processors of information;
2. Prior knowledge is activated and new knowledge is built on it;
3. Knowledge is acquired in a meaningful context;
4. Learners have opportunities for elaboration and organisation of knowledge.

These principles are indeed reflective of the importance of contextual relevance for students who are presented with an abstract concept like genomics (Chinn and Malhotra, 2002; Tobias and Hake, 1988). The dialogue amongst students and between student and instructor revealed that the principles were indeed effectively utilised.

Analysis of the instructor's role in VSG

The instructor in a student-centred learning environment takes on new roles that are crucial in maintaining an interaction and collaboration amongst students. Technology-based learning communities like VSG where learning is dependent upon a socially interactive and collaborative experience are guided by a social constructivist approach to teaching and learning (Blanton et al., 1998; Duffy and Cunningham, 1996; Jonassen and Reeves, 1996; Maor and Taylor, 1995; Tobin, 1993). Student cognition via such an approach takes place within a social context and collaboration is an essential component. In this environment, the instructor functions in several capacities: pedagogy, social interaction, management and technology (Bonk et al., 2001). Initial analysis of the instructor's contributions toward student team discussions revealed that the role as facilitator/motivator was nearly as prevalent as the pedagogical role. This is markedly distinct from what occurs in face-to-face teaching. The transfer of 'pedagogical authority' during discussions was facilitated through the collaborative nature of the modes of interaction between students and instructor. Figure 14 illustrates the modes of interaction between the instructor and the students (a coloured version of this figure is available in the Colour Section).

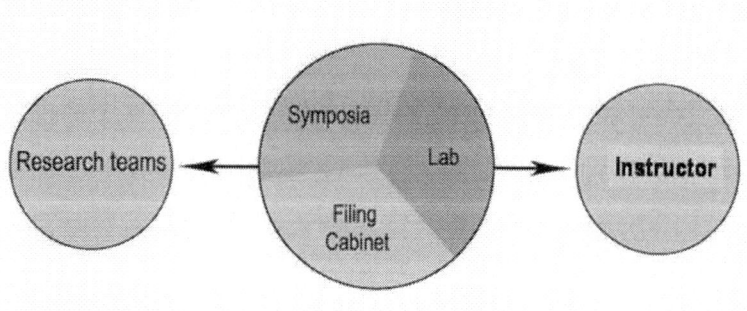

Figure 14. *Modes of interaction between the students and the instructor in VSG. A coloured version of this figure is available in the Colour Section.*

The facilitative nature of these interactive spaces was reflected in the dialogue between the instructor and the students, as visual cues were integrated with conceptual discussions. The situative learning approach also enabled the instructor to adapt to the individual needs of the teams in order to promote their metacognition of the scientific process.

SUMMARY AND CONCLUSIONS

As we continue to witness the rapid progression of post-genomic technology and the consequential exponential increase in the quantity and diversity of information, educators must effectively integrate highly networked concepts into the curriculum without overloading the cognitive capabilities of the learner. Wandersee (2000) describes 'meaning-making' in biology as the understanding of a concept, which often involves visualization. Visual representations can indeed enhance the learning process if they are strategically designed and effectively utilised with both metacognitive and metavisual scaffolding. Whether the student is viewing a static visualization in the textbook or a 3D interactive representation on a computer, the learning goals must be appropriately supported by the mode of representation. To achieve these pedagogical goals, I would venture to propose that university researcher-educators develop an awareness of their own metavisual thinking to engage in a distillation of the layers of 'assumed visual knowledge' necessitated by their practice. Such a process can facilitate the generation of cognitively more intuitive representations for not only students but the research audience as well.

Several exemplars are indeed indicative of this intuitive direction taken by biologists. SequenceJuxtaposer (Slack et al., 2004) was developed to facilitate visualization of large-scale genomic sequence comparisons in context. An information visualization technique called 'accordion drawing' was applied to enable 'stretching out' of genomic sequence regions when detailed views were required, and 'contraction' for a 'global view' of the entire genome. The creation of the visualization takes into account the cognitive load required to maintain a mental model of sequence navigation history; the accordion approach reduces this load by allowing various scales of visualization within one computer browser frame (Slack et al., 2004). The creation of an entirely new 'language' for the visualization of complex data found its basis in collaboration between artist and scientist (Keefe et al., 2004). To distill the enormous amount of data obtained from magnetic resonance imaging (MRI) scans of mouse spinal cords, concepts from oil painting were applied to represent multivalued data with multiple layers of varying 'brush strokes'. The metavisually cognisant integration of these borrowed expressive art techniques facilitated the simultaneous representation and comprehension of enormous amounts of data.

The 'classroom' is no longer a spatially and temporally fixed venue for teaching, which provides both challenges and opportunities for the educator. Similarly, as the biological playing field of 'omics' research continues to expand and evolve, the educator faces the challenge of integrating complex networks of information into multidisciplinary contexts. The potential of visual representations in facilitating the understanding of the elegance of these newly discovered systems can indeed be realised not merely through the utilisation of the visualization itself, but by the careful process behind its creation, to communicate our pedagogical goals.

ACKNOWLEDGMENTS

I am grateful for the support and opportunity provided through an Innovative Teaching and Educational Technology (ITET) Fellowship from the Office of the Pro Vice Chancellor (Education) of The University of New South Wales, which launched me on this multi-dimensional journey. I am indebted to the UNSW Faculty of Science for its funding and support of the VSG project, and I wish to acknowledge the kind support of my colleagues at the UNSW College of Fine Arts.

I would like to acknowledge the valuable input of Jessica Ang, whose honours thesis under my supervision examined the learning assessments of the VSG project. Discussions with my colleagues Barbara Tversky and Jan Plass, and my honours student Karen See have indeed been inspiring during the writing of this chapter. I am also grateful to the Carnegie Foundation for the Advancement of Teaching, at which I had the opportunity to engage in many stimulating discussions with colleagues during my tenure as a Carnegie Scholar with the Carnegie Academy for the Scholarship of Teaching and Learning. My appreciation and kudos go to the editor, John Gilbert, for his infinite patience and his insightful vision. Finally, I express my heartfelt thanks to all of my students, past and present, and especially to the VSG participants, who have challenged and enhanced my thinking in embracing new uncharted territories.

REFERENCES

Allen, B. S. (1991). Virtualities. *Educational Media and Technology: The Year in Review, 17*, 47-53.

Ang, J. (2003). *Visualizing the Science of Genomics*. Unpublished BSc. Honours, The University of New South Wales, Sydney.

Bennett, R. (1999). *Omnium*, from www.omnium.unsw.edu.au

Ben-Zvi, R., Eylon, B., & Silberstein, J. (1988). Theories, principles and laws. *Education in Chemistry (May)*, 89-92.

Benson, D.A., Karsch-Mizrachi, I., Lipman D.J., Ostell, J., Rapp, B.A., Wheeler, D.L. (2000). GenBank. *Nucleic Acids Research, 28,* 15-18

Bernal, A., Ear, U., & Kyrpides, N. (2001). Genomes OnLine Database (GOLD): a monitor of genome projects world-wide. *Nucleic Acids Research, 29*, 126-127.

Biggs, J. B., & Collis, K. F. (1982). *Evaluating the Quality of Learning: the SOLO Taxonomy*. New York: Academic Press.

Blanton, W. E., Moorman, G., & Trathern, W. (1998). Telecommunications and teacher education: a social constructivist review. *Review of Educational Research, 23*, 235-275.

Bonk, C., Kirkley, J., Hara, N., & Dennen, V. (2001). Finding the instructor in post secondary online learning: pedagogical, social, managerial, and technological location. In J. Stephenson (Ed.), *Teaching & learning online: Pedagogies for new technologies.* (pp. 76-98). London: Kogan Page.

Boyle, A. P., & Boyle, J. A. (2003). Visualization of aligned genomic open reading frame data. *Biochemistry and Molecular Biology Education, 31*(1), 64-68.

Brooks Jr., F. P. (1996). The computer scientist as a toolsmith II. *Communications of the ACM, 39*(3), 61-68.

Brown, J. S., Collins, A., & Duguid, P. (1989). Situated cognition and the culture of learning. *Educational Researcher, 18*, 32-42.

Campbell, A. M. (2004, 2004). *Genome consortium for active teaching*. Retrieved May 23, 2004, from http://www.bio.davidson.edu/projects/GCAT/gcat.html

Campbell, A. M., & Heyer, L. J. (2003). *Discovering genomics, proteomics, and bioinformatics.* San Francisco: Pearson Education, Inc. (Benjamin Cummings).

Chandler, P., & Sweller, J. (1991). Cognitive load theory and the format of instruction. *Cognition and Instruction, 8*, 293-332.

Charlin, B., Mann, K., & Hansen, P. (1998). The many faces of problem-based learning: A framework for understanding and comparison. *Medical Teacher, 20*(4), 323-330.

Chinn, C. A., & Malhotra, B. A. (2002). Epistemologically authentic inquiry in schools: A theoretical framework for evaluating inquiry tasks. *Science Education, 86*(2), 175-218.

Christopherson, J. T. (1997). *The growing need for visual literacy at the university.* Paper presented at the Visionquest: Journeys toward visual literacy; 28th Annual Conference of the International Visual Literacy Association, Cheyenne, WY.

Collins, A., Brown, J. S., & Holum, A. (1991). Cognitive apprenticeship: making thinking visible. *American Educator*(Winter), 6 - 46.

Consortium, T. G. O. (2000). Gene Ontology: tool for the unification of biology. *Nature Genetics, 25*, 25-29.

Cooper, G., Tindall-Ford, S., Chandler, P., & Sweller, J. (2001). Learning by imagining. *Journal of Experimental Psychology: Applied, 7*(1), 68-82.

Duffy, T. M., & Cunningham, D. J. (1996). Constructivism: implications for the design and delivery of instruction. In D. H. Jonassen (Ed.), *Handbook of research for educational communications and technology.* (pp. 170-198). New York: Macmillan.

Dunbar, K. (1997). How scientists really reason: scientific reasoning in real-world laboratories. In R. Sternberg & J. Davidson (Eds.), *The nature of insight.* (pp. 365-396). Cambridge, MA: MIT Press.

Edelson, D. C., & Gordin, D. (1998). Visualization for learners: a framework for adapting scientists' tools. *Computers and Geosciences, 24*(7), 607-616.

Ferk, V., Vrtacnik, M., Blejec, A., & Gril, A. (2003). Students' understanding of molecular structure representations. *International Journal of Science Education, 25*(10), 1227-1245.

Fleischmann, R., Adams, M., White, O., Clayton, R., Kirkness, E., Kerlavage, A., et al. (1995). Whole-genome random sequencing and assembly of *Haemophilus influenzae Rd. Science, 269*(5223), 496 - 512.

Gilbert, J. K. (2002). *Moving between the modes of representation of a model in science education: some theoretical and pedagogic implications.* Paper presented at the Philosophical, Psychological, Linguistic Foundations for Language and Society Literacy, University of Victoria, Canada, September 12 - 15, 2002.

Glaser, F., Pupko, T., Paz, I., Bell, R. E., Bechor-Shental, D., Martz, E., et al. (2003). ConSurf: identification of functional regions in proteins by surface-mapping of phylogenetic information. *Bioinformatics, 19*, 163-164.

Greeno, J. (1998). The situativity of knowing, learning, and research. *American Psychologist, 53*(1), 5-26.

Greeno, J. G., Collins, A. M., & Resnick, L. B. (1996). Cognition and learning. In D. Berliner & R. Calfee (Eds.), *Handbook of educational psychology.* (pp. 15-46). New York: Macmillan.

Herrington, J., & Herrington, A. (1998). Authentic assessment and multimedia: how university students respond to a model of authentic assessment. *HIgher Education Research and Development, 17*(3), 305 - 322.

Jonassen, D. H., & Reeves, T. C. (1996). Learning with technology: using computers as cognitive tools. In D. H. Jonassen (Ed.), *Handbook of research for educational communications and technology.* (pp. 693-720). New York: Macmillan.

Keefe, D. F., Kirby, R. M., & Laidlaw, D. H. (2004). Painting and Visualization. In C. R. Johnson & C. Hansen (Eds.), *Visualization Handbook 2004*: Elsevier.

Keig, P. F., & Rubba, P. A. (1993). Translation of representations of the structure of matter and its relationship to reasoning, gender, spatial reasoning, and specific prior knowledge. *Journal of Research in Science Teaching, 30*(8), 883-903.

Kirby, J. (1993). Collaborative and competitive effects of verbal and spatial processes. *Learning and Instruction, 3*, 201-214.

Kolb, D. A. (1984). *Experiential Learning: experience as the source of learning and development.* New Jersey: Prentice-Hall.

Kozma, R. (2000). Students collaborating with computer models and physical experiments. In C. Hoadley (Ed.), *Computer support for collaborative learning.* (pp. 314-322). Mahwah, NJ: Erlbaum.

Kozma, R. (2003). The material features of multiple representations and their cognitive and social afforances for science understanding. *Learning and Instruction, 13,* 205-226.

Kozma, R. B., & Russell, J. (1997). Multimedia and understanding: Expert and novice responses to different representations of chemical phenomena. *Journal of Research in Science Teaching, 34,* 949-968.

Krypides, N. (August 31, 2004). *Genomes OnLine Database.* Retrieved September 1, 2004, from http://www.genomesonline.org/

Kyrpides, N. (1999). Genomes OnLine Database (GOLD): a monitor of complete and ongoing genome projects worldwide. *Bioinformatics, 15,* 773-774.

Lander, E., Linton, L., Birren, B., Nusbaum, C., Zody, M., Baldwin, J., et al. (2001). Initial sequencing and analysis of the human genome. *Nature, 409,* 860 - 921.

Levie, H. W., & Lentz, R. (1982). Effects of text illustrations: A review of research. *Educational Communication and Technology Journal, 30,* 195-232.

Levin, J. R., Anglin, G. J., & Carney, R. N. (1987). On empirically validating functions of pictures in prose. In D. M. Willows & H. A. Houghton (Eds.), *The psychology of illustration.* (Vol. 1, pp. 51-86). New York: Springer.

Maor, D., & Taylor, P. C. (1995). Teacher epistemology and scientific inquiry in a computerised classroom environment. *Journal of Research in Science Teaching, 32,* 839-854.

Martz, E. (2002). Protein Explorer: Easy yet powerful macromolecular visualization. *Trends in Biochemical Sciences, 27*(February), 107-109.

Mayer, R. E. (1997). Multimedia learning: Are we asking the right questions? *Educational Psychologist, 32,* 1-19.

Meyer, C. A. (1992). What's the difference between authentic and performance assessment? *Educational Leadership, 49,* 39 - 40.

Paivio, A. (1986). *Mental representations: A dual coding approach.* Oxford, England: Oxford University Press.

Palloff, R. M., & Pratt, K. (1999). *Building Learning Communities in Cyberspace.* San Francisco: Jossey-Bass.

Parker, J. D., Ziemba, R. E., Cahan, S. H., & Rissing, S. W. (2004). An hypothesis-driven, molecular phylogenetics exercise for college biology students. *Biochemistry and Molecular Biology Education, 32*(2), 108-114.

Reeves, T. C., & Okey, J. R. (1996). Alternative assessment for constructivist learning environments. In B. G. Wilson (Ed.), *Constructivist learning environments: case studies in instructional design* (pp. 191 - 202). Englewood Cliffs, NJ: Educational Technology Publications.

Resnick, L. (1988). Learning in school and out. *Educational Researcher, 16*(9), 13-20.

Richardson, D. C., & Richardson, J. S. (2002). Teaching molecular 3-D literacy. *Biochemistry and Molecular Biology Education, 30*(1), 21-26.

Schnotz, W., & Bannert, M. (2003). Construction and interference in learning from multiple representation. *Learning and Instruction, 13,* 141-156.

Seddon, G. M., Eniaiyeju, P. A., & Chia, L. H. L. (1985). The factor structure for mental rotations of three-dimensional structures represented in diagrams. *Research in Science and Technological Education, 3*(1), 29-42.

See, K., & Takayama, K. (unpublished).

Shaffer, C., & Anundsen, K. (1993). *Creating Community Anywhere*. New York: Jeremy P. Tarcher/Perigee Books.

Shepard, R. N. (1978). Externalization of mental images and the act of creation. In B. S. Randhawa & W. E. Coffman (Eds.), *Visual learning, thinking, and communication* (pp. 133-189). New York: Academic Press.

Shi, L. (1998, January 7, 2002). *DNA microarray (genome chip)- monitoring the genome on a chip*. Retrieved May 23, 2004, 2004, from http://www.gene-chips.com/

Slack, J., Hildebrand, K., Munzner, T., & St. John, K. (2004). *SequenceJuxtaposer: Fluid navigation for large-scale sequence comparison in context*. Paper presented at the German Conference on Bioinformatics, October 4 - 6, 2004, Bielefeld, Germany.

Sringam, C., & Geer, R. (2000). *An investigation of an instrument for analysis of student-led electronic discussions*. Paper presented at the Learning to Choose, ASCILITE 2000 Conference, Coffs Harbour, NSW, Australia.

Sweller, J. (1999). *Instructional design in technical areas*. Camberwell, Victoria: ACER Press.

Tobias, S., & Hake, R. R. (1988). Professors as physics students: What can they teach us? *American Journal of Physics, 56*, 786-794.

Tobin, K. (1993). Constructivist perspectives on teacher learning. In K. Tobin (Ed.), *The practice of constructivism in science education*. Hillsdale, NJ: Lawrence.

Tuckey, H., Selvaratnam, M., & Bradley, J. (1991). Identification and rectification of student difficulties concerning three-dimensional structures, rotations, and reflection. *Journal of Chemical Education, 68*(6), 460-464.

Vollmeyer, R., Burns, B., & Holyoak, K. (1996). The impact of goal specificity on strategy use and the acquisition of problem structure. *Cognitive Science, 20*, 75-100.

Vrtacnik, M., Ferk, V., Dolnicar, D., Zupancic-Brouwer, N., & Sajovec, M. (2000a). The impact of visualizationn on the quality of chemistry knowledge. *Informatica, 24*, 497-503.

Vrtacnik, M., Sajovec, M., Dolnicar, D., Pucko-Razdevsek, C., Glazar, S. A., & Zupancic-Brouwer, N. (2000b). An interactive multimedia tutorial teaching unit and its effects on the students perception and understanding of chemical concepts. *Westminster Studies in Education, 23*, 91-105.

Wandersee, J. H. (2000). Language, analogy, and biology. In K. M. Fisher, J. H. Wandersee & D. E. Moody (Eds.), *Mapping Biology Knowledge*. Dordrecht, The Netherlands: Kluwer Academic Publishers.

Watson, J. D. (1968). *The double helix*. New York: New American Library.

Watson, J. D., & Crick, F. H. C. (1953). A structure for Deoxyribose Nucleic Acid. *Nature, 171*, 737-738.

Wiggins, G. (1993). *Assessing student performance: Exploring the purpose and limits of testing*. San Francisco, CA: Jossey-Bass.

Wu, H.-K., & Shah, P. (2004). Exploring visuospatial thinking in chemistry learning. *Science Education, 88*, 465-492.

STEPHEN J REYNOLDS[1], JULIA K JOHNSON[2], MICHAEL D PIBURN[2], DEBRA E LEEDY[2], JOSHUA A COYAN[2], MELANIE M BUSCH[2]

CHAPTER 12

VISUALIZATION IN UNDERGRADUATE GEOLOGY COURSES

[1]Department of Geological Sciences, Arizona State University, Tempe, AZ, USA; [2]Arizona State University, Tempe, USA

Abstract. Visualization is an essential skill in undergraduate geology courses as it is for expert geologists. Geology students and geologists must visualize the shape of the land from topographic maps, the three-dimensional geometry of geologic structures from limited exposures, and the geologic history recorded in sequences of layers and in natural landscapes. Interactive animations have proven successful in helping college students visualize the three-dimensional nature of geology. They permit interactions that are not possible with traditional, paper-based materials, are deliverable via the Internet, and can be imbedded in modules that embrace constructivist pedagogy.

INTRODUCTION

Geology is an exceptionally visual science. To solve geologic problems, geologists must visualize events and processes across a broad range of space and time. Geologists may consider and visualize processes at the microscopic and submicroscopic scale, such as deformation of crystal lattices. At the other extreme, some geologists address phenomena at the scale of a continent, such as the interaction of tectonic plates on the Earth's surface. Most geologic events occur over such long time scales that they cannot be witnessed within a single lifetime. These include the uplift and erosion of mountains, the carving of canyons, and the creation of ocean basins left in the wake of drifting continents. Some geologic events, such as earthquakes and volcanic eruptions, may occur so rapidly that, in order to visualize them, geologists must record and play back the events in slow motion. In all these extremes and the middle ground in between, geologists use maps, sketches, animations, and synthetically produced three-dimensional perspectives to place problems in a visualizable context.

Visualization is important to most scientists (Shepard, 1988) and is an essential part of learning in undergraduate science courses (Lord, 1985). As a general studies course taken by many nonmajors, geology provides an opportunity for explicit instruction and practice in spatial visualization. In fact, many geology faculty consider the development of three-dimensional thinking skills to be one of the most important goals of undergraduate geology courses, both for the major and nonmajor.

In this paper, we summarize the important role of visualization in geology and undergraduate geology courses. We then present examples of interactive computer-based visualizations that have been shown to help students improve their visualization abilities, both in geologic contexts and in more general venues. We close with a discussion of an emerging technologies and how it may bridge the gap between 2D and 3D representations and further help students improve their overall visualization abilities.

ROLE OF VISUALIZATION IN GEOLOGY COURSES

In introductory geology courses (Reynolds et al., 2001), students are immediately confronted with learning to visualize three-dimensional topography from two-dimensional topographic maps (Barsam and Simitus, 1984; Eley 1988, 1993; Schofield and Kirby, 1994). Such maps contain an array of squiggly contour lines, each representing a certain and constant elevation on the land surface (Fig. 1). Students learn that concentric contours, with a roughly circular shape, define a hill, whereas more linear contours define a slope, cliff, or valley. They also learn that closely spaced contours represent steep slopes or cliffs, and widely spaced contour represent more subdued topography. To properly comprehend the terrain, students must build a mental picture of these features in their mind (Eley, 1993). Expert geologists are able to glance at such maps, quickly construct a mental representation, and move seamlessly into describing ridges and stream valleys as if the geologist were actually out in the natural landscape. Topographic maps are used extensively in undergraduate geology courses to locate and navigate in the field, to imagine perspectives from any point on the map, to determine the steepness of slopes and suitability for land use, and as a base for showing geology, flood potential, or mineral resources.

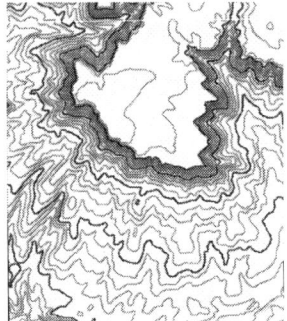 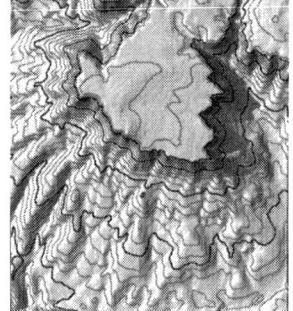

Figure 1. *Two scenes from a QTVR object movie where students can adjust the shading. Map views of a contour map (left) and its shaded-relief equivalent (right).*

Introductory geology courses also strive to teach students to visualize the three-dimensional nature of geologic structures (Davis and Reynolds, 1996; Kali and Orion, 1996). The Earth's land surface and upper crust are dominated by different rock types that occur in layers or more irregular rock bodies, such as lenses and somewhat cylindrical shapes (like a volcanic conduit). Most rock layers are horizontal and fairly continuous when first formed, but can become tilted, buckled into folds, or offset by movement on faults. Several such events may affect a single area, resulting in potentially complex geometries, as layers are tilted and then faulted. The full extent of such rock units is never exposed, because their upper parts may be eroded away and their lower parts may remain buried and hidden in the subsurface. Geology students start with the limited exposures of such features and then try to visualize their buried and eroded away original extents. Only after doing this can the student begin to reconstruct the sequence of events that formed the geometrically complex array of rocks.

Geologists also visualize how changes in the environment millions of years ago resulted in the accumulation of a sequence of different rock types. For example, seas advancing across a land surface may successively deposit a layer of beach sand followed by a layer of offshore mud. When geologists see such a sequence exposed in the side of a cliff, they envision this ancient sequence of environmental changes. Beginning geology students have much difficulty visualizing such events and how the ancient environments could have been so different from the modern ones. Our research has uncovered major misconceptions that students have about this aspect of geology, such as not being able to recognize that the rock layers exposed in a canyon must be older than the canyon itself.

To reconstruct geologic history, geologists use another type of visualization to try to imagine how a landscape evolves over time. Landscape evolution, like many other geologic processes, occurs over too long of time scales to observe directly. As a result, geologists select a number of different but related geologic localities and then mentally arrange them into a step-by-step sequence that represents how the landscape is interpreted to have changed over time. This "trading location for time" (Frodeman, 1996) permits geologists to visualize how a large plateau may be successively eroded into smaller buttes and pinnacles, or how a river may downcut through rock layers time, creating a wide chasm such as the Grand Canyon of Arizona.

TEACHING AND ASSESSING VISUALIZATION IN UNDERGRADUATE GEOLOGY COURSES

As part of several research and curriculum-development projects, we have explored the role of visualization in undergraduate geology courses by assessing students' different types of visualization abilities and by creating curricular materials to specifically improve these abilities. Our studies have been situated in two main courses at a large university in the southwestern U.S.: (1) an introductory geology course taken by nearly 2,500 students per year as a general studies requirement, and (2) a smaller, upper-level structural geology course taken by students striving for a geology major or minor. In both settings, our design of materials was guided by a constructivist pedagogy (Driver et al., 1994) that encouraged students to explore first, prior to formal introduction of terms and concepts (i.e., a learning cycle). Most visualization training was via interactive QuickTime Virtual Reality movies (animations) that students accessed via the Internet. Some movies were embedded into web-based modules and others were accessed from HTML lists. We piloted the materials in their intended settings, commonly revising them after observing how students interacted with the materials or after brief interviews with small numbers of students. For example, we noted early on that students would stop and read two lines of text on a web page, but would more likely skip over the text entirely if its length exceeded two lines. In some cases, we were able to pilot the materials in the morning lab sections and quickly revise (i.e., shorten) wording for the afternoon lab sections and immediately observe the improvement.

We developed two types of QuickTime Virtual Reality (QTVR), each of which has certain uses and advantages. QTVR *panorama* movies are a single image that students can scroll and zoom to gain different perspectives. Such panoramas traditionally represent a 360-degree view stitched together from a number of images, but in our case we use panorama movies as a way to present maps and shaded-relief images in a scrollable, zoomable format. QTVR *object* movies can be represented as a matrix of images in rows and columns. When the user clicks and drags horizontally in an object movie they are moving from one image to another within a row. Clicking and dragging vertically moves from one row to another along a column. In many QTVR object movies, an object is photographed or rendered from different perspectives to create the illusion of being able to spin or rotate the virtual object. Many of our movies are of this type. In other cases, however, we have used QTVR object movies to depict other types of actions, such as cutting into a geologic terrain, displacing a fault, eroding down across a geologic structure, or changing the shading on a landscape. Such actions are simply not possible with traditional paper-based materials. An advantage of both types of QTVR movies is that they require student involvement – nothing happens in the movie unless the student clicks and drags.

To teach students how to visualize topographic maps and construct a mental 3D representation from such 2D maps, we created an array of QTVR object movies. For many of these movies, we used a simplified contour map and draped it over a virtual terrain modeled with digital topographic data from the U.S. Geological Survey. In one type of movie, we fixed the position of the camera and terrain and changed the direction of illumination, rendering a sequence of images with increasing amounts of topographic shading. In the resulting QTVR object movies, students are able to observe a gradation from a flat 2D contour map to a shaded 3D-appearing terrain by clicking and dragging in the movie (Fig. 1). We also created object movies that allow students to spin and tilt a contour-draped terrain (Fig. 2). In a third type of topographic movie, students can raise the level of a virtual water surface to coincide step by step with each contour, reinforcing the important principle that each contour connects all the points of a certain elevation. Other movies permit students to progressively slice into the sides of a solid terrain to explore the meaning of topographic profiles.

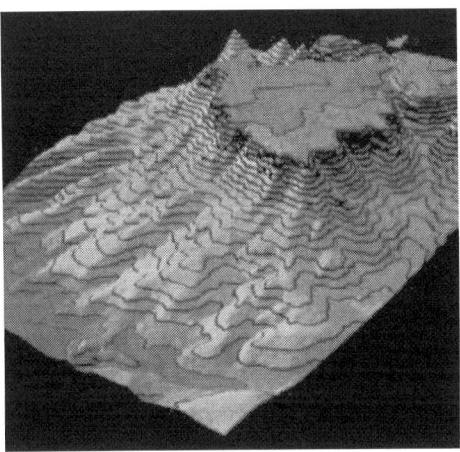

Figure 2. *Scene from a QTVR object movie of the same topographic terrain as Figure 1. Using this movie, students can spin and tilt the contour-draped terrain to gain different perspectives.*

To assess student abilities in visualizing topography and using topographic maps to solve problems, we developed several types of assessment instruments (Piburn et al., 2002, in press). We administered visual, figure-based pretests and post tests with both multiple-choice and constructed-response questions. We then interviewed small numbers of students to understand their reasoning, abilities, and misconceptions. We used this information to develop a reliable topographic visualization instrument being used in our current research (Clark et al., 2004). In our upper level course, we developed QTVR movies in which small numbers were placed on a virtual topographic terrain and assessed how closely students could locate these on a corresponding paper topographic map.

To help students visualize the three-dimensional geometry of geologic structures, we developed a module based around interactive movies of 3D geologic blocks. These movies allow students to rotate geologic blocks containing layers, folds, faults, and other geologic features (Fig. 3). For most movies, we rendered the blocks as both opaque and partially transparent, allowing us to create movies in which students could change the transparency of the block while rotating it. When the blocks are partially transparent, students can better observe the geometry of the contained geologic structures, something students traditionally have difficulty doing (Kali and Orion, 1996). In other movies, students can progressively cut into or erode down through a block, either in its opaque or partially transparent form. Movies with geologic faults allow students to displace opposite sides of the block and observe how such movement is reflected in the resulting configuration of layers. The geologic blocks module, when used in conjunction with the topographic materials, has been shown to improve student performance on content-based tests compared to a control group that did not use the interactive movies. Piburn et al. (2002, in press) documented that use of these materials eliminated the difference in performance by gender on a standard test of spatial visualization (Ekstrom et al., 1976).

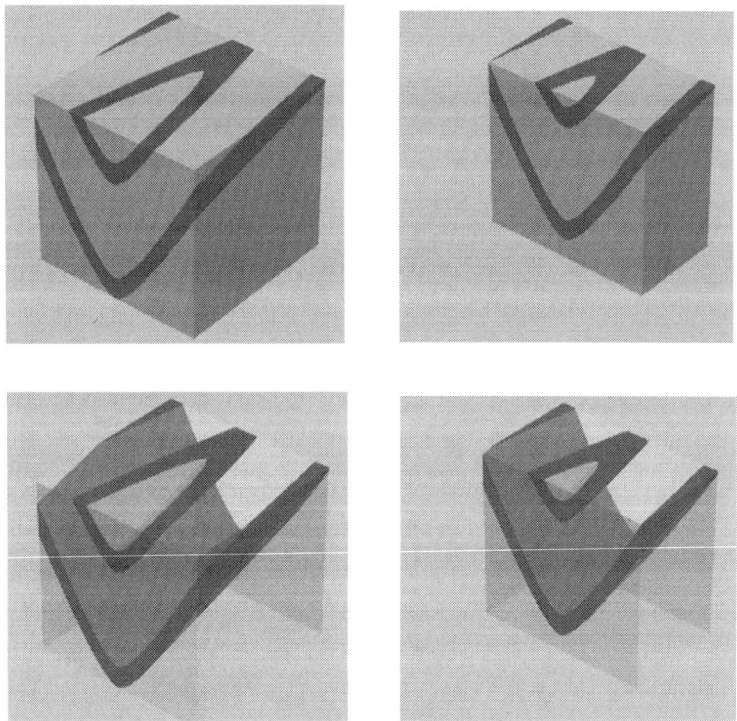

Figure 3. *Scenes from a QTVR object movie in which a user can cut into the block by dragging horizontally (the row of images) or make the block partially transparent by dragging vertically (the bottom row of images).*

Building on their experience with interactive 3D geologic blocks, students next use QTVR object movies to explore how geologic structures interact with topography. In one suite of movies, the students can adjust the tilt of a plane that represents the orientation of a rock layer in the subsurface (Fig. 4). The students observe and map the intersection between the plane and the topography to investigate the expression on the landscape of horizontal versus tilted versus vertical layers. Then students observe photographs of real landscapes that illustrate these same types of interactions of layers and topography.

Figure 4. *One scene from a QTVR object movie of a user tiltable, partially transparent plane intersecting a contour-draped terrain.*

Students extend this knowledge by observing a layer crossing a virtual landscape, mapping the trace of the layer on a corresponding contour map, and determining from this trace the orientation of the layer (Fig. 5). Next they use QTVR panorama and object movies in which geologic maps, which show the distribution of rock units and geologic structures exposed on the surface, are draped upon virtual terrains built with digital topographic data of the same area (Fig. 6). On such maps, students use the manner in which rock boundaries interact with topography to assess the orientation of that boundary.

Figure 5. *Map view of a geologic map draped over digital topography and shaded. This image is from a QTVR panorama movie that allows users to zoom in and out and scroll around in the image.*

Figure 6. *Three-dimensional perspective of a tilted virtual geologic layer intersecting topography.*

To help students envision how changing environments may deposit a stack of rock layers, we developed a series of QTVR animations with which students can interactively change the virtual environment. A student may make virtual seas advance across the landscape depositing beach sands followed by offshore muds (Fig. 7), or they can make the seas retreat so that offshore muds are successively covered over by beach sands and then river deposits. After students explore how a sequence of rock layers is thusly formed, they can observe how this sequence will be exposed in the walls of a canyon as it is eroded deeper and deeper (Fig. 8).

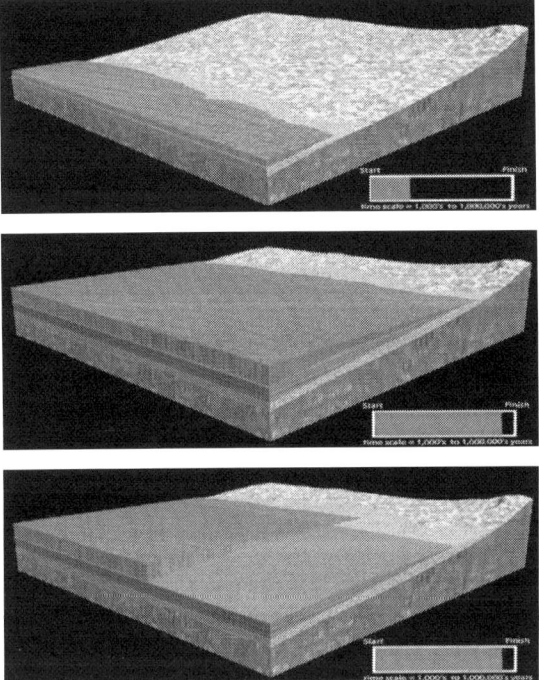

Figure 7. *Three scenes from a QTVR object movie in which the sea advances onto the land from left to right, depositing a layer of beach sand followed by a layer of mud. In the last image, the water has been partially hidden to reveal the sediments on the seafloor.*

To help students learn the strategy of "trading location for time," we developed interactive animations that illustrate how landscapes evolve over time. One movie, for example, shows how a rock layer is progressively eroded from a broad plateau to a smaller, flat-topped mesa, and to a tall, thin rock pinnacle (Fig. 9). This is supplemented with actual landscape photographs of features that correspond to each stage in the evolution of the landscape.

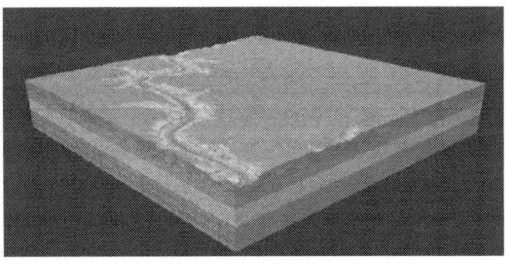
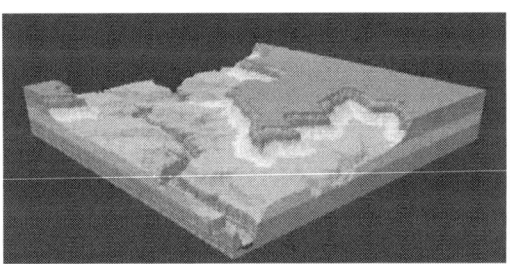

Figure 8. Three scenes from a QTVR object movie in which a sequence of layers are deposited and then progressively eroded and exposed in a canyon.

After completing these virtual exercises, students go into the field to locate features and navigate on a topographic map, to map the boundaries between geologic units on that map, and to determine the orientation of the layers and their boundaries. From these maps that the students create, they construct topographic profiles and geologic cross sections that integrate all that they have learned from their virtual and field-based studies.

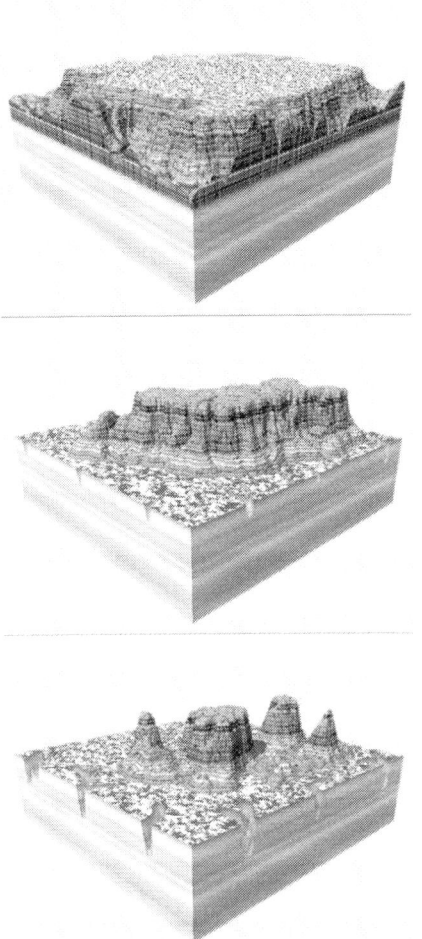

Figure 9. Three scenes from a QTVR object movie in which erosion successively forms a broad plateau, flat-topped mesa, and pinnacles.

RESULTS FROM CLASSROOM USE

We have a number of measures of how successful these materials are in the classroom. The most detailed assessment was done using control and experimental groups in intact sections of the laboratory of physical geology, a general sciences course for nonmajors. Both groups were pre- and post-tested using tests of general spatial ability (spatial visualization and spatial orientation) and a geospatial test.

The geospatial test focused on the more spatial aspects of the course, such as visualizing and using topographic maps and geologic cross sections. The experimental group completed two modules, one on visualizing topography and another on interactive 3D geologic blocks. Both modules were constructed around interactive QTVR object movies that were embedded in a learning-cycle approach. The results of this experiment were that both groups improved in spatial visualization and geospatial measures, but the experimental group improved more, as measured with normative gains (Piburn et al., in press). Also, the experiment equalized the performance of males and females in a case where the performance of males was initially superior to that of females (Piburn et al., in press).

In a separate study, Clark et al. (2004) used student interviews and pre- and post-tests addressing student understanding and strategies using topographic maps. These results demonstrated that students' abilities at visualizing and using topographic maps improved as a result of using the interactive animations.

On a less quantitative level, our experience with using the entire array of visualizations has been very successful in preparing geology majors for field studies. From years of teaching field geology, both with and without the visualizations, we conclude that the visualizations help students better visualize topography and better understand the geometry of geologic structures, and how these structures interact with topography. Using the interactive visualizations in the laboratory, when integrated with complementary field exercises, makes the students better at locating themselves in the field, visualizing real geologic structures from clues on the land surface, and portraying these geologic structures by drawing geologic boundaries on a topographic map (i.e. making a geologic map). Every year students tell us how much the interactive animations have helped them solve real geologic problems out in the field. We receive similar reports from colleagues who use our materials to prepare university and high school-students from Arizona to Papua New Guinea.

THE GEOWALL – AN EMERGING TECHNOLOGY

New opportunities have arisen from increases in the power of personal computers, decreases in the costs of such computers and of digital projectors, and improvements in the capabilities of visualization software. Perhaps the best example of this, for the geologic sciences, is the invention of the GeoWall, an inexpensive stereo 3D projection system (GeoWall Consortium, 2004). A GeoWall consists of an off-the-shelf personal computer with a graphics board capable of outputting different video signals to two monitor ports. The video stream from each port is sent to a digital projector, with one projector displaying a left-eye view and the other a right-eye view. Polarizing filters are placed in front of each projector and students wear inexpensive polarized glasses matched to the polarization directions of the projectors, so that the student's left eye sees the image from the left-eye projector and the right eye sees the image from the right-eye projector. There are three types of geologic software currently available for the GeoWall. The simplest type of stereo images are stereo pairs, consisting of two still images taken from slightly different perspectives to impart a stereo effect. Stereo pairs can be captured in the field using matched digital cameras with an appropriate spacing or can be created digitally be rendering a virtual scene from two adjacent perspectives. Such stereo pairs have been used to bring natural landscapes into the classroom and to provide students stereo views of virtual terrains, such as draped geologic maps.

A second type of software treats virtual terrains as objects that can be spun, tilted, and zoomed into and out of with a click of a mouse. Such terrains include digital topography draped with topographic maps, geologic maps, satellite images, and various types of geophysical data. In addition, these programs can be used to display the distribution of earthquakes in map view or on a globe of the entire world. They also are suited for displaying crystal lattices, molecular structure, fossils, and internal anatomy of plants, animals, and cells.

The third type of software allow users to walk or fly over the virtual terrain. Different textures, such as aerial photographs and geologic maps, can be draped over the terrain and users can typically switch from one draped map to another as they walk across the virtual landscape. When done in stereo on the GeoWall, these provide students with a virtual experience that is more authentic to actual field studies than other virtual activities we have seen.

There have been almost no detailed assessments of GeoWall-related activities, such as using control and experimental groups. Anecdotal experience for various GeoWall users, as well as our own personal observations of student use of GeoWalls, suggests that it is an effective tool for helping students bridge the gap between 2D representations of 3D objects and actual observation in three dimensions. We suspect that seeing objects in true stereo on the GeoWall lessens the students cognitive load compared to mentally creating 3D representations from two dimensional data. This avenue of research is sorely needed and hopefully will be addressed in the near future.

REFERENCES

Barsam, H.F. and Simitus, Z.M. (1984). Computer-based graphics for terrain visualization training. Human Factors, 26(6), 659-665.

Clark, Douglas, Reynolds, et al. (2004) Interpreting topographic maps: strategies and assumptions of university students: Proceedings of the NARST 2004 Annual Meeting (Vancouver, BC, Canada): National Association for Research in Science Teaching (NARST) April 1-3, 2004, 97 p. Available online at: http://courses.ed.asu.edu/clark/Papers/NARST_2004/205043_ASU_Topo.pdf

Davis, G.H., & Reynolds, S.J. (1996) *Structural Geology of Rocks and Regions* (2nd ed). New York, John Wiley & Sons, Inc., 776 pp.

Driver, R., Asoko, H., Leach, J., Mortimer, E. F. and Scott, P. (1994) Constructing scientific knowledge in the classroom. *Educational Researcher, 23, (7), 5-12*

Ekstrom, R., French, J., Harman, H. and Dermen, D. (1976) Manual for Kit of Factor Referenced Cognitive Tests. Princeton, NJ, *Educational Testing Service*.

Eley, M.G. (1988). Determining the shapes of landsurfaces from topographic maps. *Ergonomics, 31(3), 355-376.*

Eley, M.G. (1993) The differential susceptibility of topographic map interpretation to influence from training. *Applied Cognitive Psychology*, 7, 23-41.

Frodeman, R.L. (1996) Envisioning the outcrop. *Journal of Geoscience Education, 44, 417-427.*

GeoWall Consortium, 2004, Web Site: *http://geowall.org*

Kali, Y. and Orion, N. (1996) Spatial abilities of high-school students and the perception of geologic structures. *Journal of Research in Science Teaching*, 33, 369-391.

Lord, T. (1985). Enhancing the visuo-spatial aptitude of students. *Journal of Research in Science Teaching, 22(5), 395-405.*

Piburn, M.D., Reynolds, S.J., Leedy, D.E., McAuliffe, C., Birk, J.E., and Johnson, J.K. (2002) The Hidden Earth: Visualization of geologic features and their subsurface geometry: Paper accompanying presentation to national meeting of National Association of Research in Science Teaching (NARST), New Orleans, LA, *47 p.* with CD-ROM.

Piburn, M., Reynolds, S., McAuliffe, C., Leedy, D., Birk, J. & Johnson, J. (in press). The role of visualization in learning from computer-based images. *International Journal of Science Education.*

Reynolds, S.J., Johnson, J.K., and Stump, E. (2001). *Observing and interpreting geology: A laboratory manual for introductory geology*: Tucson, Terra Chroma, Inc., 148 p.

Schofield, N. J., & Kirby, J. R. (1994). Position Location on Topographical Maps: Effects of Task Factors, Training, and Strategies. *Cognition and Instruction, 12 (1), 35-60.*

Shepard, R. (1988). The Imagination of the scientist. In K. Egan & D. Nadaner (Eds.), *Imagination and education* (pp. 153-185). New York: Teachers' College Press.

SECTION D

ASSESSING THE DEVELOPMENT OF VISUALIZATION SKILLS

VESNA FERK SAVEC[1], MARGARETA VRTACNIK[2]
AND JOHN K GILBERT[3]

CHAPTER 13

EVALUATING THE EDUCATIONAL VALUE OF MOLECULAR STRUCTURE REPRESENTATIONS

[1] [2]Department of Chemistry and Informatics, University of Ljubljana, Slovenia; [3]Institute of Education, The University of Reading, U.K.

Abstract. An investigation examined the value of various representations (e.g. concrete three-dimensional models, virtual computer models, static two-dimensional computer models, stereo-chemical formulas) in supporting the achievement by students of an effective perception of molecular structures. Additionally, the usefulness was studied of concrete three-dimensional models, virtual computer molecular models, and their combination, as help tools for students in solving spatial chemistry tasks involving three-dimensional perception, rotation and reflection. Altogether 477 students from secondary schools (age: 18-19 years) took part in the investigation. For purpose of the inquiry a set of four Molecular Visualization Tests was developed. Information about students` manner of thinking while solving spatial tasks was initially gained with a questionnaire and then examined in depth with a structured interview. The data was processed by methods suitable for the respective quantitative and qualitative approaches taken. The results suggest that the information sources which serve as a foundation for students` perception of molecular structure decrease in value from concrete models, to virtual models, to static computer models. Students` perception of three-dimensional structure was better when a stereo-chemical formula was used in comparison to that supported by a computer image. The results indicate that both molecular models types used as help-tools can ease the solving of chemistry tasks that require three-dimensional thinking. Virtual computer models seem to be as effective as concrete models, but the combined usage of both can cause splits in students' attention and therefore seems to be less appropriate.

INTRODUCTION

The ancient Chinese proverb says that one picture is worth a thousand words. Comenius introduced this old wisdom into holistic pedagogy in his Latin textbook 'Orbis sensualium pictus' (1658), The World in Pictures. Since the time of Comenius visualization of concepts has remained a central concept in pedagogy and didactics, and illustration has become an indispensable component in teaching materials (Molitor, Ballstaedt, and Mandl, 1989).

Johnstone (1991) indicated that the nature of scientific concepts and the threefold manner of representing science (at the macro, sub-micro and symbolic levels) make science difficult to learn. Visualization elements, i.e. structural models, play an important role by supporting students when connecting the three levels of concept representations (Barke and Wirbs, 2002).

Hand-held molecular models have been used for educational purposes since Dalton used them to illustrate his ideas (Hardwicke, 1995). However, the massive interest in the value of molecular structure representations in chemical education has risen since the 1970s. Numerous articles have appeared that both gave instructions for the production of cheap self-made concrete / material (*3d*) molecular models and which pointed to their potential in chemical education, especially at school level (e.g. McGrew, 1972; Roberts and Traynham 1976; Chapman, 1978; Battino, 1983; Birk and Foster, 1989; Eggeton, Williamson, Lovesless and Grimes, 1990; Hanson, 1995). The usefulness of different versions of such models in teaching has been investigated. Thus, Goodstein and Howe (1978) and Yamana (1989) found that, by virtue of the fact that they can be physically manipulated, they are effective in teaching students about molecular structure, whilst Gabel and Sherwood (1980) showed that high school chemistry students who manipulated such models achieved better assessment grades.

The use of representations increased rapidly with the advent of computer-based molecular models, what we will term 'virtual models', and molecular modelling packages. Enquiries focused solely on the use of such packages have shown them to be educationally worthwhile. Hyde, Shaw, Jackson, and Woods (1995) developed and evaluated a package which integrated tutorial instruction with the use of an interactive molecular modelling programme. University students' self-assessment indicated that this integration helps them understand spatial chemistry. Barnea (1997) reported that high-school students who learned about molecular structure and its relation to macroscopic properties with the use of computer-based models gained better insight into the concept of 'model'. They could also explain more phenomena with the aid of variety of types of model itself than did a control group. Ealy's (1999) research on undergraduate students showed that 87% of the students believed that an appreciation of the three-dimensional nature of models and of the structure of molecules were gained by the use of the program 'Spartan'. 38% of them said that it made abstract ideas more concrete, and, perhaps most significantly, 37% said that work with the computer was fun and the program easy to use. Canning and Cox (2001) reported that 85% of their undergraduate students thought that the use of the visualization software 'RasMol' in class and in projects enhanced their understanding of the structural nature of biological molecules and of the role of non-covalent interactions in protein structure.

One important form of enquiry has been into the comparative effects of using *2d* models, virtual models, and *3d* models, on student learning of the structures of organic compounds. The results show that a mixture of model types is the most effective approach. Copolo and Hounshell (1995) taught organic structures to high school students using one of four manners of molecular representations: (1) two-dimensional textbook representations only (2) virtual models only (the software was Molecular Editor), (3) concrete *3d* models (ball-and-stick) models only , and (4) a combination of virtual models and ball-and-stick models. They concluded that both the *3d* and virtual models used separately were effective tools for the teaching of molecular structures and isomers. However, students using both forms scored significantly higher on the retention test of isomeric identification when compared to the other groups. Dori and Barak (2001) based their research on the assumption that both concrete *3d* models and virtual models have advantages and disadvantages. The authors report that high-school students in their experimental group, where a mixture of representational types was used, both gained a better understanding of the concept of 'model' and were more capable of defining and implementing newly acquired concepts, such as isomerism and functional group. Their experimental group students were more capable of transforming two-dimensional representations of molecules, provided by either in symbolic form or as a structural formula (a type of *2d* representation), into three-dimensional representations, and vice versa. They recommend incorporating both virtual and *3d* models in chemistry teaching/learning as a means to foster model perception and spatial understanding of molecular structure. Wu, Krajcik, and Soloway (2001) developed the visualization tool 'eChem', which allows students to build virtual molecular models and view multiple representations simultaneously, and evaluated its usefulness. The results show that the visualization tool in combination with ball-and-sticks models enabled high-school students to develop a better conceptual understanding of chemical representations. Furthermore, eChem improved students' ability to translate *3d* representations to *2d* representations, which was consequently exhibited in an improved understanding of the concepts of isomerism and polarity.

The usefulness of dynamic virtual models – computer animations – has also been investigated. Williamson and Abraham (1995) studied the effect of them on college students' mental models of chemical phenomena. Animations were used in two treatment situations: as a supplement to large group lectures; as both the lecture supplement and as part of an assigned individual activity. Both treatments increased the students' understanding. Sanger and Greenbowe (2000) investigated the effect on students' learning about the nature of current flow in electrolyte solutions both of computer animations of microscopic chemical processes and also of instruction-based demonstrations. They found that instruction was effective in dispelling student misconceptions but that the use of computer animations did not appear to affect student test responses. Russell, Kozma, Jones, and Wykoff (1997) developed the computer based teaching material '4M: CHEM', in which they synchronized macroscopic representations (photo, video), microscopic representations (animations), and symbolic representations (chemical equations), in an attempt to enhance the teaching and learning of chemical concepts. They report that college students achieved significantly better results on the post-test after use of their educational material.

Other researchers have focused on the mental operations involved in the manipulation of representations. Seddon and Eniaiyuju (1986), Seddon and Shubber (1985a), and Tuckey and Selvaratnam (1991) found that high school students' success in tasks involving the mental rotation of *2d* representations improved when they were trained with instructional programs, e.g. using videotapes that demonstrated rotations. Surprisingly, research has also shown that high-school students who used *3d* models at the same time as watching instructional videotapes were less effective in such tasks (Seddon and Moore, 1986). The authors assume the result to be the consequence of students' attention being divided between the two resources. A study by Seddon and Shubber (1985b) showed that high-school students' efficiencies in conducting mental rotations depended on the use of colour in the representations with which they were presented. The use of some colour did cause an improvement in the students' results, but very colourful schemes actually decreased students' efficiency in performing rotations.

PROBLEM DEFINITION AND THE SCOPE OF A STUDY

In summary then, molecular structures can be presented to students with the use of many different types of representations in both educational materials (e.g. textbooks, video-tapes, CD-ROMs, tutorials on the web, etc.) and in actual teaching situations. Some of these types are three-dimensional (i.e. concrete/material models), others are pseudo-three dimensional (i.e. virtual models), but the majority of those currently used are two-dimensional (e.g. photos of three-dimensional models, static computer models, schematic representations, stereo-chemical formulas). It is often presumed by educators that students can recognize the three-dimensional structure of a molecule on the basis of all the different representations of it, and, even more significantly, that students are able to conduct different three-dimensional mental manipulations with these mental images. On the contrary, a study by Ferk, Vrtacnik, Blejec, and Gril (2003) showed that even the most 'simple' mental process – the perception of three-dimensional molecular structure – depends on the type of two-dimensional molecular structure representation used. The present chapter goes a step further, in dealing with the value of the *3d* representations and virtual representations as help tools for students. Their tasks were to perceive *3d* molecular structure and to solve problems involving basic three-dimensional operations (e.g. rotation, reflection) on the basis of two fundamental types of molecular structure representation (static computer model, stereochemical formula).

RESEARCH QUESTIONS

Three research questions were posed on students' perceptions of three-dimensional molecular structure representations:
- 1st research question: Does the type of molecular structure representation used have an impact on the quality of students' perception of its three-dimensional structure?
- 2nd research question: Do different types of representations, when used as help tools, support students' perception of three-dimensional molecular structure?
- 3rd research question: Is the usage of a certain kind of representation more efficient than others in supporting students' perception of three-dimensional molecular structure?

Three further research questions were posed about students' three-dimensional mental operations with molecules:
- 4th research question: Does the type of molecular structure representation have an impact on students' efficiency in solving spatial chemistry problems involving rotation or/and reflection?
- 5th research question: Do different types of molecular models, when used as help tools, support students in solving spatial chemistry problems involving rotation or/and reflection?

- 6th research question: Are traditional three-dimensional models, virtual models, and a combination of them, equally useful to students in solving spatial tasks with molecular models?

METHODS

The research presented here took place in several phases. An outline of the main phases of the investigation (Phases A-C) and of the terminology used are briefly presented in an Advance Organiser section. This is done to provide the reader with the overview of the investigation, with more detail of the instruments, sample, procedures for data collection, and data analysis, following in later sections.

Advance Organiser

Phase A: Development and selection of the research instruments used

A central instrument used in the investigation was a set of 'Molecular Visualization Tests' (*MVTs*); Ferk, 2003), with which students' efficiency in solving simple spatial tasks with molecules (represented in different manners) was studied. The *MVTs* were constructed to require students to solve spatial tasks purely mentally (*MVT1*), or with various help tools available to them during task solving (*MVT2s*). Because three different types of help tools were available to students, three sub-types types of *MVT2s* emerged: *MVT2-3d*, in which three-dimensional (*3d*) models were available; *MVT2-v*, in which virtual models (*v*) were available; *MVT2-3dv*, in which three-dimensional models and virtual models were available at the same time.

To gain additional information about students' mental processes during task-solving, a brief 'Questionnaire' and a more detailed 'Structured Interview' were developed (Ferk, 2003). 'Questionnaire' was administered to students immediately after testing with *MVTs* and was used only as a guideline for the development of a more detailed qualitative instrument - 'Structured Interview'. All the instruments were pretested on a small sample of students and adjusted in the light of that pilot enquiry.

In collaboration with psychologists, standardised tests of students' perceptional ('Patterns Test'; Pogacnik, 1998a) and spatial abilities ('Rotations Test'; Pogacnik, 1998b), fluid intelligence ('Series Test'; Pogacnik, 1994) and visual memory ('BIS4-OG1', 'BIS4-FM1', 'BIS4-WE1'; Jaeger, Suess, and Beauducel, 1997) were identified. Students' mental capabilities, as identified using these tests, were taken into account when selecting the sample for interview as described in the Phase C. Details of the correlations between students' abilities, gender, and achievements on the *MVTs*, are given in Ferk (2003).

Phase B: Preparatory work before data collection

This involved the selection of the student sample, the acquisition of necessary written agreements for the research to take place, and the preparation of computer classrooms for the testing (Scheme 1).

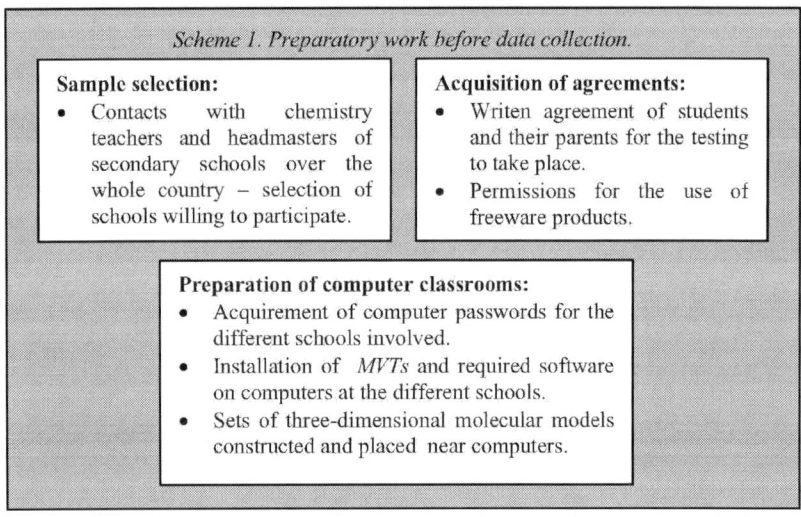

Phase C: Data collection and analysis

Firstly, data was collected using *MVTs*, 'Questionnaire' and ability tests. Then, on the basis of the analysis of students' achievements on *MVTs*, the results on the 'ability' tests, and on students' readiness to participate further in the investigation, a smaller sample of students was selected and interviewed (Scheme 2).

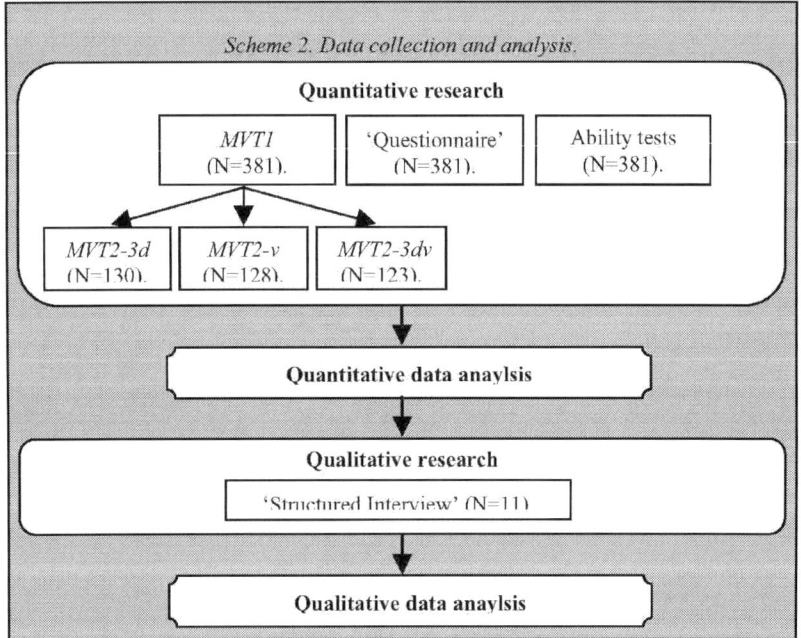

Instruments

The tests developed and associated help tools are summarised in Table 1.

Table 1. Types of MVTs according to availability of help-tools.

MVT types	Help tools available during the tests
MVT1	/
MVT2-3d	Plastic three-dimensional molecular models (*3d*)
MVT2-v	Computer pseudo-three-dimensional molecular models (*v*)
MVT2-3dv	Plastic three-dimensional molecular models and computer pseudo-three-dimensional molecular models (*3dv*)

Each *MVT* consisted of eight tasks which involve spatial operations, e.g. perception of a three-dimensional molecular structure, rotation of a molecule, reflection of a molecule, and a combination of all those operations. The mental operations involved in particular task type are evident from its name, e.g. 'Perception'; 'Perception and Rotation'; 'Perception and Reflection'; 'Perception, Rotation and Reflection' (Table 2).

Table 2. Structure of the MVTs.

Task type	Students' task in particular task types
'Perception'	Recognise a three-dimensional molecular structure on the basis of static computer molecular model (Task 1) and stereo-chemical formula (Task 4) with / without application of different help tools.
'Perception and Rotation'	Recognise a three-dimensional molecular structure on the basis of a static computer molecular model (Task 7) and stereo-chemical formula (Task 2), and rotate it around labelled axis in the system shown of coordinates with / without application of different help tools.
'Perception and Reflection'	Recognise a three-dimensional molecular structure on the basis of a static computer molecular model (Task 3) and stereo-chemical formula (Task 6), and derive its mirror image with / without application of different help tools.
'Perception, Rotation and Reflection'	Recognise a three-dimensional molecular structure on the basis of a static computer molecular model (Task 5) and stereo-chemical formula (Task 8), and use this mental image in the following operations of reflection and rotation.

As indicated in Table 2, each of the task types consisted of two tasks with the same instructions but with different visual materials included. The molecular structures represented were of simple molecules with up to four carbon atoms, presented as static computer molecular models and as stereo-chemical formula. Further details the *MVTs* are given in the Colour Section.

The 'Structured Interview' was developed to gain access to students' recall of their mental processes during the solution of the spatial operation tasks under different circumstances. It was based on the analysis of the students' responses to the *MVTs* and to the 'Questionnaire'. It consisted of forty-seven questions in five groups, respectively concerning: (1) the construction of molecular models and the writing of stereo-chemical formulas, (2) perception of molecular structures, (3) rotation of molecular structures, (4) reflection of molecular structures, and (5) rotation and reflection of molecular structures. Further details of the 'Structured Interview' are given in the Appendix (see Colour Section).

Sample

Altogether 477 students from the third- and fourth- grade of eleven general upper level secondary schools from different Slovenian regions participated in the investigation. [1]

- 94 participants took part in the pilot-phase of the investigation, when the instruments (*MVT1*, *MVT2*s, 'Questionnaire', 'Structured Interview') were refined.
- A further 381 students took *MVT1*, an *MVT2*, 'Questionnaire' and the 'ability' tests.
- From these 381 students, nine students were selected to be interviewed. To identify a variety of students in terms of their personal characteristics and achievements, the criteria for selection were: achievements on the *MVTs*, gender, 'ability', and willingness to participate in the further investigation. In that manner an approximately 50:50 male: female balance was achieved and poorer, average, and better results on *MVT1* and *MVT2* were represented within a range of 'ability'. Besides those nine students, two students who had not participated in the main section of the investigation (therefore they presumably have no experiences in solving similar tasks) were also selected for interview to track down students' thoughts when they are solving this type of task for the very first time - in case that their accession to task solving would reveal some differences.

Data Collection

Testing was conducted in the school year 2001/2002 in the computer classrooms at the secondary schools involved with up to 15 students being tested at the same time. This was done so that each student had an individual computer. After an introduction by the first author (V.F.S.), students were randomly divided into three groups of the same size. Firstly, all students - regardless of the group – took *MVT1* (20 minutes were allowed for this). During the following 20 minutes, students in each of the groups solved the problems using only one variation of *MVT2* (e.g. using *3d*, *v*, or *3dv*, as help tools). Students wrote their answers to *MVTs* in printed answer booklets. After taking the *MVTs* students were asked to answer the 'Questionnaire' (10 minutes). After a short break, the psychologist conducted the 'ability' testing (45 minutes was needed).

The first author interviewed students one by one according to the prepared scheme approximately 6 months later. The questions were asked during the task-solving process to gain additional information on students' way of thinking, their understanding of the mental processes required by particular tasks, their perception of molecular structures, and the usefulness of representational forms. Their comments and behavior (gestures, construction of *3d* representations, and their manipulation, etc.) were captured by video camera. The interviewer additionally wrote notes. The duration of the interviews varied from 56 – 106 minutes. In this manner 990 minutes (16.5 hours) of videotape materials were collected. The first author prepared 107 pages of transcription from the videotapes, with repeated viewing taking place to capture additional details.

Data Analysis

Analysis of MVTs

Students' written responses to *MVT1*, *MVT2*s and 'Questionnaire' were and entered into computer files and processed with the programmes MS Excel and SPSS (Table 3).

Table 3. *Statistical methods used in processing data from MVT1 and the MVT2s.*

Statistical methods used	Purpose
Absolute and relative frequencies of students' answers	To find out proportions of certain answers.
t-test and ANOVA (Post Hoc Test, LSD)	To find out whether the differences in success in solving *MTV* among groups of students were statistically significant.

Analysis of 'Structured Interviews'

The large amount of data which was obtained by transcribing the structured interviews was first of all analysed to identify the natural units of meaning produced by each question. In the next step, units with similar meaning were amalgamated. Those two steps were repeated several times during a one-month period to obtain the most appropriate final version. In this way, for each of the questions in the interview, codes were ascribed to each of the final natural units of meaning, enabling a coding table to be established.

To assure a high reliability of categorisation, all the transcriptions were re-evaluated, this time with the coding table, approximately one month after the first analysis. Altogether 95 % of repeated evaluation (reliability) was achieved. Subsequently both evaluations were contrasted at points where differences occurred and, after consideration, the more appropriate one was chosen.

RESULTS AND DISCUSSION

1st research question: Does the type of molecular structure representation used have an impact on the quality of students' perception of its three-dimensional structure?

The results for the *MVT1* task section on 'Perception' indicated that students' perception of three-dimensional structure was statistically significant better when a stereo-chemical formula was used to present a molecular structure in comparison to that provided by a computer image (Table 4).

Table 4. Results of 3d, v and 3dv students' groups on Task 1 and Task 4 of MVT1.

Students' groups according to help tools in *MVT2*	*MVT1* (no help tools)				t	p
	Task 1 (computer model)		Task 4 (stereo-chemical f.)			
	Mean	SD	Mean	SD		
3d (N=130)	0.86	0.14	0.91	0.17	-2.41*	0.018
v (N=128)	0.84	0.15	0.88	0.24	-2.27*	0.025
3dv (N=123)	0.85	0.15	0.93	0.15	-4.73**	0.000
Total Mean (N=381)	0.85	0.15	0.91	0.19	-5.27**	0.000

** Significant at the confidence level 0.01.   * Significant at the confidence level 0.05.

This issue was examined in depth in structured interviews. Some students explained that the notations used in stereo-chemical formulas defined the positions of atoms in a molecule clearly for them. On the other hand, they were not certain about the positions of some atoms in a molecule presented by a static computer image. Some typical answers were:

> 'The stereo-chemical formula is my favourite representation, because it clearly defines the positions of atoms. Someone has done the work for you and you don't have to think which atoms are in the front, which are behind.' (Gle41)

> 'Task solving with stereochemical formula was easier, because the positions of all atoms are defined.' (Gle13)

Interestingly, on the contrary, some students reported that they could more easily visualize molecular structure on the basis of a static computer image, because it has a more 'concrete' representation, than from the abstract notation with dots and lines used in stereo-chemical formulas. Some typical answers were:

> 'The static image better presents the appearance of the molecule, but it is difficult to define which atoms are in front and which are behind. From the stereochemical formula the positions of atoms are clear, but I can hardly imagine the molecule – there are just dots and lines.' (Gle11)

> 'I can imagine the molecules' appearance from a static image, but no so well on the basis of stereo-chemical formula.' (Glc29)

Overall, the results from the interviews suggest that the information sources which serve as a foundation for students' perception of molecular structure decrease in value from three-dimensional molecular models, to pseudo-three-dimensional molecular models, to a static computer model.

The perception of molecular structure from a *three-dimensional representation* is based on: direct observation, rotation of the molecule, and palpability of the three-dimensional representation. When perceiving molecular structure from *virtual computer representations*, only the first two of the above mentioned information sources are available (direct observation and rotation of molecules), for palpability is not possible. Information about molecular structure is only obtainable from the *static molecular representation* by direct observation. As the consequence of that, the positions of some atoms were not clear and students had to make assumptions about their positions. Students' opinion suggested that the virtual representations are a bridge between three-dimensional molecular representations and two-dimensional representations. This is because, while they can be regarded as a set of static computer models, they can be directed by moving the computer mouse as requested by the user to create the impression of a three-dimensional representation. The broader range of information obtainable from three-dimensional representations implies that they are useful to more students, because they facilitate the engagement of different learning styles. Some typical answers were:

> 'From the static image I came to the solution on the basis of assumptions and mental imagination. Task solving with stereochemical formula was easier, because the positions of all atoms are defined.' (Glc13)

> 'The positions of some atoms are not clear from virtual computer models – when the virtual computer model is still, it is the same as the static computer model. The three-dimensional structure is seen from it when it is rotated. Plastic models are better because they can be held by hand and are more concrete.' (Glc23)

All students shared the opinion that it is crucial to know the meaning of the symbols used in stereo-chemical formulas in order to be able to recognise spatial relations among atoms in a molecule so represented. This statement can be supported when comparing the current study, where the meanings of notations used in stereo-chemical formulas were written on the blackboard and were available to students during the testing with *MVTs*, with the study by Ferk et al. (2003), where the meanings of notations were not rehearsed with students prior to the testing and not directly available during that testing. Although, the tasks were similar, different conditions of testing resulted in students' success on tasks of the 'Perception' type on the basis of static computer image being quite similar in both studies (current study *MVT1*: f%=85%; study by Ferk et al., 2003: f%=89%). However, when stereo-chemical formulas were used as initial representations, bigger differences were observed (current study *MVT1*: f%=91%; study by Ferk et al., 2003: f%=82%). This means that, although students should be familiar with the meanings of notations in stereo-chemical formulas by the end of general upper secondary school, this cannot be taken for granted and can still be a source of problems to some students.

2nd research question: Do different types of molecular models as help tools support students' perception of three-dimensional molecular structure?

Help tools led to a statistically significant improvement in the students' results on *MVTs* in task section 'Perception', when structure was perceived on the basis of a computer image only (Table 5), but not from the use of a stereo-chemical formula only (Table 6).

Table 5. Results of 3d, v, and 3dv students' groups on Task 1 of MVT1 and MVT2.

Students' groups according to help tools in *MVT2*	Task 1 (computer model)				t	p
	MVT1 (no help tools)		*MVT2* (with help tools)			
	Mean	SD	Mean	SD		
3d (N=130)	0.86	0.14	0.92	0.11	-4.50**	0.000
v (N=128)	0.84	0.15	0.91	0.14	-5.66**	0.000
3dv (N=123)	0.85	0.15	0.92	0.11	-5.17**	0.000
Total Mean (N=381)	0.85	0.15	0.92	0.12	-8.87**	0.000

** Significant at the confidence level 0.01. * Significant at the confidence level 0.05.

Table 6. Results of 3d, v, and 3dv students' groups on Task 4 of MVT1 and MVT2.

Students' groups according to help tools in *MVT2*	Task 4 (stereo-chemical formula)				t	p
	MVT1 (no help tools)		*MVT2* (with help tools)			
	Mean	SD	Mean	SD		
3d (N=130)	0.91	0.17	0.93	0.13	-1.48	0.143
v (N=128)	0.88	0.24	0.92	0.18	-1.91	0.058
3dv (N=123)	0.93	0.15	0.89	0.19	2.09*	0.038
Total Mean (N=381)	0.91	0.19	0.91	0.17	-0.74	0.460

** Significant at the confidence level 0.01. * Significant at the confidence level 0.05.

The results obtained suggest that the use of a stereo-chemical formula was already sufficiently well known to the majority of students that the introduction of help tools did not contribute to a further improvement of their results. On the contrary, help tools that clarified the positions of some atoms in the static computer models do result in the improvement of student achievement. Never the less, the transition from the abstract notation of a stereo-chemical formula into three-dimensional perception still causes trouble to some students. They can be confused even more when they have to deal with more than one type of help tool at the time (e.g. with *3dv*, Table 6).

It is also important to note that students' results, when based on different molecular structure representations and with the use of help tools, became very similar. No statistically significant differences in students' results were observed in these circumstances (Table 7) which was not the case when help-tools were not available (Table 4).

Table 7. Results of 3d, v, and 3dv students' groups on Task 1 and Task 4 of MVT2.

Students' groups according to help tools in *MVT2*	MVT2 (with help tools)				t	p
	Task 1 (computer model)		Task 4 (stereo-chemical f.)			
	Mean	SD	Mean	SD		
3d (N=130)	0.92	0.11	0.93	0.13	-0.57	0.567
v (N=128)	0.91	0.14	0.92	0.18	-0.37	0.713
3dv (N=123)	0.92	0.11	0.89	0.19	1.62	0.108
Total Mean (N=381)	0.92	0.12	0.91	0.17	0.49	0.624

** Significant at the confidence level 0.01.   * Significant at the confidence level 0.05.

This is supported by students' comments in interview, where they explained that stereo-chemical formulas defined the positions of the atoms clearly so that they did not need the help tools. On the other hand, some positions of the atoms were not clear from the static computer representation and in such cases the help-tools helped students to overcome their uncertainties regarding three-dimensional structure. Some typical answers were:

> 'It would be nice to have help tools available with the static image. Basically, I am more certain with concrete model than with an image, because the atoms' positions are clear. This is confirmation of the solution.' (Glc20)

> 'The help tools are not necessary with stereochemical formulas. I don't have a clue how they could additionally help me, because everything is clear.' (Glc9)

> 'When solving a task with stereochemical formulas I don't need help tools. My manner of thinking would change, if I were to use molecular models in addition. I am forced to imagine the molecule when using a three-dimensional model, on the other hand the meaning of symbols can be simply coped without requiring any imagination to solve the task, if you understand the symbols in stereochemical formula.' (Glc28)

3rd research question: Is the usage of a certain kind of molecular model more efficient than others in supporting students' perception of three-dimensional molecular structure?

The comparison of *MVT* achievements on task section 'Perception' of students groups using different help tools showed no statistically important differences (Table 8). This indicates that three-dimensional representations, virtual representations, and a combination of both types, are equally useful to students in supporting perception of the molecular structure.

Table 8. ANOVA for 3d, v and 3dv students' groups on Tasks 1 and Task 4 of MVT1 and MVT2.

MVTs	Sources of variance	df	Sum of Squares	Mean Square	F	p
MVT1 Task 1	Between Groups	2	0.04	0.02	1.03	0.357
	Within Groups	378	8.31	0.02		
	Total	380	8.36			
MVT2 Task 1	Between Groups	2	0.01	0.00	0.17	0.841
	Within Groups	378	5.76	0.02		
	Total	380	5.77			
MVT1 Task 4	Between Groups	2	0.15	0.08	2.06	0.129
	Within Groups	378	13.86	0.04		
	Total	380	14.01			
MVT2 Task 4	Between Groups	2	0.08	0.04	1.53	0.219
	Within Groups	378	10.44	0.03		
	Total	380	10.52			

** Significant at the confidence level 0.01.   * Significant at the confidence level 0.05.

However, the majority of students said in interview that they prefer more concrete (*3d*) representations to virtual representations. This was so for the following two reasons: (1) spatial relations are more naturally seen from them, (2) they are easier to manipulate. However, there were also exceptions: one student believed that spatial relations are better perceived from computer virtual models; another student said that it doesn't matter which help-tool is used. Some typical answers were as follows:

> 'The positions of some atoms are not clear from virtual computer models – when the virtual computer model is still, it is the same as the static computer model. The three-dimensional structure is seen from it when it is rotated. Plastic models are better because they can be held and are more concrete.' (Glc23)

> 'If I could choose between concrete and virtual help tools, I would rather take concrete model, because virtual are difficult to manipulate.' (Glc13)

> 'Virtual models are better help tools because they enrich the static computer image with three-dimensional properties. When using concrete model, transference from three-dimensions to two dimensions (static image of the model) is additionally needed.' (Glc42)

> 'I would solve tasks in the same way with virtual and concrete help tools.' (Glc29)

4th research question: Does the type of molecular structure representation have an impact on the efficiency of solving spatial chemistry problems involving rotation or/and reflection?

In *MVTs* task sections 'Perception and Rotation' (Table 9) and 'Perception, Rotation and Reflection' (Table 10), students achieved better results when a static computer model was used in the task instruction instead of symbolic notation of a molecular structure.

Table 9. Results of 3d, v, and 3dv students' groups on task section 'Perception and Rotation' of MVT1.

Students' groups according to help tools in *MVT2*	*MVT1* (no help tools)				t	p
	Task 7 (computer model)		Task 2 (stereo-chemical f.)			
	Mean	SD	Mean	SD		
3d (N=130)	0.82	0.39	0.55	0.50	-5.41**	0.000
v (N=128)	0.84	0.37	0.52	0.50	-5.66**	0.000
3dv (N=123)	0.84	0.37	0.46	0.50	-7.14**	0.000
Total Mean (N=381)	0.83	0.38	0.51	0.50	-10.49**	0.000

** Significant at the confidence level 0.01.   * Significant at the confidence level 0.05.

Table 10. Results of 3d, v, and 3dv students' groups on task section 'Perception, Rotation and Reflection' of MVT1.

Students' groups according to help tools in *MVT2*	*MVT1* (no help tools)				t	p
	Task 5 (computer model)		Task 8 (stereo-chemical f.)			
	Mean	SD	Mean	SD		
3d (N=130)	0.40	0.49	0.18	0.39	4.21**	0.000
v (N=128)	0.35	0.48	0.19	0.39	3.10**	0.002
3dv (N=123)	0.34	0.48	0.16	0.37	3.46**	0.001
Total Mean (N=381)	0.36	0.48	0.18	0.38	6.22**	0.000

** Significant at the confidence level 0.01.   * Significant at the confidence level 0.05.

At first sight these quantitative results seem contradictory to those found for the task section on 'Perception', but the further qualitative approach suggested that they are in accordance one with another. The interviews indicate that static computer images are a better foundation for the visualization of a three-dimensional molecular structure, which was crucial in these task sections, than stereo-chemical formulas. Some typical answers were as follows:

> 'Task solving on the basis of a static image is easier. I cannot imagine the molecule on the basis of stereochemical formula even though I can read the positions of atoms. When I want to rotate the molecule mentally, it is essential that I imagine it three-dimensionally. All I see from stereochemical formula is dots and lines and it is up to me to imagine them three-dimensionally. However, this is not the case with a static image – the three-dimensional structure is already given. In task section 'Perception' the three-

dimensional imagination is not necessary for task solving – stereochemical formulas are better because they are more evident. In task section 'Perception and Rotation' three-dimensional imagination is crucial - the static image is a better initial representation. I believe there is a need for three-dimensional imagination that determines which initial representation is more appropriate for solving certain kinds of task.' (Gle11)

'A static image is better for this kind of task than stereochemical formula. I have to translate the meaning of symbols used in stereochemical formula all the time to be able to solve the task, but when using a static image the three-dimensional structure can be seen directly.' (Gle41)

Additionally, when students were dealing with combination of the processes of rotation and reflection, some of them pointed to the amount of information to be processed. The majority of students said that the great amount of information required to solve this kind of task is more easily processed with static computer models than with stereo-chemical formulas. Some typical answers were as follows:

'Task solving is much easier with a static image, because there is less information to be processed at the same time, than when stereo-chemical formulas are used. The same statement can be extended to the use of help tools, they are more a necessary supplement for task solving on the basis of stereochemical formula. Stereochemical formula are more difficult to imagine and I need something to concretise the mental image, with the use of help tools less information has to be simultaneously processed.' (Gle3)

'It is much more difficult to think on the basis of dots and lines than on the basis of balls. The colours of balls are helpful. Stereochemical formula representation is suitable for simple tasks, but when more information has to be simultaneously processed, it is easier to rely on concrete things, e.g. it is useful that the balls in a model are coloured. The dots and lines used in stereochemical formula don't mean anything to me in the sense of presenting the three-dimensional object.' (Gle29)

On the other hand, one student believed that the great amount of information is more easily processed when the initial representation is a stereo-chemical formula. His explanation was:

'It is even more difficult to solve the task on the basis of a static image in comparison to a stereo-chemical formula. The main advantage of stereochemical formula is that the positions of atoms are clearly defined and I don't have to put additional effort into guessing about the positions of atoms, as is the case when a static image is used.' (Gle24)

The preferences expressed for task sections 'Perception and Rotation' and 'Perception, Rotation and Reflection' were different to those expressed for 'Perception and Reflection'. Students were on average more successful on the basis of stereo-chemical formulas in comparison to static computer representations (Table 11).

Table 11. Results of 3d, v and 3dv students' groups on task section 'Perception and Reflection' of MVT1.

Students' groups according to help tools in MVT2	MVT1 (no help tools)				t	p
	Task 3 (computer model)		Task 6 (stereo-chemical f.)			
	Mean	SD	Mean	SD		
3d (N=130)	0.65	0.48	0.73	0.45	-1.61	0.109
v (N=128)	0.56	0.50	0.73	0.44	-2.98**	0.003
3dv (N=123)	0.67	0.47	0.68	0.47	-0.15	0.880
Total Mean (N=381)	0.63	0.48	0.72	0.45	-2.82**	0.005

** Significant at the confidence level 0.01. * Significant at the confidence level 0.05.

This is probably due to students' proneness to use mathematical rules with stereo-chemical formulas, whilst sticking to mental conceptualization when using static computer models. The application of a rule, from the viewpoint of the mental processes involved, is less demanding in comparison to creating one's own mental mirror image. An example of a typical answer was:

> 'When I imagine mathematically, the distance from the mirror to the object has to be the same on both sides of the mirror. It is a pity that I didn't solve prior tasks with reflection (based on computer image) by the use of a mathematical rule.' (Glc3)

These results suggest the need to make cross-curriculum linkages between mathematics and chemistry when teaching reflection, e.g. by using different molecular structure representations as objects when learning reflection.

5th research question: Do different types of molecular models as help tools support students' in solving spatial chemistry problems involving rotation or/and reflection?

When help tools were available in solving the task sections 'Perception and Rotation' (Table 12, Table 13) and 'Perception, Rotation and Reflection' (Table 14, Table 15) student achievement improved significantly.

Table 12. Results of 3d, v and 3dv students' groups on Task 7 (task section 'Perception and Rotation' of MVT1 and MVT2).

Students' groups according to help tools in MVT2	Task 7 (computer model)				t	p
	MVT1 (no help tools)		MVT2 (with help tools)			
	Mean	SD	Mean	SD		
3d (N=130)	0.82	0.39	0.97	0.17	-4.58**	0.000
v (N=128)	0.84	0.37	0.96	0.19	-3.56**	0.001
3dv (N=123)	0.84	0.37	0.91	0.29	-1.75	0.083
Total Mean (N=381)	0.83	0.38	0.95	0.22	-5.55**	0.000

** Significant at the confidence level 0.01. * Significant at the confidence level 0.05.

Table 13. Results of 3d, v and 3dv students groups on Task 2 (task section 'Perception and Rotation' of MVT1 and MVT2).

Students' groups according to help tools in *MVT2*	Task 2 (stereo-chemical formula)				t	p
	MVT1 (no help tools)		*MVT2* (with help tools)			
	Mean	SD	Mean	SD		
3d (N=130)	0.55	0.50	0.92	0.27	-8.08**	0.000
v (N=128)	0.52	0.50	0.87	0.34	-6.86**	0.000
3dv (N=123)	0.46	0.50	0.80	0.40	-7.57**	0.000
Total Mean (N=381)	0.51	0.50	0.87	0.34	-12.98**	0.000

** Significant at the confidence level 0.01. * Significant at the confidence level 0.05.

Table 14. Results of 3d, v and 3d students' groups on Task 5 (task section 'Perception, Rotation and Reflection' of MVT1 and MVT2).

Students' groups according to help tools in *MVT2*	Task 5 (computer model)				t	p
	MVT1 (no help tools)		*MVT2* (with help tools)			
	Mean	SD	Mean	SD		
3d (N=130)	0.40	0.49	0.54	0.50	-2.85**	0.005
v(N=128)	0.35	0.48	0.55	0.50	-3.71**	0.000
3dv (N=123)	0.34	0.48	0.54	0.50	-3.60**	0.000
Total Mean (N=381)	0.36	0.48	0.55	0.50	-5.90**	0.000

** Significant at the confidence level 0.01. * Significant at the confidence level 0.05.

Table 15. Results of 3d, v and 3dv students' groups on Task 8 (task section 'Perception, Rotation and Reflection' of MVT1 and MVT2).

Students' groups according to help tools in *MVT2*	Task 8 (stereo-chemical formula)				t	p
	MVT1 (no help tools)		*MVT2* (with help tools)			
	Mean	SD	Mean	SD		
3d (N=130)	0.18	0.39	0.55	0.50	-6.49**	0.000
v (N=128)	0.19	0.39	0.59	0.49	-8.17**	0.000
3dv (N=123)	0.16	0.37	0.54	0.50	-7.55**	0.000
Total Mean (N=381)	0.18	0.38	0.56	0.50	-12.71**	0.000

** Significant at the confidence level 0.01. * Significant at the confidence level 0.05.

The use of help tools leads to the differences in results between the two initial types of representation becoming smaller (compare Table 9 with Table 16, and Table 10 with Table 17).

Table 16. Results of 3d, v and 3dv students' groups on task section 'Perception and Rotation' of MVT2.

Students' groups according to help tools in *MVT2*	*MVT2* (with help tools)				t	p
	Task 7 (computer model)		Task 2 (stereo-chemical f.)			
	Mean	SD	Mean	SD		
3d (N=130)	0.97	0.17	0.92	0.27	-1.92	0.057
v (N=128)	0.96	0.19	0.87	0.34	-3.10**	0.002
3dv (N=123)	0.91	0.29	0.80	0.40	-2.66**	0.009
Total Mean (N=381)	0.95	0.22	0.87	0.34	-4.45**	0.000

** Significant at the confidence level 0.01.   * Significant at the confidence level 0.05.

Table 17. Results of 3d, v, and 3dv students' groups on Task 5 and Task 8 of MVT2.

Students' groups according to help tools in *MVT2*	*MVT2* (with help tools)				t	p
	Task 5 (computer model)		Task 8 (stereo-chemical f.)			
	Mean	SD	Mean	SD		
3d (N=130)	0.54	0.50	0.55	0.50	-0.27	0.790
v (N=128)	0.55	0.50	0.59	0.49	-0.56	0.574
3d v(N=123)	0.54	0.50	0.54	0.50	0.15	0.885
Total Mean (N=381)	0.55	0.50	0.56	0.50	-0.40	0.687

** Significant at the confidence level 0.01.   * Significant at the confidence level 0.05.

The improvement was greater when stereo-chemical formulas were initially used. It can be assumed that the help tools, on the one hand, partially eliminated uncertainties about the positions of some atoms in the static computer molecular model, and on the other hand, enabled a visualized three-dimensional structure to be perceived from a stereo-chemical formula. Some typical student explanations in the interviews were:

> 'When using stereochemical formula I don't need additional help tools, because the positions of bonds are clear. I need help tools when using a static image to check the positions of some atoms.' (Gle9)

> 'Help tools were useful regardless of the type of initial structure representation. It is difficult to imagine the molecule on the basis of stereochemical formula and additionally the rotation of all symbols; therefore the help tools were more necessary with stereochemical formula.' (Gle41)

> 'It is much easier to solve the task with help tools when a static image is the used as representation. When a stereochemical formula is the initial representation, there is no drastic difference in task solving; the conceptualisation of molecular structure is better. Anyway, once you have checked that the concrete model is the same as the initial molecule, it is easier to compare concrete model with possible answers and in this way

avoid the multiple mental transference from three-dimensions to two-dimensions and the other way around, as is the case when comparing the possible answers with initial stereo-chemical formula.' (Gle20)

It can be concluded that help tools also supported students in performing three-dimensional mental operations. Regardless of the fact that they were useful to the majority of students, we still have to keep in mind that they could evoke confusion in some students. A typical example of the answer illustrating this latter point was:

> 'Help- tools are useless, regardless of the type of initial representation. I think, they even make task solving harder, because you have to think of more things when using them.' (Gle23)

However, the use of concrete models as help tools actually decreased students' average results in task section 'Perception and Reflection' (Table 18, Table 19).

Table 18. Results of 3d, v and 3dv students' groups on Task 3 MVT1 and MVT2.

Students' groups according to help tools in *MVT2*	Task 3 (computer model)				t	p
	MVT1 (no help tools)		MVT2 (with help tools)			
	Mean	SD	Mean	SD		
3d (N=130)	0.65	0.48	0.60	0.49	0.83	0.407
v (N=128)	0.56	0.50	0.52	0.50	0.69	0.494
3dv (N=123)	0.67	0.47	0.54	0.50	2.66**	0.009
Total Mean (N=381)	0.63	0.48	0.56	0.50	2.27*	0.024

** Significant at the confidence level 0.01. * Significant at the confidence level 0.05.

Table 19. Results of 3d, v and 3dv students' groups on Task 6 of MVT1 and MVT2.

Students' groups according to help tools in *MVT2*	Task 6 (stereo-chemical formula)				t	p
	MVT1 (no help tools)		MVT2 (with help tools)			
	Mean	SD	Mean	SD		
3d (N=130)	0.73	0.45	0.52	0.50	4.08**	0.000
v (N=128)	0.73	0.44	0.59	0.49	2.72**	0.007
3dv (N=123)	0.68	0.47	0.53	0.50	3.34**	0.001
Total Mean (N=381)	0.72	0.45	0.55	0.50	5.85**	0.000

** Significant at the confidence level 0.01. * Significant at the confidence level 0.05.

These interesting outcomes were examined further in the interviews. This showed that the majority of students believed the concrete models are not useful as help tools in the solving the problems of this task section. Some even stated that they distracted them, which might explain the decrease of student achievement. An additional help tool offered to students in the interview was a mirror. Students found the mirror to be an excellent help tool, but some needed guidance when using it for the first time. Some typical students' comments:

'I don't need help tools. When you know the positions of atoms before reflection, then the positions after reflection are clear as well. Models can help me only to determine the positions of atoms about which I am not sure from the static image.' (Glc9)

'I would be lost without help tools - I can not imagine whole appearance of the molecule in my mind. Models are welcome, although the best help tool is the mirror.' (Glc3)

'We don't need molecular models as help tools in this task section, because it is not possible to derive the mirror image with their rotation. Even more, molecular models can mislead us into making rotation instead of reflection. With the use of the mirror, I can directly get the mirror images.' (Glc24)

The last of the above students' comments points to a crucial issue. The most probably reason for the decrease of students efficiency when *3d* models were used as help tools was that some students were mislead, undertaking rotation instead of reflection. This issue was confirmed with the fact that not all students were sure about the difference between rotation and reflection. Some typical students' explanations of the operation reflection were:

'We see the opposite image in the mirror. No, that is rotation. I confused rotation and reflection... When conducting reflection, the object will be transferred to the other side of the mirror, but spatial configuration will remain the same. It is essential that one understands the concept of rotation to solve these tasks.' (Glc9)

'Reflection is transference of the object around one axis only, but when conducting rotation we can move the object around all axes.' (Glc13)

'Reflection means deriving mirror images, similarly to seeing your face in the mirror. What was above and below will remain at the same position, but the left and the right side will exchange. The part of the molecule that was closer to the mirror will have to be closer to the other side of the mirror when reflected.' (Glc23)

Probably because an understanding of the nature of reflection seems to be crucial for in the solving of this task section with the use of help tools, no statistically significant differences in students' results were revealed for the use of different forms of representations (Table 20).

Table 20. Results of 3d, v and 3dv students' groups on the task section 'Perception and Reflection' of MVT2.

Students' groups according to help tools in *MVT2*	*MVT2* (with help tools)				t	p
	Task 3 (computer model)		Task 6 (stereo-chemical f.)			
	Mean	SD	Mean	SD		
3d (N=130)	0.60	0.49	0.52	0.50	1.27	0.205
v(N=128)	0.52	0.50	0.59	0.49	-1.22	0.226
3dv (N=123)	0.54	0.50	0.53	0.50	0.29	0.774
Total Mean (N=381)	0.56	0.50	0.55	0.50	0.23	0.816

** Significant at the confidence level 0.01. * Significant at the confidence level 0.05.

Interviews additionally revealed diverse preferences for certain initial representations when solving these tasks. The majority of students believed that the static computer representation is more appropriate for solving this task section, because it gives a better perception of the spatial structure of the molecule. Some others thought that the stereo-chemical formula is more evident in defining the positions of atoms and is therefore more appropriate. One student believed it is not important which initial representation is used in this task type. Some typical answers were as follows:

> 'It is easier to solve this task with static images than with stereo-chemical formula. Probably because static images look more realistic and the colours of the atoms support our memory more than symbols.' (Gle11)

> 'It is even more difficult to solve the task on the basis of static image than on the basis of a stereo-chemical formula. The positions of atoms are evident from stereochemical formula, but when the molecule is presented by the static image, I have to limit myself to assumptions about atoms' positions additionally to making reflection.' (Gle24)

> 'I am not able to solve such tasks without the use of help tools (models and the mirror) regardless of the type of initial representation.' (Gle23)

6th research question: Are traditional three-dimensional models, virtual models, and a combination of them equally useful to students in solving spatial tasks with molecular models?

In the task section 'Perception and Rotation', no statistically significant differences in *MVT* achievements between the groups using concrete and virtual models were found (Table 21, Table 22).

Table 21. ANOVA for 3d, v and 3dv students' groups on task section 'Perception and Rotation' of MVT1 and MVT2.

MVTs	Sources of variance	df	Sum of Squares	Mean Square	F	p
MVT1 Task 7	Between Groups	2	0.04	0.02	0.14	0.873
	Within Groups	378	53.87	0.14		
	Total	380	53.91			
MVT2 Task 7	Between Groups	2	0.25	0.13	2.55	0.079
	Within Groups	378	18.70	0.05		
	Total	380	18.95			
MVT1 Task 2	Between Groups	2	0.56	0.28	1.12	0.327
	Within Groups	378	94.66	0.25		
	Total	380	95.22			
MVT2 Task 2	Between Groups	2	0.88	0.442	3.86*	0.022
	Within Groups	378	43.29	0.115		
	Total	380	44.17			

** Significant at the confidence level 0.01. * Significant at the confidence level 0.05.

Table 22. Results of a detailed analysis of MVT2 Task 2 with ANOVA..

I (type of model)	J (type of model)	Mean Difference (I-J)	p
3d	v	0.06	0.186
	3dv	0.12**	0.006
v	3d	-0.06	0.186
	3dv	0.06	0.146
3dv	3d	-0.12**	0.006
	3dv	-0.06	0.146

** Significant at the confidence level 0.01.   * Significant at the confidence level 0.05.

However, differences in the achievements of the group using only concrete models and the group which had both concrete and virtual models available while solving Task 2 of *MVT2*, were found (Table 21, Table 22). The availability of a second kind of help tool evidently made task solving less efficient. Such an outcome is probably not due to the overload on students' working memory, because the help tools were used one at a time, but is the consequence of divided attention during task solving. This outcome indicates that the assumption that, when both kinds of help-tools are available, students would benefit from using one alongside the other as each has its advantages and disadvantages is not justified. To overcome the issue of split attention, it is probably necessary to allow students enough time to deal with each type of molecular model separately and without haste.

Regardless of the fact that no statistically significant differences between *MVT* achievements were observed between the group using concrete models and the group using virtual models (Table 21, Table 22), in the interview students indicated that they were aware of both the advantages and the disadvantages of each of the forms of representation when used as help tools. Some students preferred three-dimensional form, either because they are said to give a better spatial impression of the molecule, or because they are easier to manipulate, or for both reasons. Other students preferred virtual models because they enabled the spatial structure of the molecule to be more readily perceived. This has probably to do with students' way of processing information: their cognitive style. Some typical students' answers were:

'I think that the process of translation of a concrete model into a two-dimensional image is the most difficult. The final solution is therefore easier to get with help of virtual model. Otherwise, it is easier to make rotation with concrete model, because it is easier to follow the changes during rotation.' (Glc24)

'When a static computer model is the initial representation, it is better to use virtual models as help tools, because they are similar. On the basis of stereochemical formula, it is easier to use concrete model, because it gives a three-dimensional appearance.' (Glc11)

'It is easier to use *3d* as help tools, because when rotating *3dv* you are not sure how much you have turned the molecule. We perceive virtual space differently; when objects are in stillness it cannot be even regarded as space.' (Glc23)

'I like virtual models more than concrete models. When we get used to them, the positions of atoms are more evident than from concrete model. The background already defines the planes in the virtual molecule. The black colour of the background gives the depth to the virtual space.' (Gle9)

The statistically important differences between groups of students using different help tools were not observed in the task section 'Perception and Reflection' (Table 23).

Table 23. ANOVA for 3d, v and 3dv students' groups on the task section 'Perception and Reflection' of MVT1 and MVT2.

MVTs	Sources of variance	df	Sum of Squares	Mean Square	F	p
MVT1 Task 6	Between Groups	2	0.21	0.10	0.50	0.604
	Within Groups	378	77.18	0.20		
	Total	380	77.39			
MVT2 Task 6	Between Groups	2	0.40	0.20	0.80	0.45
	Within Groups	378	93.96	0.25		
	Total	380	94.36			
MVT1 Task 3	Between Groups	2	0.86	0.43	1.85	0.159
	Within Groups	378	88.22	0.23		
	Total	380	89.08			
MVT2 Task 3	Between Groups	2	0.40	0.20	0.81	0.444
	Within Groups	378	93.63	0.25		
	Total	380	94.04			

** Significant at the confidence level 0.01.   * Significant at the confidence level 0.05.

Interestingly, students' comments were quite different from those given for the section above. The majority of students said that they would use neither the three-dimensional nor the virtual models, because they are not helpful. Two students would rather use the three-dimensional representation because it better enables a better visualization of molecular structure to take place than the pseudo-three dimensional representation. One student would use a help tool but doesn't care which one. Some typical answers were as follows:

'I don't need models as help tools, because I am not able to make reflection with them. I would have to construct a new molecular model to make good use of models. The role of help tools is in visualizing the molecule.' (Gle11)

'With the use of molecular models as help tools we are prone to make rotation instead of reflection. Only the mirror has helped me in solving this task, because it gives directly the right image.' (Gle13)

'I try to imagine the mirror image on the concrete model. That atom will be closer to the mirror, that hydrogen will be orientated towards me... It is much easier to imagine reflection on the basis of concrete models; virtual models cannot help me.' (Gle29)

'With the use of molecular models as help tools we are prone to make rotation instead of reflection. Only the mirror has helped me in solving this task, because it gives directly the right image.' (Glc13)

'I try to imagine the mirror image on the concrete model. That atom will be closer to the mirror, that hydrogen will be orientated towards me... It is much easier to imagine reflection on the basis of concrete models; virtual models cannot help me.' (Glc29)

The statistically important differences between groups of students using different help tools were also not observed in task section 'Perception, Rotation and Reflection' (Table 24).

Table 24. ANOVA for 3d, v and 3dv students' groups on task section 'Perception, Rotation and Reflection' of MVT1 and MVT2.

MVTs	Sources of variance	df	Sum of Squares	Mean Square	F	p
MVT1 Task 5	Between Groups	2	0.25	0.13	0.54	0.585
	Within Groups	378	88.04	0.23		
	Total	380	88.29			
MVT2 Task 5	Between Groups	2	0.02	0.01	0.04	0.966
	Within Groups	378	94.43	0.25		
	Total	380	94.45			
MVT1 Task 8	Between Groups	2	0.05	0.02	0.16	0.855
	Within Groups	378	55.82	0.15		
	Total	380	55.86			
MVT2 Task 8	Between Groups	2	0.16	0.08	0.32	0.727
	Within Groups	378	93.76	0.25		
	Total	380	93.92			

** Significant at the confidence level 0.01.   * Significant at the confidence level 0.05.

When students were asked in the interview whether they preferred certain kinds of help tools while solving this type of task, three students said that concrete models are better for presenting spatial structure, whilst one student thought the opposite. Four students believed that both help tools have their advantages in the context of spatial representation. Three students would decide for the three-dimensional molecular representation because it is easier to manipulate. One student wouldn't use either of the help tools, because in her opinion both evoke confusion. Some typical answers were as follows:

'I rather used virtual models as help tools when a static image was the initial representation, because in concrete models the atoms' size and colours are slightly different. But, I can orientate better with the use of concrete models then virtual. This is probably due to problems that I have with manipulation of virtual models, but also because molecular structure can be directly perceived from concrete model; unlike virtual, which has to be rotated to give a three-dimensional appearance.' (Glc11)

CONCLUSIONS AND IMPLICATIONS

The main findings of the study and their implications for chemistry education are concerned with:

The form of molecular structure representation used
- Three-dimensional molecular structure seems to be clearly defined by stereo-chemical formulas, but knowing the meaning of the symbols used is essential! Static computer images do not provide unequivocal information about the positions of all atoms in a molecule. However, students reported that they could more easily visualize molecular structure on the basis of a static computer image than from the abstract notation used in stereo-chemical formulas. Apparently, the molecular structure representation form initially met can be the limiting factor for students' successful in solving chemistry tasks.
- The most important criteria which distinguish the value of static computer models from stereo-chemical formulas are, according to students' opinion: their ability to support a comprehension of the overall space occupied by the molecule, their ability to clarity the positions of individual atoms, their capacity to 'chunk' the amount of information that has to be to be processed during the manipulation of molecular structures. Consequently, a majority of students believed that when a lot of information has to be processed to solve a task (e.g. complex mental operations), it is easier to use static computer models than stereo-chemical formulas.

The use of molecular models as help tools
- Concrete and virtual models supported students in overcoming the deficiencies of both applied two-dimensional molecular structure representations and also in performing three-dimensional mental operations. Therefore, their availability as 'visualizers' is recommended for widespread use in science classrooms. However, it should not be forgotten that their use can also confuse some students.
- Virtual computer models are as effective as concrete models as help tools in supporting students' performance of three-dimensional tasks. However, the majority of students preferred three-dimensional representations due to their three-dimensional nature and palpability. Due to a human proneness to rely on concrete objects, concrete *3d* models should be available to students, at least when new ideas and concepts are being introduced. A transition to the use of pseudo three-dimensional (virtual) analogs can be made later on in the learning of a topic. A simultaneous use of both types of representation is not advisable, in order to avoid causing students to suffer from split attention, which could lower their problem-solving efficiency.

An understanding of the concepts and processes underlying three-dimensional tasks
- It has been shown that students' lack of understanding of the concepts of rotation and reflection can be a limiting factor in solving three-dimensional

tasks. Therefore, adequate attention should be given to developing the applicable understanding of basic operations in three-dimensional space.

This study has suggested that a range of further research is needed. Studies could profitably:

- Extend the investigation described into the primary and university level of education.
- Adapt *MVT* for diagnostic purposes in the classroom environment. This would provide feedback information to teachers about students' accuracy of *3d* perception, understanding of spatial operations (rotation, reflection), and skill in spatial manipulation.
- Develop of an additional *MVT2* test by including stereo-projections as help tools and evaluate their value in comparison to concrete and virtual models.
- Introduce structural formulas as another type of initial molecular presentation into *MVT2* and evaluate it in the classroom environment.
- Develop and evaluation of a molecular models web-base with exercises for students' improvement of *3d* perception, understanding of spatial operations (rotation, reflection) and skilful in spatial manipulation.
- Conduct an investigation based on long-term classroom observations to study the 'objective' frequency and strategies of molecular models use in school practise and their consequences for students' understanding of chemical concepts and processes.
- Develop and evaluation a web-base of molecular models with didactic suggestions and hints for their use in accordance with the curriculum.
- Conduct investigations into students' eye-movements when perceiving information on molecular structure from different molecular structure representations and during solving spatial tasks. This would provide additional evidence about the positions of atoms that are easy/difficult for students to recognise and, from that, suggest how molecular structure could be presented efficiently.

NOTES

1. General upper secondary education is provided in upper level secondary schools (slo. gimnazija) and lasts 4 years (15-19 years of age). Upper level secondary schools end with the general state examination (slo. matura), which is a condition for entering the tertiary level of education.

REFERENCES

Barke, H. D., & Wirbs, H. (2002). Structural units and chemical formulae. *Chemistry Education: Research and Practice in Europe, 3* (2), 185-200.
Barnea, N. (1997). The use of computer-based analog models to improve visualization and chemical understanding. In J. K. Gilbert (Ed.), *Exploring Models and Modelling in Science and Technology Education* (pp. 145-161). Reading: University of Reading, Faculty of Education and Community Studies.
Battino, R. (1983). Giant atomic and molecular models and other lecture demonstration devices designed for concrete operational students. *Journal of Chemical Education, 60* (6), 485-488.
Birk, J. P., & Foster, J. (1989). Molecular models for the do-it-yourselfer. *Journal of Chemical Education, 66* (12), 1015-1018.

Canning, D. R., & Cox, J. R. (2001). Teaching the structural nature of biological molecules: molecular visualization in the classroom and in the hands of students. *Chemistry Education and Practice in Europe*, 2 (2), 109-122.
Chapman, V. L. (1978). Inexpensive space-filling molecular models useful for VSPR and symmetry studies. *Journal of Chemical Education*, 55 (12), 798-799.
Comenius, J. A. (1896). *Orbis sensualium pictus*. Wien: Freytag. (Original work published 1658)
Copolo, C. F., & Hounshell, P. B. (1995). Using three-dimensional models to teach molecular structures in high school chemistry. *Journal of Science Education and Technology*, 4 (4), 295-305.
Dori, Y. J., & Barak, M. (2001). Virtual and physical molecular modeling: fostering model perception and spatial understanding. *Educational Technology and Society*, 4 (1), 61-74.
Ealy, J. B. (1999). A student evaluation of molecular modeling in first year college chemistry. *Journal of Science Education and Technology*, 8 (4), 309-321.
Eggeton, G. L., Williamson, J. J., Lovesless C. E., & Grimes, B. C. (1990). *Creative student-made molecular models*. Journal of Chemical Education, 67 (12), 1028.
Ferk, V., Vrtacnik, M., Blejec A. & Gril A. (2003). Students' understanding of molecular structure representations. *International Journal of Science Education*, 25 (10), 1227-1245.
Ferk, V. (2003). *The Significance of Different Kinds of Molecular Models in Teaching and Learning of Chemistry: Doctoral Dissertation*. Ljubljana: University of Ljubljana, Faculty of Natural Sciences and Engineering, Department of Chemical Education and Informatics.
Gabel, D., & Sherwood, R. (1980). The effect of student manipulation of molecular models on chemistry achievement according to Piagetian level. *Journal of Research in Science Teaching*, 17 (1), 75-81.
Goodstein, M., & Howe, A. (1978). The use of concrete methods in secondary chemistry instruction. *Journal of Research in Science Teaching*, 15 (5), 361-366.
Hanson, R. M. (1995). *Molecular origami: precision scale models from paper*. Sausalito: University Science Books.
Hardwicke, A. J. (1995). Using molecular models to teach chemistry. Part I: modelling molecules. *School Science Review*, 77 (278), 59-64.
Hyde, R. T., Shaw, P. N., Jackson, D. E., & Woods, K. (1995). Integration of molecular modelling algorithms with tutorial instruction. *Journal of Chemical Education*, 72 (8), 699–702.
Jaeger, A. O., Suess, H. M., & Beauducel A. (1997). *Berliner Intelligenzstruktur Test - Form 4: Handanweisung*. Hogrefe: Verlag fuer Psychologie.
Johnstone, A. H. (1991). Why is science difficult to learn? Things are seldom what they seem. *Journal of Computer Assisted Learning*, 7 (2), 75-83.
McGrew, L. A. (1972) Stereoscopic projection in the chemistry classroom. *Journal of Chemical Education*, 49 (3), 195-199.
Molitor, S., Ballstaedt, S. P., & Mandl, H. (1989). Problems in knowledge acquisition from text and pictures. In H. Mandl & J. R. Levin (Eds.), *Knowledge aquisition from text und pictures* (pp. 3-35), Amsterdam: North-Holland.
Pogacnik, V. (1998a). *Test hitrosti percepcije 'Vzorci' ('Patterns' Test)*. Ljubljana: Center za psihodiagnosticna sredstva.
Pogacnik, V. (1998b). *Spacialni test 'Rotacije' ('Rotations' Test)*. Ljubljana: Center za psihodiagnosticna sredstva.
Pogacnik, V. (1994). *Test 'Nizi' ('Series' Test)*. Ljubljana: Center za psihodiagnosticna sredstva.
Roberts, R. M., & Traynham, J. G. (1976). Molecular geometry: as easy as blowing up balloons. *Journal of Chemical Education*, 53 (4), 233-234.
Russell, J. W, Kozma, R. B., Jones, T., & Wykoff, J. (1997). Use of simultaneous-synchronized macroscopic, microscopic, and symbolic representations to enhance the teaching and learning of chemical concepts. *Journal of Chemical Education*, 74 (3), 330-334.
Sanger, M. J., & Greenbowe, T. J. (2000). Addressing student misconceptions concerning electron flow in aqueous solutions with instruction including computer animations and conceptual change strategies. *International Journal of Science Education*, 22 (5), 521-37.
Seddon, G. M., & Eniaiyeju, P. A. (1986). The understanding of pictorial depth cues, and the ability to visualise the rotation of three-dimensional structures in diagrams. *Research in Science and Technological Education*, 4 (1), 29-37.
Seddon, G. M., & Moore, R.G. (1986). An unexpected effect in the use of models for teaching the visualization of rotation in molecular structures. *European Journal of Science Education*, 8 (1), 79-86.

Seddon, G. M., & Shubber, K. E. (1985a). Learning the visualization of three-dimensional spatial relationships in diagrams at different ages in Bahrain. *Research in Science and Technological Education, 3* (2), 97-108.

Seddon, G. M., & Shubber, K. E. (1985b). The effects of colour in teaching the visualization of rotations in diagrams of three-dimensional structures. *British Educational Research Journal, 11* (3), 227-239.

Tuckey, H., & Selvaratnam, M. (1991). Identification and rectification of students' difficulties concerning three-dimensional structures, rotation, and reflection. *Journal of Chemical Education, 68* (6), 460-464.

Williamson, V. M., & Abraham, M. R. (1995). The effects of computer animation on the particulate mental models of college chemistry students. *Journal of Research in Science Teaching, 32* (5), 521-534.

Wu, H. K., Krajcik, J. S., & Soloway, E. (2001). Promoting understanding of chemical representations : students' use of a visualization tool in the classroom. *Journal of Research in Science Teaching, 38* (7), 821-842.

Yamana, S. (1989). An easy constructed bicapped trigonal prism model. *Journal of Chemical Education, 66* (12), 1021-1022.

ACKNOWLEDGEMENTS

The authors would like to express the gratitude to the participating students and their schools (Skofijska gimnazija, Ljubljana; Gimnazija Bezigrad, Ljubljana; Gimnazija Kranj, Kranj; Gimnazija Koper, Koper; Gimnazija Lava, Celje; Gimnazija Ledina, Ljubljana; Gimnazija Litija, Litija; Gimnazija Novo Mesto, Novo Mesto in Slovenia), and to the Slovenian Ministry of Education, Science and Sport for the financial support of the research work.

Please turn to pp. xxxiii–xxxix for the Appendices to Chapter 13.

JOEL RUSSELL[1] AND ROBERT KOZMA[2]

CHAPTER 14

ASSESSING LEARNING FROM THE USE OF MULTIMEDIA CHEMICAL VISUALIZTION SOFTWARE

[1]*Department of Chemistry, Oakland University, Rochester, MI;* [2]*Center for Technology in Learning, SRI International, Menlo Park, CA*

Abstract: This chapter extends the use of cognitive and "situative" theories of learning described in our earlier chapter to discuss the design of five chemistry multimedia visualization projects. All five projects are shown to enhance the learning of chemistry concepts and development of scientific investigative process skills. Two projects emphasize the social processes associated with scientific investigation with bench laboratory components; two others without laboratory components could be easily utilized in ways that develop such social processes; and the fifth is shown to enhance visualization process skills although it was designed based upon the cognitive theory of multimedia learning. In order to assess the developing levels of visualization skills of novice chemistry students supported by all five visualization projects, non-traditional testing items must be utilized. Samples of several multimedia testing items addressing both conceptual and process skills are discussed and used in studies of the efficacy of the most cognitive-theory based project. A study using the project with the closest alignment to "situative" theory illustrates how the rubric for representational competence levels discussed in our earlier chapter is applied to show changes in visualization abilities of students. The chapter concludes with brief suggestions for future development of visualization software, visualization-based instructional activities and testing activities.

VISUALIZATIONS IN CHEMISTRY

In an earlier chapter (Kozma & Russell, this volume) we established the desirability for inclusion of chemistry curricular components specifically designed to address the acquisition of chemical visualization skills which we defined as representational competence. These skills include the ability to utilize chemical symbols, chemical equations, various types of structural diagrams, diverse graphical formats including spectral plots and computer models, and nanoscale animations as appropriate for solutions of problems or tasks and the investigation and understanding of phenomena and concepts. Two theoretical bases for development of visualization skills were discussed – cognitive and situative. Cognitive theory (Mayer, 2003, 2002, 2001) addresses the transformation of these external symbolic representations to internal mental representations – mental models. Situative theory (Greeno, 1998; Roth, 1998, 2001) focuses on learning science as an investigative process and the use of social discourse and representations to support this process. In the earlier chapter we showed that representational competence levels of novice students can be enhanced through classroom, laboratory, and individual use of molecular modeling and computer simulations and animations. In this chapter we extend this discussion to several multimedia chemical visualization software packages. We include packages with both cognitive and situative goals and shall assess their designs and efficacies from one or both perspectives. Critical to such analyses are the development of new assessment tools to probe visualization skills and students' abilities to apply such skills.

The goals of this chapter are:
- to review five multimedia chemical visualization packages with respect to their design characteristics and learning goals,
- to discuss special assessment tools and techniques useful for measuring visualization skill levels,
- to summarize some of our studies of the efficacy of our multimedia visualization programs, *Synchronized Multiple Visualizations of Chemistry* (*SMV:Chem*) (Russell, Kozma, Becker, & Susskind, 2000), and , *ChemSense* (Schank & Kozma, 2002) using these new assessment tools, and
- to suggest directions for future instructional improvements and research.

This chapter is divided into four major sections addressing each of these four goals. We include a sufficient number of black and white figures in the printed text to allow logical development of our goals without access to a computer or the Internet. However, in the Colour Section and on the website http://www2.oakland.edu/users/russell, we provide color versions of all figures, examples of all *SMV:Chem* experiments discussed in the text in a new web-based format, examples of *SMV:Chem*-based homework assignments in the web format, and an example of visualization skills assessment tools used in our study of the efficacy of *SMV:Chem* in classroom and out-of-class use. Whenever possible we provide URLs for viewing samples of the multimedia software packages.

MULTIMEDIA VISUALIZATION PROJECTS

A number of large-scale multimedia projects have been developed in the U.S., often with funding from the National Science Foundation, to use the capabilities of new technologies to improve chemistry instruction and learning. In this section we review several of the major projects and feature our project, *SMV:Chem*.

We use the two theoretical approaches described above—the cognitive and situative theories—as structural schema for analyzing these projects. Some projects explicitly cited one or the other of these theories in developing their software tools and instructional approaches; others did not. Nonetheless, it appears to us that some of these projects emphasize the learning of chemical concepts, principles, and procedures, while others emphasize the learning of the scientific process of chemical investigation. As we mentioned above, these two types of educational goals are not mutually exclusive; they can be complementary. Indeed some of these projects incorporate aspects of both. As multimedia projects, they also often incorporate several types of visualizations—models, simulations, and/or animations.

Synchronized Multiple Visualizations of Chemistry (SMV:Chem)

SMV:Chem is a chemical visualization software program designed to assist students to develop their abilities to use appropriate visualizations including nanoscale models and animations as well as various symbolic representations as they seek to understand and explain chemical concepts and solve chemical problems (Russell & Geno, 2000; Russell, 2004). Version 1.0 (MAC and Windows 95 & 98) was distributed in 2000-2002 by John Wiley & Sons (Russell, Kozma, Becker, & Susskind, 2000). Version 2.0 (MAC and Windows 2000, ME, XP, and NT) is available from the authors (russell@oakland.edu). A sample subset of the web-based Version 3.0 can be accessed via the Internet at http://www2.oakland.edu/users/russell. *SMV:Chem* was based upon a prototype, *4M:Chem – MultiMedia and Mental Models in Chemistry* (Russell & Kozma, 1994; Russell, Kozma, Jones, Wykoff, Marx, & Davis, 1997).

SMV:Chem and *4M:Chem* were designed for use in the classroom to allow instructors to show examples of real experiments and ways to represent the experimental phenomena with molecular-scale animations, graphs, molecular models, and equations and for use by students outside the classroom to enhance their chemical visualization skills and conceptual understanding. A secondary classroom goal was to promote an interactive classroom environment that stimulated class discussion with the pace and directions of usage controlled by the instructor's assessment of class responses. Both programs used a consistent screen design with experimental videos, nanoscale animations, graphical representations, and other representations such as chemical equations, text, data collection sheets, and molecular models in respective quadrants of the screen. All experiments discussed are included on the CD-ROM in the new web-based format.

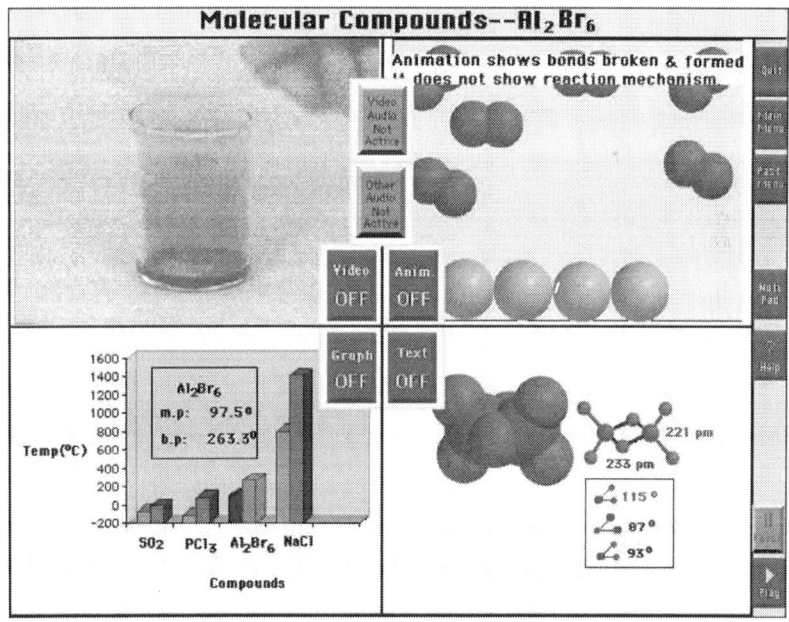

Figure -1. *Window selection and program control screen for a SMV:Chem experiment.*

Once a specific experiment has been chosen a screen such as in Figure 1 appears to allow selection of one to four windows using the buttons at the center of the screen and program control using buttons on the right vertical toolbar. There are separate audio tracks for all windows that can be activated or deactivated. In Figure 1 the visualizations in the lower two windows are static although in most experiments the graphs are dynamic. Using the Pause/Play buttons windows can be frozen at any point and restarted.

The experiment depicted in Figure 2 shows the effect of decreasing the volume on an equilibrium mixture of two gases, $N_2O_4 \rightleftharpoons 2NO_2$. This particular example was selected since at room temperature a sample can be prepared with near equal amounts of both gases but only the NO_2 can be observed by the eye. The UV-visible absorption spectrum shows a broad absorption centered at 400 nm with its long wavelength tail extending to 650 nm giving the sample a red-brown color. N_2O_4 has a sharper absorption band at 335 nm which does not extend past 400 nm making it invisible to the human eye.

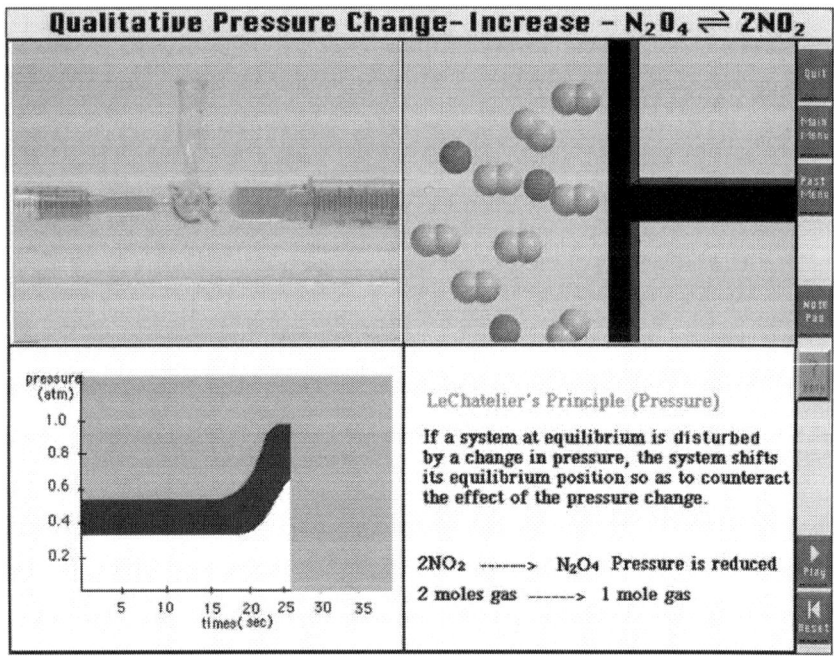

Figure -2. *SMV:Chem screen for LeChatelier's principle experiment showing effect of a change in volume on the equilibrium compositions of the sample. Viewed on the computer screen the color of the gas in the 1 cm square tube left of the stopcock is definitely lighter than the color in the gas in the 2 cm square tube right of the stopcock.*

Users are encouraged to view first only the video window and to replay the experiment until they have a through understanding of how the experiment shows a change in equilibrium composition that produces more N_2O_4 resulting from decreasing the volume. At the start of the video the three-way stopcock is positioned 45^0 counterclockwise from its position in Figure 2 such that there is an open path between the two sides of the apparatus. The stopcock is rotated clockwise to the position shown which isolates the two sides. At this point the partial pressures of NO_2 are the same in both the left 1.00 cm square tube and the right 2.00 cm square tube. The sample appears darker on the right since there are twice as many NO_2 molecules in the viewing path as you look through the tube. The left syringe is pushed in until reaching the point where the volume in the isolated left side is exactly half its initial volume. If there was no change in the equilibrium composition, by Boyle's law the partial pressure of NO_2 would have doubled. With double the pressure and half the path length the color of NO_2 gas on both sides should be identical. The observation that after this change in volume the color on the left is lighter than the color on the right shows the composition has shifted to produce more colorless N_2O_4 leaving less red-brown NO_2.

If the experiment is rerun activating both the video and animation windows, the animation shows initially five red-brown spheres and seven coupled gray spheres all in constant motion. Prior to the piston moving in from the right to cut the volume (area) in half the distribution of species is seen to alternate between five red-brown monomers/seven gray dimers and seven monomers/six dimers as the dimers decompose and the monomers dimerize showing the dynamic property of a chemical equilibrium. After the volume was cut in half, the animation shows the composition alternating between three monomers/nine dimers and one monomer/ten dimers.

Class discussion or homework questions might ask students to consider what is shown in the animation that is not visible in the video of the experiment, which properties of a chemical equilibrium are correctly and which incorrectly shown by the animation, and how LeChatelier's principle could be used to explain the observed and portrayed shift in composition. The animation shows the dimers that form upon reaction of two monomers even though these species represent the unseen colorless N_2O_4 molecules. The animation correctly shows the dynamic properties of a chemical equilibrium but shows a fluctuation in composition where equilibrium compositions are constant. This discrepancy is usually quickly attributed by students to the small numbers of species shown in this animation. The increase in pressure produced by the decrease in volume can be partially offset by forming more N_2O_4 since such formation reduces the total number of gas molecules present in the sample. On rare occasions some student will note that the shift in composition in the animation appears very dramatic while that in the video was barely visible. This is a perfect lead-in to viewing the graphical representation which is quantitative rather than qualitative as the animation. After discussing the graph instructors can show the video, animation, and graph simultaneously and ask if the equilibrium shift could be accurately portrayed by the animation and how many molecules would be required.

Animations used for the eleven *SMV:Chem* gas equilibrium experiments with NO_2 and N_2O_4 all use animations with red-brown spheres for NO_2 and coupled gray spheres for N_2O_4. This choice of models rather than use of space filling models of these molecules with the usual color coding for atoms with blue for N and red for O is consistent with the adaptation of Mayer's Coherence Principle to visual objects (See Mayer 2001, 2002 and the discussion of cognitive theory in the previous chapter). This principle states that extraneous objects should be excluded. The primary point to all equilibrium animations is to show the dynamic property of chemical equilibria. We did do a pilot test using the red-blue space filling models for these molecules. In that case reactants and products could only be distinguished by their shapes. When asked to view animations using both types of models no students noticed the dimerization and dissociation reactions occurring with the space filling models but many made this observation with the red-brown spheres and gray coupled spheres.

In designing the animation and graphical representations we took advantage of color to connect species with the two gases, brown for NO_2 and light gray for N_2O_4. Our earlier work showed that most students first use surface features of representations to answer questions concerning all types of visualizations. Color was shown to be a most important property when comparing two visualizations. The sample used in this experiment was initially 38% NO_2 and after the volume was cut in half became 28% NO_2. The experiment in Figure 2 has been paused just as the final volume was reached and the new final equilibrium established. At the end of the experiment the 28% and 38% figures are added to the graph as are two lines to show where the two vertical boundaries of the brown NO_2 band would have been with no shift in equilibrium composition.

Figure 3 shows a screen capture for another *SMV:Chem* experiment designed to allow the determination of an empirical formula. A glass tube of the shape shown in the animation window is weighed, a sample of red mercury(II) oxide added, and the tube and solid weighed. The tube is clamped with its U-bend in a water-ice bath and the closed end with the red solid heated. Upon heating the amount of solid decreases and droplets of liquid mercury are observed to form in the tube. Once the solid at the end of the tube has disappeared the tube is removed form the bath, dried, and weighed. When used in class or for homework this experiment is first run with the video and text windows active. As mass measurements are made the resulting data is entered into the table shown in the text window.

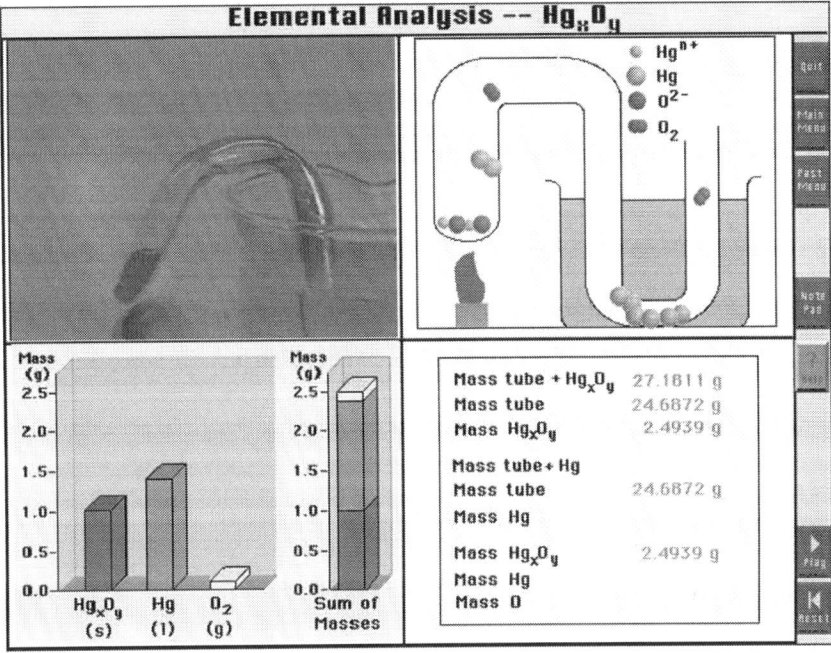

Figure 3. *SMV:Chem screen from experiment to determine the empirical formula of a red mercury oxygen compound.*

The animation for this experiment includes both models of nanoscale species (Hg atoms, O_2 molecules, and Hg^{2+} and O^{2-} ions) and macroscopic objects (sample tube, water-ice bath, and Bunsen burner). During meetings of the advisory board for the *4M:Chem* prototype project when shown animations mixing objects of such difference size scales in a diagram the science educators were concerned that would promote or reinforce students' misconceptions. All board members who were scientists thought that it was obvious the intent and students would not be confused. Over the years that we have used such animations no student has questioned in class or indicated on homework that he/she did not recognize the scale differences of the components of the animation. In fact, we have used this animation to discuss the distinction between nanoscale representations of chemical phenomena and macroscopic observations. The mercury ion's legend on the animation is shown with unknown charge since the primary objective of the experiment was to find the empirical formula of the red solid. Although the animation does start with equal numbers of mercury and oxygen ions thus signifying an HgO empirical formula, you can count on students to use numbers wherever available and to determine the empirical formula from the numeric data rather than by simple examination of the animation.

The graphical representation for this experiment is rich in data and allows discussion in class or questions on homework of many topics. For example, the experiment could be paused at the position shown in Figure 3 and the class asked what would be the weight of the sample in the tube at that point if it were removed from the bath and dried and if the empirical formula could be determined with this data. Other potential questions are – What does the right graph show? How will the right graph appear at the end of the experiment (students have already seen the video of the experiment)? How does the graph show conservation of mass? This last question is, of course, another version of the first question that provides the concept but does not direct attention to the right graph. When observed on the computer screen you would note the use of silver or gray to represent mercury in the animation and graph, the use of white in the graph to represent the colorless oxygen gas released into the atmosphere, and the red color in the graph to show the original red solid.

The final data shows there were 2.3112 g of Hg and 0.1827 g of O in the original 2.4939 g of the red solid. This data corresponds to 0.011522 moles of Hg and 0.01142 moles of O clearly giving an empirical formula of HgO. Using the data in this experiment, students could be asked to find the theoretical and percent yield of Hg knowing the empirical formula is HgO. They would get 2.3097 g theoretical yield and 100.06 percent yield. If asked to explain how they could get a percent yield above 100%, the most common responses are that there are impurities in the sample or the tube was not dry. Seldom do students respond that within the experimental uncertainty 100.06 is 100%. It is worth noting that these data, as all data in the *SMV:Chem* experiments, were those recorded in the laboratory as the experiments were filmed. No video results were adjusted to give "good" data.

We have made use of audio tracks throughout *SMV:Chem* experiments consistent with Mayer's principles of multimedia learning. Using the audio option the software satisfies the multimedia principle of using words and pictures and the Contiguity, Modality, and Redundancy Principles of using words and picture simultaneously without repeating the words in an on-screen text. The style of the narration is conversational rather than a more formal scientific style thus satisfying the Personalization Principle. We address the Interactivity Principle by providing buttons that allows students conveniently to pause, restart, and replay experiments and to select visualization viewing combinations from single, all possible duets and triplets, to all four windows.

Although *SMV:Chem* was designed to help students gain deeper understanding of basic chemical concepts and phenomena, its use of alternative nanoscale and symbolic representations was also designed to promote student use of such representations when solving chemical problems or considering chemical phenomena. Attainment of the cognitive goals as noted above is supported through consideration of Mayer's guidelines for multimedia instruction. In the assessment section below we show sample questions that measure attainment of both deeper conceptual understanding and abilities to utilize alternative forms of representations to understand chemical phenomena – process skills characteristic of investigative science encompassed by situative theory.

Connected Chemistry

Like *SMV:Chem*, *Connected Chemistry* is an example of a multimedia project that is targeting the learning of difficult chemistry concepts (Stieff & Wilensky, 2003). The software is a collection of computer-based simulations of closed chemical systems that students can interact with in several ways. Each simulation focuses on a chemical concept such as factors effecting rates of reaction or LeChatelier's principle. *Connected Chemistry* is written in NetLogo, a multi-agent modeling language. The interface window of *Connected Chemistry* for the "Simple Kinetic 2" module is shown in Figure 4. The interface consists of three fundamental components: a graphics window for a nanoscale representation of the system, a plotting window for graphs of macroscopic properties of the system, and an area for setting system parameters and starting, pausing, restarting, and resetting the simulation.

This module in Figure 4 simulates the equilibrium, $2\ NO_2(g) \rightleftharpoons N_2O_4(g)$, and is designed to explore LeChatelier's principle. The graphics window displays visualizations for two types of molecules, a three atom linear molecule shown on the computer screen as white-green-white circles and a six atom molecule visualized with two central red circles and two white circles attached to each red circle representing NO_2 and N_2O_4 respectively. (Note: these models do not reflect the geometric structures of bent NO_2 and N_2O_4 whose lowest energy structure has mutually perpendicular NO_2 groups.) Unless the user pauses the simulation, all molecules in the graphics window are in rapid motion. The plotting window shows the number of reactants, NO_2, and products, N_2O_4, versus time. The graph for the number of N_2O_4 product molecules is shown in red corresponding to the red color used for the nitrogen atoms in the N_2O_4 models in the graphic window. The color green is used for the NO_2 graph in the plotting window and nitrogen atoms in the NO_2 models in the graphics window. The use of red and green for nitrogen atoms in the N_2O_4 and NO_2 models in the graphics window allows a quick visual estimate of relative numbers of these species once the simulation is paused. Note in the figure in the text the difficultly in making such an estimate based on model shapes since the red and green colors appear the same in the black and white print.

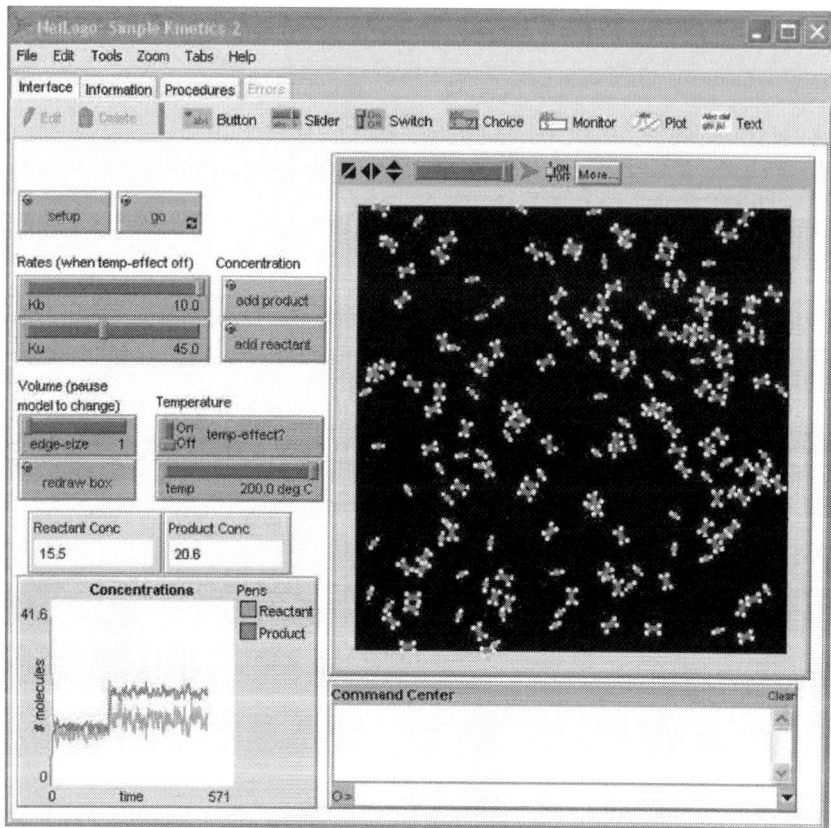

Figure 4. *Connected Chemistry interface window for Simple Kinetics 2 module.*
http://ccl.northwestern.edu/netlogo/models/SimpleKinetics2 (Wilensky, 2002)

For the conditions of this particular simulation the user chose to have initially only reactants present. The reactant numbers rapidly dropped and product numbers increased until they were fluctuating with about equal numbers of each as shown by the first horizontal region of the graph in the plotting window. At this point the user stopped the simulation and added extra NO_2 molecules causing the spike in the plot. Once the user restarted the simulation, reactant numbers dropped but leveled off at a new equilibrium higher than the old one. The product numbers increased but reached equilibrium at a higher value than the new reactant equilibrium value. This shows the LeChatelier's principle effect for changing concentration. The boxes at the top of the plotting window show the numbers of reactant and products at any time. These numbers are for some arbitrary volume to signify a concentration rather than the actual number of molecules shown in the two dimensional graphics area.

The system parameter area contains three sliders to set values for the forward rate constant, K_b, the reverse rate constant, K_u, and the temperature. Three other variables can be adjusted to add reactants, add products, and make the box smaller by making its walls thicker. The "setup" box is used to reset the graphic and plotting windows for the initial conditions selected for system parameters. Clicking the "go" box starts, pauses, and restarts the simulation. The tabs above the top tool bar open an "information" window and a "procedures" window. The information window contains a description of the chemical concept behind the model and suggestions on how students may wish to modify the model. The procedures window shows the programming code for the model which students can modify to test different models.

Stieff and Wilensky give six examples of use of *Connected Chemistry* in structured interviews with academically-talented upper division science majors that show them confronting and changing former misconceptions of LeChatelier's principle. A primary advantage of this software is the ease with which changes can be made in a wide range of system parameters and the consequences of each change rapidly observed. This should be most advantageous for instructors to the classroom where they can use the software to confront students' misconceptions. For example the effect of increasing the reactant concentration on a system at equilibrium shown in Figure 4 is obvious since the initial equilibrium had equal amounts of reactants and products. This initial equilibrium condition was achieved by adjustment of the relative values of K_b and K_u. Experienced instructors will recognize the experimental conditions that best show specific results that confront misconceptions or most obviously show the concept under discussion. For *Connected Chemistry* to be effective with less academically-talented students and high school or first year college students, we speculate these students will need a well designed guided inquiry approach whether in the classroom or for out-of-class assignments.

In an effort to make the simulation as close to reality as possible and take advantage of NetLogo's ability to show rule-driven motions of hundreds of objects, when the simulation is running the graphics window is a blur of motion. The slider above the graphics window allows the speed of the simulation to be reduced. Only when the simulation is paused can features of any molecules be observed and then there are so many objects that only crude conclusions can be drawn such as there appears to be many more product molecules than reactant molecules. All quantitative conclusions from this simulation are derived from the graph in the plotting window and the counters for concentrations. The nanoscale display in the graphics window should help students tie the observable macroscopic properties shown in the plotting window to the underlying molecular behavior and is probably the focal point for student attention. The option for students to modify the code and test alternative models available in *Connected Chemistry* is likely to be seldom used due to time constraints on both students and instructors.

Analyzing *Connected Chemistry* using cognitive theory and Mayer's principles of multimedia instruction, it comes out very high on the Interactivity Principle, if that principle is expanded to control more than rates of presentation. The absence of audio and restriction of words to a minor role in the separate information window take the other eight principles outside the realm of the software and rely on the instructor's verbal narration. This supports our conclusion stated above that *Connected Chemistry* requires a skilled instructor and a well-designed guided inquiry organization when implementing the module.

Connected Chemistry allows users to plan experiments to test various hypotheses or to simply explore the results of variations of several experimental parameters in an attempt to formulate hypotheses and to use the simulation to quickly see the results. In this light, users are learning chemistry in a true investigative mode and on a time scale much compressed from what would be required with hands-on laboratory experiments. In order for *Connected Chemistry* to be more aligned with situative theory, it would have to engage students in the investigation of real chemical phenomena and the social processes associated with these investigations.

Molecular Workbench

As with *Connected Chemistry*, the goal of *Molecular Workbench* is to use advanced computational techniques and visualizations to help students develop appropriate mental models of different chemical systems and concepts. *Molecular Workbench* provides a variety of real-time, interactive simulations of chemical phenomena by adding sets of rules describing chemical reactions to a molecular dynamics modeling system (Xie & Tinker, in press). Simulations for a wide variety of topics are available at http://workbench.concord.org/modeler/index.html. All simulations are calculated and displayed for a two-dimensional molecular dynamics model with the potential energy for forming molecules at 0 K taken as the sum of electronic energies for each bond and adjacent pair of bonds with the total potential energy the sum of two- and three- center terms. Added to the potential energy are terms for intermolecular forces for van der Waals, electrostatic, bond stretching, and bond angle bending. Motions of all species in the simulation are calculated with Newton's equations of motion with the parameterized potentials. To account for chemical reactivity at each step in the simulation the energy of each bond is compared with a table of energies required to break particular bonds and the energies of nearby unbound atoms to determine if a bond could form as discussed in the example below. The software has the flexibility to allow users to set most initial parameters including the atoms, their positions, velocities, and bonds as well as all potential energies parameters. To simplify use, an initial simulation with preset values for these parameters is provided for each topic. Many topics allow users to add or subtract thermal energy and rerun the simulations.

As an example of one Molecular Workbench simulation, Figure 5 shows a snapshot of the computer screen as the simulation progresses for a homogeneous catalysis. This screen is for the catalyzed mechanism:

$$C + A_2 \rightarrow CA + A^* \qquad (1)$$
$$C + B_2 \rightarrow CB + B^* \qquad (2)$$
$$A^* + CB \rightarrow AB + C \qquad (3)$$
$$B^* + CA \rightarrow AB + C \qquad (4)$$

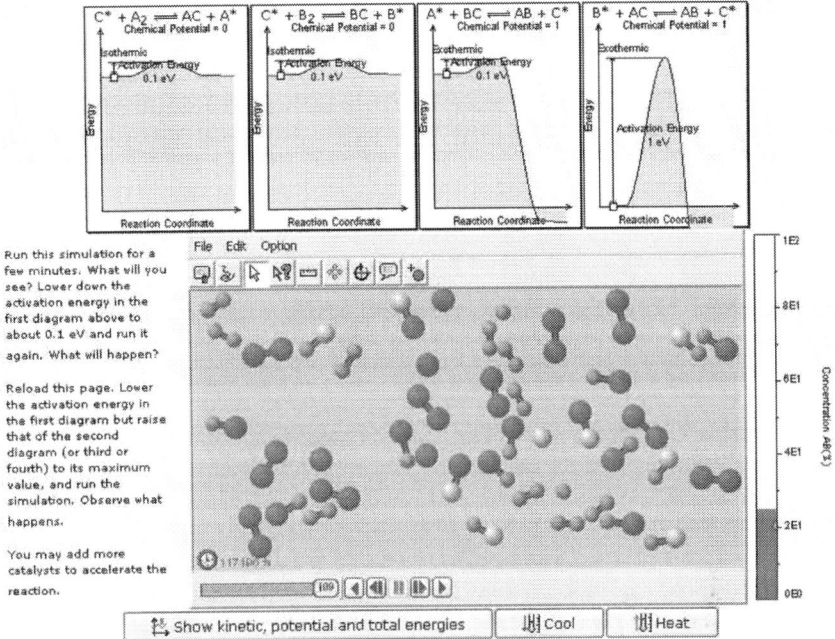

Figure 5. *Molecular Workbench simulation of homogeneous catalysis for $A_2 + B_2 \rightarrow 2\,AB$ with catalyst C. Models: A – small dark sphere, B – large dark sphere, C – intermediate light sphere*

To simulate this mechanism the activation energies of the dissociations of A_2 and B_2 and for the reactions $A^* + B_2$ and $B^* + A_2$ are set high so they will not occur at the temperature of the simulation. The screen in Figure 5 shows that five parameters can be easily adjusted – the four activation energies by moving the square at the end of the arrows for the activation energies up or down on the four energy diagrams at the top of the figure and the temperature using the cool and heat buttons at the bottom of the figure. In the particular simulation shown in Figure 5 the activation energy for step (4) was made high cutting off this step. Note at the point of the simulation shown the product AB was formed by step (3) and every species shown in this mechanism is present. Observation of the simulation shows that CA and CB molecules do dissociate upon some collisions with other molecules. If CA molecules did not dissociate, then CA would accumulate since step (4) is not possible. The reaction would proceed until all C was bound in CA. This end state could be formed by lowering the temperature such that collisions of CA molecules don't have enough energy to break the CA bond. The original simulation for this topic shown on a prior screen had the activation energies for all four of these steps set low such that the reaction proceeds to give only AB and free catalyst C since the numbers of A_2 and B_2 were initially equal.

An initial study of *Molecular Workbench* found that it did support student understanding. Pallant and Tinker (2004) found that when students used *Molecular*

Workbench they accurately recalled arrangements of the different states of matter, and could reason about atomic interactions. These results were independent of gender and they held for a number of different classroom contexts. Additionally, a close evaluation of students' responses about the bulk properties of atoms and molecules revealed that many fewer students had misconceptions following the intervention as compared to their responses on the pre-test. Follow-up interviews indicated that students were able to transfer their understanding of phases of matter to new contexts, suggesting that the knowledge they had acquired was robust. Xie and Tinker (2004) claim their work with other simulation systems (Pallant & Tinker, 2004) indicates maximum learning occurs when students experiment with the simulations with some instructor guidance. Instructors might ask students to determine if products will form if any one of the four activation energies is set to its upper limit with the other three set low, perform the simulations and explain what they observe. Instructors could follow-up by asking students to predict if they expect the same results at high and low temperatures, to try those experiments, and explain the results.

Just as with *Connected Chemistry*, *Molecular Workbench* scores high on the Mayer Interactivity Principle but the other Mayer multimedia instruction principles do not apply due to the lack of audio tracks. *Molecular Workbench* simulations promote investigative science learning. The catalysis example in Figure 5 allows users to explore five parameters associated with the proposed mechanism and rapidly observe results of changes in any of these parameters. This catalysis simulation is much more sophisticated than the states of matter simulations used in the Pallant and Tinker (2004) study noted above. If maximum learning using the states of matter simulation required instructor guidance, a more directed guided inquiry approach would likely be needed for students to understand the full consequences of various relative activation energies for the four steps in the proposed mechanism. This example does allow students to explore the mechanism in ways impossible in the laboratory since each simple change in any activation energy would mean use of a different catalyst. This point should be made specific by the instructor as well as noting the assumptions about the reactivities of A_2, B_2, and AB with other species implicit in this four step mechanism. As with *Connected Chemistry*, *Molecular Workbench* would have to connect these simulations with the physical and social processes of scientific investigation if it were to be strongly aligned with situative theory.

ChemSense

While *Connected Chemistry* and *Molecular Workbench* support student learning of chemical concepts and principles, *ChemSense* is designed to support student inquiry in the wet lab. The *ChemSense* environment offers an ensemble of tools that enable students to create their own representations of chemical phenomena (Schank & Kozma, 2002). The basic premise of the *ChemSense* design is that these tools will be used within a social context (students collaboratively create representations) to investigate, analyze, and discuss chemical phenomena (students conducting wet lab experiments). Students use the tools in the *ChemSense* to express their chemical understanding in a jointly shared representational space and do this within a classroom and task context which includes physical lab equipment and data collection probeware.

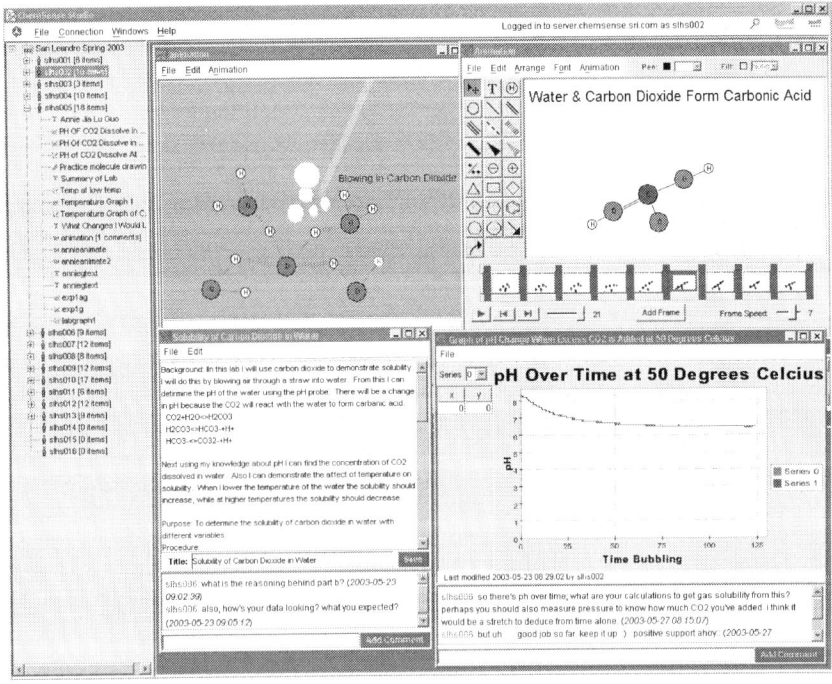

Figure 6. *The ChemSense screen for work by high school students. Upper left shows picture students drew to depict experiment of exhaling breath through a straw into water with pH initially adjusted to > 8.0. Upper right shows frame in student animation showing formation of H_2CO_3 and its partial ionization to H^+ and HCO_3^-. Graph is pH measured with pH probe showing formation of acidic solution as CO_2 dissolves in water.*

Figure 6 shows the basic layout of *ChemSense*. The environment contains a set of tools – drawing, animation, graphing, and text tools – for creating and viewing representations, and a commenting feature for peer review. A web interface, called the *ChemSense Gallery*, is also available for viewing and commenting on work, and managing groups and accounts. Several examples of student-generated items can be seen in Figure 6. At the top of the window are two animations that students created to show the process of a gas – specifically, carbon dioxide – dissolving in water. To create these animations, students constructed individual frames that stepped through the breaking of bonds between the carbon, oxygen, and hydrogen, and the subsequent formation of carbonic acid, H_2CO_3 (aq). Below and to the right of the animations is a display of a dynamic graph that shows student-collected data from an experiment on the change in pH of the solution as carbon dioxide is dissolved over time. This data was collected at the lab bench through the use of probeware (developed by Pasco Inc., Roseville, CA) and imported into *ChemSense*. Below the graph is a comment area where other students can submit and view comments and questions. (Every item in the workspace has its own personal comment area.) To the left of the graph is a student-constructed description of the lab purpose, procedure, and findings.

In its use, *ChemSense* is designed to shape the way students think and talk while using representations to describe, explain, and argue about physical phenomena in terms of underlying chemical entities and processes. For example, in considering the dissolving of carbon dioxide into water, students can use the animation tool to build a representation of this dynamic process. As they create their animation they are confronted with a set of decisions about how to represent their understanding of the dissolving process – What does a carbon dioxide molecule look like? What is the structure of carbonic acid? What happens to the water and carbon dioxide structures as they meet? Which atoms are involved? How many are there of each kind? Discussions related to these decisions can engage students in the deep understanding of chemical phenomena in terms of underlying molecular structures and processes.

The *ChemSense* environment also gives students the means to coordinate these representations with observable phenomena using Probeware. Probeware allows students to directly import graphical or tabular data into their representations from bench top investigations and other inquiry-based activities (Krajcik et al., 1998; Roth & Bowen, 1999). For example, students collect wet-lab data such as temperature or dissolved oxygen content and import the data into *ChemSense*. They are then able to create and run two representations – a nanoscopic level representation showing the underlying process, and a tool-generated representation showing the change in observable properties. The goal of using two representations that show parallel changes at the nanoscopic and physical levels is to get students to question what is happening at the nanoscale that determines the emergent properties of what they see on the lab bench.

ChemSense is used in the context of specially designed curriculum units and investigative activities that scaffold student use of interconnected forms of visual and discursive representations and ask students to describe, explain, and argue about the chemical experiments they are conducting on the lab bench. In addition,, the environment allows students to peer-review each other's work by way of commentary. For example, a teacher may include as part of an activity a section towards the end of a unit that asks students to "review the work of two other lab groups and ask two questions related to the chemistry in their representation." As part of their "assignment", each lab group is responsible for providing critical feedback on other students' work. Used appropriately, this function further supports the possibility for students to collectively arrive at new understandings of scientific concepts by asking students to probe other students' thinking (Bell & Linn, 2000; Brown & Campione, 1996; Greeno, 1998; Kozma, 2000a; Linn, Bell, & Hsi, 1998; Pea, 1992, 1994; Scardamalia & Bereiter, 1994).

Analyzed from a situative perspective, *ChemSense* supports students as they engage in investigations and use visualizations while working with each other to understand the chemical phenomena they are observing in terms of their underlying molecular structures and processes. As with *Connected Chemistry and Molecular*

Workbench, ChemSense relies heavily on the skill of the teacher to structure the laboratory as an inquiry experience and to organize the class as a knowledge-building community. An observational study of students using *ChemSense* (Schank & Kozma, 2002) illustrating how the software can help students incorporate visualizations into their investigations and deliberations over their findings is discussed below.

ChemDiscovery

Like *ChemSense*, *ChemDiscovery* (formerly called *ChemQuest*) also supports chemistry inquiry. *ChemDiscovery* provides a technology-based, inquiry-oriented learning environment (Agapova, Jones, Ushakov, Ratcliffe, & Martin, 2002). Rather than using lectures and worksheets as the primary instructional vehicles, *ChemDiscovery* features interactive web pages linked to activities, databases, and design studios and coordinates with hands-on laboratory activities. Students work in pairs or small cooperative learning teams with instructors functioning most of the time as coordinators and facilitators rather than as spigots for knowledge. Agapova et al. list the following set of synergistic learning strategies that comprise the instructional philosophy of *ChemDiscovery*:

- Approaching content with relevant contexts
- Visualizing and modeling the molecular level of matter
- Engaging in inquiry through authentic science and design activities
- Discovering knowledge in a step-by-step manner
- Learning independently and in cooperative learning groups
- Self-constructed meaningful learning (from computer feedback to problem-solving and problem-constructing strategies
- Accepting responsibility for the environment
- Designing individual learning pathways through the course

ChemDiscovery consists of eight projects (i.e., quests) that can be used individually to supplement a traditional classroom or as a set to replace most of the traditional curriculum. These quests include topics such as design elemental substances, design chemical reactions between elements, and design and explore chemical systems. There are three entry points for each project: a set of learning goals for project activities or two contextual motivational tools introducing the project from environmental, scientific, and social perspectives. Each project includes design activities such as those shown in Figures 7 and 8.

After completing the design studio in which they design an atom of carbon (Figure 7), student teams construct a chemical reaction for formation of CO_2 from its elements as shown in Figure 8. They then build models of the molecules. A typical *ChemDiscovery* classroom would have student teams working simultaneously on a range of activities, either using computers for such activities as using design studios and extracting information from databases, performing hands-on laboratories, or working on designing strategies for problem solving. The multiple paths from initiation to completion of each project allow students to benefit from actions of their team as well as the others and to use learning activities and approaches suitable for varying abilities and interests. Students seek the knowledge and information to complete their activities, as needed. The goal is for students to gain a better understanding of chemical concepts and principles through investigation and accept greater responsibility for their own learning.

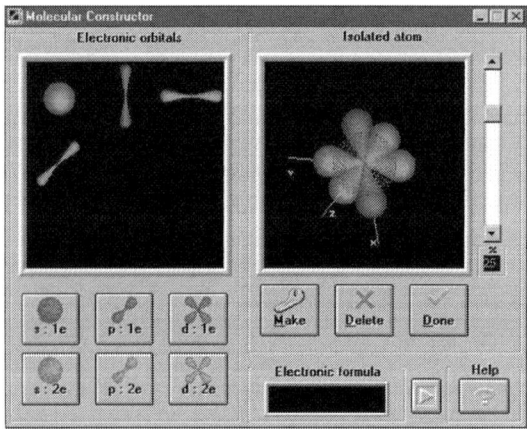

Figure 7. *ChemDiscovery design studio for construction of an atom from atomic orbitals.*

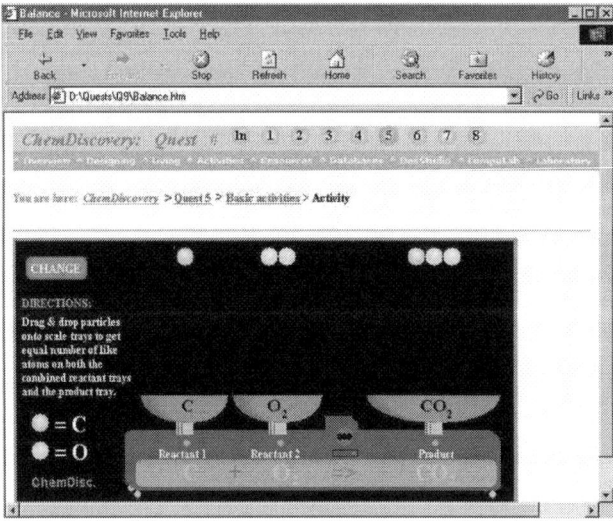

Figure 8. *ChemDiscovery design studio to model a chemical reaction between elements.*

A team of researchers observed teachers using ChemDiscovery to varying degrees in a number of classrooms. Use of ChemDiscovery shifted the classroom focus as intended from a teacher-led lecture format to student-teacher interactions. Two teachers both taught different sections of their classes using ChemDiscovery and traditional instruction. They were observed to spend 38% of their time facilitating independent study work in their ChemDiscovery classes compared to 13% in their traditional classes. Journal entries of these teachers noted that students using ChemDiscovery had become more successful learners – connected new knowledge to prior knowledge, organized and reviewed new knowledge, and monitored their own understanding.

Although *Connected Chemistry*, *Molecular Workbench*, *ChemSense*, and *ChemDiscovery* all focus on chemistry as an investigational process rather than chemistry as a set of key concepts, *ChemSense* and *ChemDiscovery* include laboratory activities. Activities, such as working in cooperative teams, designing chemical systems, and analyzing extensive databases, engage students in challenging authentic scientific activities. *ChemDiscovery* supports a situative approach to learning with focus on the physical and social characteristics of the environment. However, as with *ChemSense*, implementation of the situative approach with this project relies upon skilled instructors to guide student teams through the quests, structure their use of representation to support understanding, and make the quests authentic inquiry experiences.

ASSESSMENT OF VISUALIZATION SKILLS

Williamson and Abraham (1995) in their study of the efficacy of computer animations showing nanoscale representations of chemical phenomena used a specially designed assessment tool called the Particulate Nature of Matter Evaluation Test (Williamson, 1992). This test required drawings, written explanations, and choosing multiple choice explanations of chemical phenomena. Noting the need for specially designed tools for assessment of visualization skills, we designed three types of multimedia examination questions: explanations of computer displayed visualizations, selection of a visualization (nanoscale animation, video of experiment, dynamic graph of macroscopic property) that shows a specific feature, and matching of a target visualization with one in a different representational form (Russell & Kozma, 1996). Bowen (1998) argued that since chemists use multiple representations (macroscopic, particulate, and symbolic) to solve problems, chemistry students need to develop problem solving skills across all three levels and special types of test items are required to assess these skills. Bowen suggested examples of test items involving explanations of videos of chemical phenomena, animations of atoms and molecules during chemical or physical changes, and combinations of these with traditional test questions that feature symbolic representations. The following discussions of tools for assessing learning from use of multimedia chemical visualization software focuses on the expert-like visualization skills we believe should be developed in undergraduate chemistry curricula as discussed in the earlier chapter (Kozma & Russell, this volume).

Multimedia Test Items to Assess Conceptual Understanding and Visualization Skills

We have developed a variety of multimedia test items to use for the assessment of conceptual understanding and visualization skills (Russell & Kozma, 1996; Russell & Geno, 2000). Our earlier studies (Kozma & Russell, 1997) showed that test items requiring students to supply answers rather than choosing between possible answers are more likely to cause them to look beyond the surface features of the visualizations and produce responses based upon their views of the underlying chemistry. In the following paragraphs, we provide a number of examples of multimedia test items. All these examples require short free format responses. As we discuss these sample text items, we classify each question as a process question if it focuses on a visualization skill routinely used by chemists or as a conceptual question if its focus is on a fundamental chemical concept. Conceptual understanding may aid in answering process questions and hopefully the various visualizations will trigger the identification and application of an appropriate concept for the conceptual questions. Conceptual questions address learning from the cognitive theory perspective while process questions address learning from both cognitive and situative theory perspectives.

Figure 9 shows a series of screens taken from an animation depicting the equilibrium between liquid water and water vapor in a closed container. The animation begins with all molecules moving about in the lower layer representing liquid water. Soon molecules begin escaping into the vapor phase and returning to the liquid phase with on average three molecules shown at all times in the vapor phase. As this animation is displayed with a data projector on a large screen in front of the class the students have a printed test that shows one screen still to represent the animation. Next to this still shot is a model of water labeled H_2O.

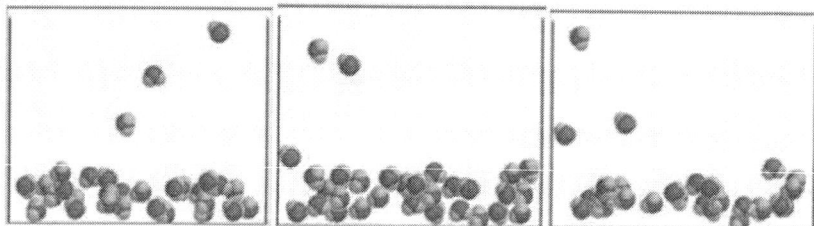

Figure 9. *Three frames from animation of equilibrium between liquid water and water vapor.*

Test items associated with this animation are:
1. Write an equation to represent the phenomena illustrated in the animation.
2. Describe how each species in the equation you wrote in part 1 is represented in the animation.
3. Does this animation show a dynamic equilibrium?
4. What could you do to this system to get more species into the upper part of the container?

These first two questions are designed to assess the student's ability to interpret a nanoscale animation and to transform this symbolic representation into the symbolic chemical equation. Although these questions require conceptual understanding of liquids and gases, they are classified as process questions with their focus on interpretation of nanoscale models and construction of an appropriate chemical equation. Students' responses might indicate their understanding of the concept of a dynamic equilibrium. The second pair of questions is strictly conceptual testing student's understanding of dynamic equilibrium and of vapor pressure and/or heats of vaporization.

The next example shows a test item that uses both a video of an experiment and a chemical equation for the reaction and asks responses in written language. The experiment starts with 100 mL of vinegar in four 250 mL volumetric flasks with attached balloons containing various masses of baking soda. The balloons are shaken to drop the baking soda into the vinegar. The masses of baking soda were chosen such that vinegar is the limiting reactant in two and baking soda the limiting reactant in the other two.

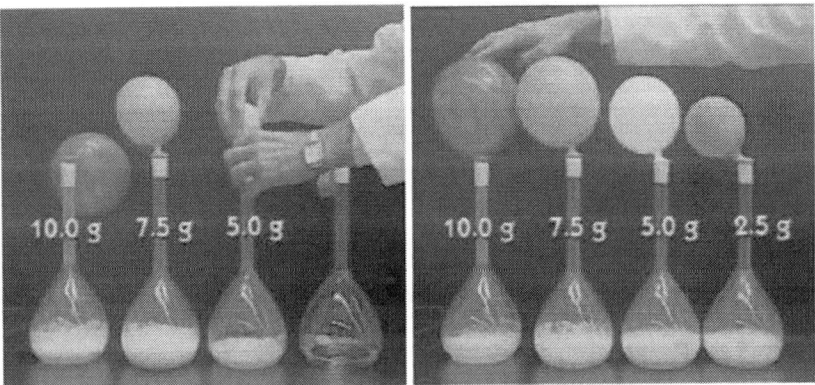

Figure 10. Specified masses of $NaHCO_3(s)$ are added to 100 mL samples of vinegar resulting in the reaction, $NaHCO_3(s) + CH_3COOH(aq) \rightarrow NaCH_3COO(aq) + H_2O + CO_2(g)$.

The printed test item shows the final frame of the video shown on the right in Figure 10. Next to this picture is printed the caption to Figure 10 giving the chemical equation. Three questions that have been used with this video are:
1. Why do the balloons inflate?
2. Does the HCO_3^- ion act as an acid or a base?
3. Why do the red and orange balloons inflate to the same volume but the yellow and blue balloons inflate to smaller volumes?

Although this experiment shows the limiting reactant concept, questions 1 and 3 can be answered by interpretation of the experiment as one forming a gas with the amount produced dependent upon the quantities of reactants available. The chemical equation confirms the formation of a gas and identifies it as CO_2. A complete answer to question 3 based only on observation of the video will state the limiting reactant concept whether or not students use the term limiting reactant. These two questions are classified as process questions. Question 2 requires understanding of Bronsted-Lowry acids and bases and is classified as conceptual.

Figure 11 shows three frames from animations showing the partial dissolving of $Cu(IO_3)_2(s)$ and total dissolving of $CuSO_4(s)$. For the partially soluble salt most of the ions remain in the solid phase shown as the regular array at the bottom of the frame. Ions in the solid continuously exchange with ions in the solution to maintain the constant saturated concentrations. The soluble salt results in all the ions moving throughout the solution.

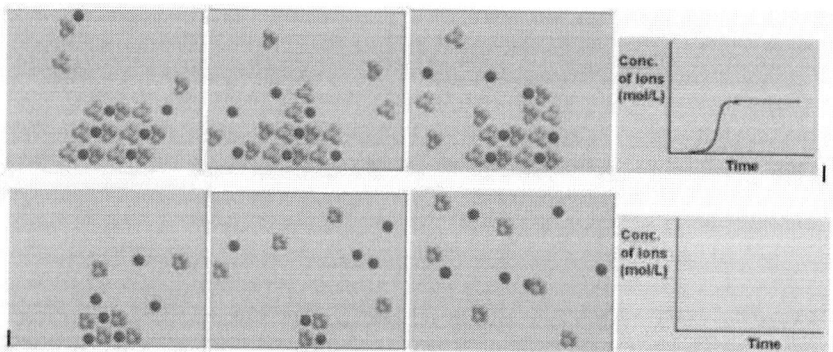

Figure 11. *Three sequential screen shots from animations showing partial dissolving of $Cu(IO_3)_2(s)$ forming a saturated solution in upper sequence and total dissolving of $CuSO_4(s)$ to give an unsaturated solution in lower sequence. Upper graph shows change in concentration of Cu^{2+} with time. Graphs are used in question 3 below.*

These two animations are played simultaneously before students answer the following three questions:
1. Explain the difference in the addition of $Cu(IO_3)_2(s)$ and of $CuSO_4(s)$ to water depicted in these animations.
2. Do either (or neither) of these examples show a dynamic equilibrium?
3. The change in concentration of Cu^{2+}(aq) for the addition of $Cu(IO_3)_2$ to water is shown on the graph (Figure 11 upper). Add to this graph a curve representing the corresponding change in concentration of IO_3^-(aq). On the other graph (Figure 11 lower) draw two curves representing the changes in the concentrations of Cu^{2+}(aq) and SO_4^{2-}(aq) for the addition of $CuSO_4(s)$ to water.

Questions 1 and 3 can be answered by observing the two animations without understanding the underlying concepts of identities (specific ions and solubility) and are thus classified as process questions. These questions can be answered based solely on the ability to interpret a nanoscale animation and to convert this representation to a graphical representation. Question 2 requires a conceptual understanding of dynamic equilibrium and is classified as conceptual.

Student responses to visualization items can be interpreted as assessing *both* student's conceptual knowledge and their ability to interpret and transform various types of visualizations. The following items show examples of responses of two students, whom we identify by the pseudonyms Sue and Joe, to three parts of the questions on water vapor equilibrium and HgO decomposition illustrated in Figures 9 and 4 respectively.

1a. Write an equation corresponding to the phenomena shown by this animation.

Sue: $H_2O(l) \rightleftharpoons H_2O(g)$
Joe: $H_2O(s) \rightarrow H_2O(g)$

Sue correctly interprets the animation as showing liquid and gaseous water and is able to express this in terms of a chemical equation showing the symbol for equilibrium. Joe interprets the water molecules moving throughout the lower layer as a solid and shows an equation for a one way reaction. Sue has the expert's ability to interpret the animation and write a chemical equation. Joe wrote an equation that corresponded to his interpretation of the animation. The interpretation of the animation and its expression as a chemical equation is classified as a visualization (process) skill.

1b. How are all species in your equation represented in the animation?
Sue: Liquid H_2O is at the bottom with the molecules close together and moving around. H_2O gas is at the top with the molecules far apart and moving very quick.
Joe: Yes, $H_2O(s)$ and $H_2O(g)$

Sue is able to express her interpretation of the animation in written language. Joe does not answer the question merely noting without identification the presence of $H_2O(s)$ and $H_2O(g)$. This is a chemical process question addressing transformation from the symbolic animation to an alternate symbolic system – language.

1c. Explain how this animation illustrates a dynamic equilibrium.
Sue: Some of the liquid molecules went into the gas form and some of the gas molecules went into the liquid form. The same number that went into the gas is the same number that went into liquid.
Joe: The substance reactant product is constantly going at equal rates and the concentrations of reactant and products stay same.

This is a purely conceptual question requiring prior knowledge which both students have. By viewing their pretests we can determine if they had this knowledge before taking this course or acquired it during the course.

Question 9 below refers to the SMV:Chem experiment for the determination of the empirical formula of Hg_xO_y. One screen shot from this experiment is shown in Figure 4.

9a. Write a chemical equation for the reaction.
Sue: $Hg_xO_y \rightarrow xHg(l) + (y/2)O_2(g)$
Joe: $HgO_2 \rightarrow Hg + O_2$

Sue used the formulas on the graph to write a balanced equation showing the physical states of the two products. Joe wrote a balanced equation for a nonexistent oxide of mercury. These responses are to the process aspect of writing chemical equations from information supplied in a different representational form.

9b. How is the law of conservation of mass illustrated?
Sue: Each bar in the graph, when added together, still equals the same amount as the starting amount.
Joe: You end up with what you started with. Nothing is lost.

Sue correctly interprets the right hand graph showing constant total mass and expresses this correctly in words. Joe states, "nothing is lost" when the animation clearly shows oxygen is lost. If Joe is referring to the right-hand graph, then "nothing is lost" could mean no mass is lost. This question requires knowledge of the concept of conservation of mass and the skill to interpret graphs and/or animations. Both students focused on the graph. Although it is possible that students could using the right-hand graph figure out what conservation of mass means without prior knowledge, we classified this question as a conceptual question.

9c. If the experiment were stopped at the position indicated in the graph and animation with the mass determined by weighing the tube and subtracting the mass of the empty tube, what mass would be determined? (Express your answer to the nearest 0.1 g.)
Sue: The mass determined would be of Hg_xO_y and Hg.
1.0 g + 1.4 g = 2.4 g
Joe: 1.4 g because everything in the tube is Hg_xOy.

Sue drew horizontal lines on the graph to estimate the masses of Hg_xO_y and Hg. Joe ignored the fact that the animation showed all the Hg formed remained in the tube. This question is a process question testing the ability to interpret the graph and the animation.

These questions illustrate how tests of this type can both assess components of students' conceptual knowledge and their abilities to perform the processes associated with doing chemistry. Some questions about chemical processes can be formulated using still images like the three frames for the liquid water and gaseous water equilibrium question, as shown in the prior section of this chapter, but the questions are much clearer if they are shown as an animation movie. All of the questions on this test involve some parts that require a transformation from one form of representation to another – a processing skill chemists must develop. Ten multiple-part questions used on the visualization tests discussed below are included on the website: http://www2.oakland.edu/users/russell. The test section on the website contains a MS-Word document that serves as the test booklet and a MS PowerPoint presentation that has been compressed with Impatica. In the compressed form all visualization movies will play as soon as the slide containing them is opened. When used in the PowerPoint form for an in-class test, the instructor can start each movie and replay them as needed.

Assessment of Representational Competence after Use of ChemSense

In a study involving the use of *ChemSense* in a high school chemistry course (Schank & Kozma, 2002), we found a significant increase in representational skills and a movement from using representation as depiction to using representations to describe, explain, and predict what was happening at the nanoscopic chemical level.

To provide a more detailed picture of how representational competence can be assessed, we present a test item example and work though the scoring of student's representational competence. [See earlier chapter (Kozma & Russell, this volume) for a discussion of "representational competence".] The item for this example is a four-step "storyboard" question asking students to draw and explain at the nanoscopic level how NaCl dissolves in water over time. This item allows students the opportunity to show their understanding of a solubility-related *process* and represent it accordingly. Students were scored on both their chemical understanding and their representational competence. Here we compare pretest and posttest responses for a given student only for representational competence. In the earlier chapter we defined representational competence and provided a five level scale for representational competence: Level 1 – representation as depiction, Level 2 – early symbolic skills, Level 3 – syntactic use of formal representations, Level 4 – semantic, social use of formal representations, and Level 5 – reflective, rhetorical use of representations.

Figure 12 shows a sample pretest response for our example test item. At pretest, the student completed all frames of the storyboard. However, instead of creating nanoscopic-level representations, the student provided a macroscopic-level drawing of the solution and a representation of the ionic lattice using the symbols "Na" and "Cl" to represent nodes in the lattice. This is evidence that the student is operating at a "surface level" representational competence—the discussion centers only on observable, macroscopic level features. The student uses representations as depictions of what they might see at the lab bench, providing only an "isomorphic, iconic depiction of the phenomenon at a point in time." This student's response at pretest received a "level 1" score.

At posttest this student demonstrated a richer, more complex representation of the underlying process. Here the student correctly used space filling molecules to represent the underlying, non-observable entities and processes, and provided an accurate description of the dissolving process except for showing atoms rather than ions as the solutes. The student shows a semantic and social use of formal representations, using these representations to explain the physical phenomena rather than simply depicting what may be seen and the response was scored at a "level 4."

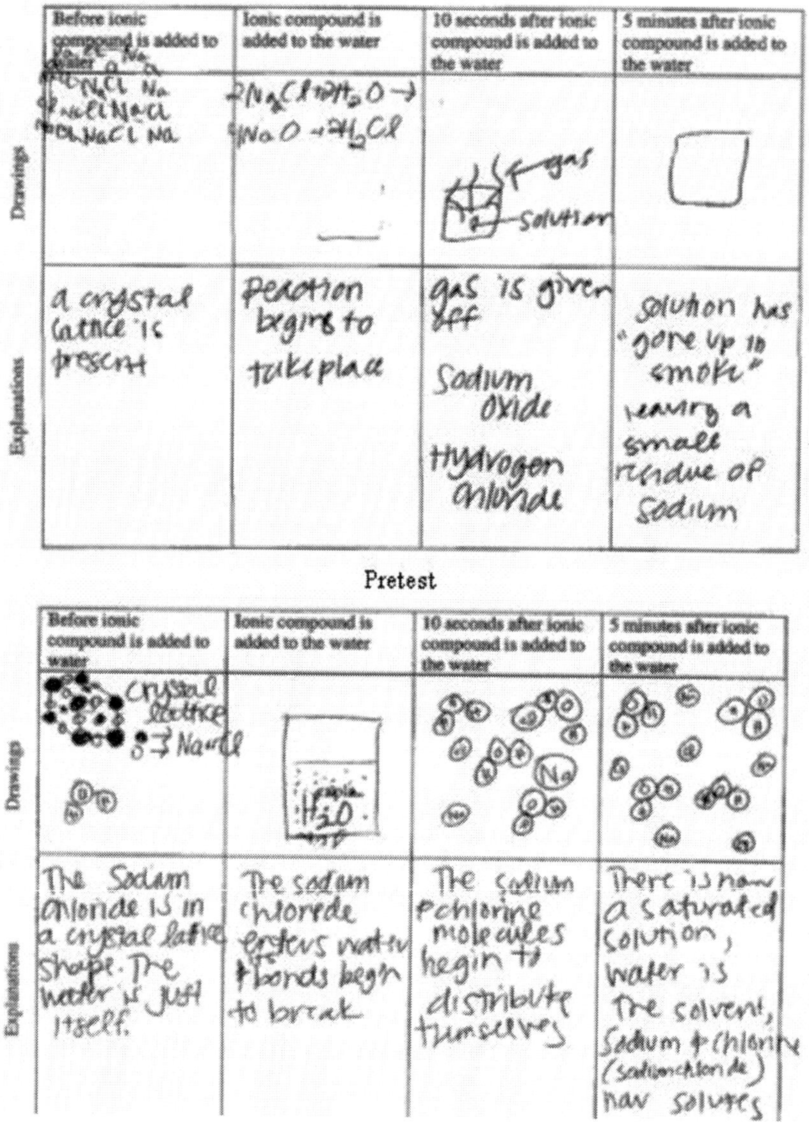

Figure 12. *Sample pretest (upper) and posttest (lower) for student using ChemSense.*

Studies of Efficacy of SMV:Chem for Enhancing Visualization Skills

We used multimedia visualization test items as discussed above as components of assessment instruments for studies of the efficacy of *SMV:Chem* in the classroom and for out-of-class assignments. Three recent studies (Russell, 2004) are summarized. The first two studies conducted during the 2001-2002 academic year involved six sections of a first semester general chemistry course taught by five different very experienced instructors. We distinguished the fall 2001 and winter/spring 2002 studies for two reasons: fall and winter semester students typically are drawn from different groups as confirmed by their pretest scores in these studies and we used extra questions on the winter/spring pre- and posttests. One section each term taught by instructors A and B used no *SMV:Chem* components and served as control sections. Instructor C taught one section in fall 2001 using eight *SMV:Chem* homework assignments with occasional use of videos and animations from *SMV:Chem* in lectures. Instructor D taught one section in the fall, winter, and spring semesters using four SMV:Chem homework assignments and made extensive use of software visualizations during lectures. A sample of one of the SMV: Chem homework assignments used in these studies as well as examples of three homework assignments in the new Version 3.0 web-based format. Additional homework assignments may be viewed at http://www2.oakland.edu/users/russell.

Prior to discussing the results of these studies, we report on a subset of the data from the Winter 2002 study that shows one measure for comparing the overall performance in first semester general chemistry of students in the control and experimental sections. This comparison also shows the correlation between the visualization skills measured on our test and performance on a standardized chemistry examination. The American Chemical Society's Examination Institute has developed conceptual questions that contain limited assessments of visualization skills within the limitations of a printed multiple-choice format exam. We have compared student performance on the 2000 ACS First Semester General Chemistry exam with both conceptual and quantitative questions with performance on tests we have constructed to assess visualization skills (Russell, 2002; Russell, 2004).

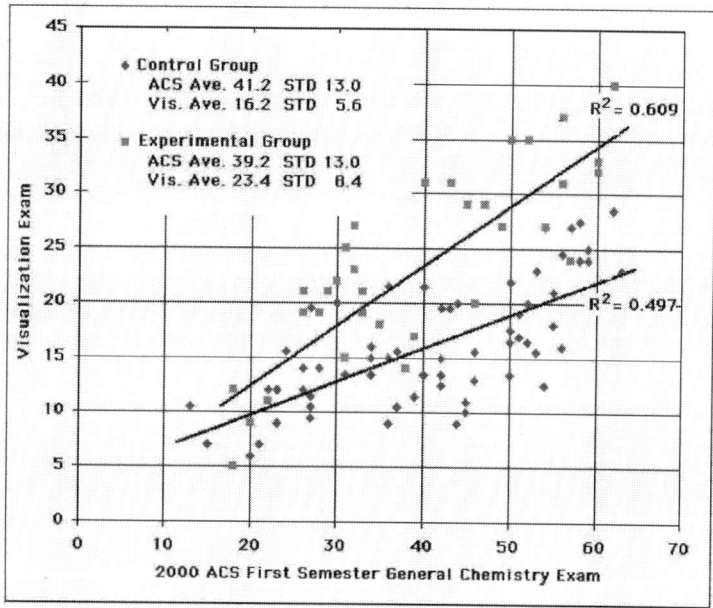

Figure 13. *Correlation of scores on visualization test and ACS 2000 First Semester General Chemistry Test. Experimental group used four SMV:Chem homework assignments and control group used none.*

Figure 13 shows correlations between scores on the ACS exam (70 points) and our ten multiple-part question visualization test (43 points) for two sections (experimental, n = 34; control, n = 62) of first semester general chemistry taught by two instructors in Winter 2002. As shown by t-tests (two-tailed, two samples with unequal variance), the differences in scores on the visualization test were significant (p = 0.0002) but insignificant on the ACS exam (p = 0.45). The data in Figure 12 show not only is the average score on the visualization test higher for the experimental group but visualization test scores are higher across the range of ACS exam scores. The data show that students with ACS scores less than 25 had similar performance levels on the visualization test. The increased slope of the linear least square line for the experiment group compared to the control group showed use of the *SMV:Chem* assignments was more effective for higher ability (as measured by the ACS exam) students. The R^2 values of the two least squares regression lines show a slight increase in the correlation between visualization test and ACS exam scores for the experimental group compared to the control group.

For all six sections in the Fall 2001 and Winter/Spring 2002 studies a pre- and posttest methodology was used with the pretest given during the first week recitation section. The posttest for the control groups was given during the last week recitation section and for the experimental groups was part of the final exam. To encourage students to provide their best efforts on the pretest and the control group posttests extra credit points toward their total course grade were awarded to students who made serious efforts to answer every question. A copy of the 2002-2003 test is included in the assessment section on the website: http://www2.oakland.edu/users/russell. For all sections in the 2001-2002 study the first six questions were used on the pre- and posttests. For the winter and spring 2002 sections, questions seven and eight were added. The fall tests were scored based upon a total of 25 points with 10 points for process questions and 15 points for conceptual questions as discussed above. The winter/spring test had 30 points with 13 process and 17 conceptual. Results for the 2001-2002 study are shown in Table 1.

Table -1. *2001-2002 sections of first semester general chemistry. Average differences in posttest and pretest scores for all questions, for process questions, and for conceptual questions with respective (standard deviations). The last three columns show t test p values for comparing each experiment group with its control group.*

Sect.	Inst.	No	Post-Pre (25 pts) Total	Post-Pre (10 pts) Process	Post-Pre (15 pts) Concept	T test p Total	T test p Process	T test p Concept
F01 Contr.	A	58	5.66 (3.14)	2.58 (1.83)	2.98 (2.160)			
F01 Exp.8	C	40	8.21 (3.87)	4.24 (2.30)	3.98 (2.68)	6×10^{-4}	3×10^{-4}	0.06
F02 Exp.4	D	33	8.82 (3.33)	3.88 (2.04)	4.94 (2.27)	2×10^{-5}	0.003	2×10^{-4}
			Post-Pre (30 pts) Total	Post-Pre (13 pts) Process	Post-Pre (17 pts) Concept			
W02 Contr.	B	63	3.35 (3.33)	2.52 (1.96)	1.03 (2.42)			
W02 Exp.4	D	30	6.95 (4.58)	2.63 (2.05)	4.32 (3.32)	7×10^{-4}	0.79	2×10^{-5}
Sp02 Exp.4	D	17	7.94 (5.72)	4.18 (2.78)	3.76 (3.91)	0.007	0.03	0.01

The differences between the average increases in pretest to posttest overall scores were statistically significant at the $p < 0.01$ level for all experimental groups compared with their control groups. The most interesting conclusion from the t test p values for the separated process and conceptual questions is that except for the eight homework fall 2001 section the pretest posttest gains between experimental and control groups were more significant for the conceptual questions. However, there were some confounding factors in this study. For example, students in the group who received eight homework assignments had used all of the visualizations that appeared on the posttest while answering different questions in the homework assignment than used on the visualization test.

The differences between the average increases in pretest to posttest overall scores were statistically significant at the $p < 0.01$ level for all experimental groups compared with their control groups. The most interesting conclusion from the t test p values for the separated process and conceptual questions is that except for the eight homework fall 2001 section the pretest posttest gains between experimental and control groups were more significant for the conceptual questions. However, there were some confounding factors in this study. For example, students in the group who received eight homework assignments had used all of the visualizations that appeared on the posttest while answering different questions in the homework assignment than used on the visualization test.

In order to eliminate a number of these confounding issues another study was performed in 2002-2003. Since the earlier study showed fairly consistent pretest scores for all sections of first semester general chemistry, the pretest was eliminated. This meant that no student had seen the posttest questions before the posttest. Both instructors for the experimental sections agreed to use five homework assignments chosen to minimize exposure to experiments covered on the posttest. Instructor C used different parts of *SMV:Chem* experiments covered on 22% of the posttest questions while instructor D used a different part of one experiment covered on 4% of the posttest questions. Questions seven and eight on the visualization test used in the 2001-2002 studies and covered in the lab and homework assignments were replaced by questions nine and ten not covered in the homework. Table 2 summaries the results of this study.

All experimental posttest scores for overall questions, process questions, and conceptual questions are significantly higher than their control sections at the $p < 0.0001$ level. The posttest score for the winter 2003 control group and fall 2002 experimental groups are essentially the same for all questions, process questions, and conceptual questions. It appears the gains in visualization skills acquired through use of *SMV:Chem* homework assignments during the first semester of general chemistry are matched by students who did not use the software after two semesters of general chemistry. Students who used the software during the second semester showed additional gains for both process and conceptual questions.

The scoring for the eight questions used in the 2002-2003 study (questions 1-6, 9,10 of visualization exam) were equally divided between questions judged as process and conceptual as discussed above. Table 2 shows that for all sections students showed nearly identical performance on process questions and conceptual questions. Thus although *SMV:Chem*'s design conforms to Mayer's principles for multimedia learning tools to enhance conceptual understanding it's use also contributes to the acquisition of skills used by chemists as they interact in work environments consistent with situative theory.

Table -2. *Average posttest scores for all questions and for process and conceptual questions for CHM 157 and 158, 1st and 2nd semesters of general chemistry, with (standard deviations). t test p values compare row and column sections for total questions.*

Sect.	Inst.	No	Posttest (38 pts) Total	Posttest (19 pts) Process	Posttest (19 pts) Concept	T test p F02 Contr.	T test p F02 Exp.	T test p W03 Contr
F02 Contr. C157	B	50	13.12 (3.92)	5.68 (2.07)	7.44 (2.66)			
F02 Exp.5 C157	C	34	19.16 (6.61)	9.27 (3.76)	9.90 (3.52)	2×10^{-5}		
W03 Contr C158	E	32	18.81 (6.49)	9.30 (3.97)	9.42 (3.01)	5×10^{-5}	0.83	
W03 Exp. 5 C158	D	29	26.84 (5.59)	13.09 (3.29)	13.72 (3.14)	4×10^{-15}	5×10^{-6}	3×10^{-6}

FUTURE DIRECTIONS OF VISUALIZATIONS IN CHEMICAL EDUCATION

Education Recommendations for the Development of Visualization Software

Over the last 10 years, there has been a significant amount of development of visualization software in chemistry. Molecular modeling software packages, such as *Spartan, Hyperchem, Gaussian+GaussView, etc.*, are being successfully used in undergraduate college and even high school chemistry courses for their visual and numerical results without requiring understanding of their theoretical basis. As reported above, several large-scale visualization packages have been specifically design for chemistry education, often with the support of the U.S. National Science Foundation. Many of these packages involve simulations and animations of molecular systems and the use of multiple representations. Results from initial studies of these packages warrant continued support for their development and scaling up.

Modeling tools usually display the structure of a single molecule. These tools are important in helping students understand the structure and properties such as dipole moments and energy states of molecules. Simulations help students understand difficult concepts related to the dynamics, rate, concentration, etc. of chemical systems that involve relatively large numbers of molecules and reactions. We feel there is an important place for simulations of single reactions. Simulations of reactions based on the electronic structure methods showing transformations of molecular (frontier) orbitals, electrostatic potentials, electron densities, etc., during reactions will aid the visualization of single reactions.

Many of the Chime-format models now available are built using data for atomic coordinates available from web-based databases: Inorganic Crystal Structure Database, http://fiz-informationsdienste.de/en/DB/icsd/; Cambridge Structural Database for small molecules, http://www.ccdc.cam.ac.uk/products/csd; Protein Data Bank, http://rcsb.org/pdg/; Nucleic Acid Database, http://ndbserver.rutgers.edu; and Crystmet database for metals and alloys, http://www.tothcanada.com.

Recommendations for Visualization-Based Instructional Activities

Our review of the research and development literature warrants the use of visualization technologies in chemistry instruction. These findings are strongest when the instructional goal is to teach concepts and principles. We identified instances in which lectures were supplemented by animations to successfully teach concepts related to equilibrium, reaction chemistry, electrochemistry, and miscibility. (See also Kozma & Russell, this volume.) We found studies in which molecular models supported students' understanding of structure and bonding. Several significant development projects, such as *SMV:Chem, Connected Chemistry*, and *Molecular Workbench*, provide instructors and instructional designers with new technological resources to bring multimedia into the chemistry classroom. However, the current state of research on chemical visualization does not allow us to say much beyond this. There are no carefully designed experiments that tell us when it is best to use animations versus still pictures or whether dynamic molecular models are better than physical ball-and-stick models. Nor can we say with much precision how these various media can be used together and when it is best to do so. For the time being, all of these practical issues are left up to the judgment of instructors and instructional designers.

The current research allows us to say even less, if the goal is to teach chemistry as a process of investigation. Situative theory would argue for the use of various representations in the context of laboratory investigations, using them to ask questions, plan experiments, carry out procedures, analyze data, and present findings. New software systems, such as *ChemSense* and *ChemDiscovery*, provide instructors with powerful tools to support students' investigations. Initial studies with these environments seem to suggest that they foster the kinds of investigative activities that are encouraged by the theory. But there is little systematic research so far that helps guide instructors and instructional designers on how to effectively integrate these tools into classroom and laboratory activities.

Recommendations for Testing Activities

Almost all of the studies that we examined used specially designed tests which measured students' understanding in new ways and measured new goals for chemistry learning, such as increasing students' abilities to use appropriately diverse forms of representations in problem solving as are used by experts. Indeed, in the experimental studies we reviewed, visualizations had the greatest effect on test items of this sort. These findings argue for the integration of still pictures, animations, models, and other visualizations into the tests and examinations of chemistry courses. There are now a variety of multimedia tools that make it easy for chemistry instructors to include these kinds of items in their regular student assessment activities.

ACKNOWLEDGEMENTS

Robert Kozma's participation in this chapter was made possible by the National Science Foundation under Grant No. 0125726. Any opinions, findings, and conclusions or recommendations expressed in this material are those of the authors and do not necessarily reflect the views of the National Science Foundation.

REFERENCES

Agapova, O., Jones, L., Ushakov, A., Ratcliffe, A., and Martin, M., 2002, Encouraging independent chemistry learning through multimedia design, *Chemical Education International*, **3**(1): AN-8. http://www.iupac.org/publications/cei/vol3/0301x0an8.html

Bell, P., and Linn, M. C., 2000, Scientific arguments as learning artifacts: Designing for learning from the web with KIE, *International Journal of Science Education* [Special Issue], **22**(8): 797-817.

Bowen, C., 1998, Item design considerations for computer-based testing of student learning in chemistry, *Journal of Chemical Education*, **75**: 1172-1174.

Brown, A., and Campione, J., 1994, Guided discovery in a community of learners, in: *Classroom lessons: Integrating cognitive theory and classroom practice*, K. McGilly, ed., MIT Press, Cambridge, MA, pp. 229-270.

Greeno, J., 1998, The situativity of knowing, learning, and research, *American Psychologist*, **53**(1): 5-26. .

Kozma, R., 2000, Representation and language: The case for representational competence in the chemistry curriculum, Paper presented at the Biennial Conference on Chemical Education, Ann Arbor, MI.

Kozma, R., and Russell, J., 1997, Multimedia and Understanding: Expert and Novice Responses to Different Representations of Chemical Phenomena, *Journal of Research in Science Teaching*, **43**(9): 949-968.

Krajcik, J., Blumenfeld, P., Marx, R., Bass, K., Fredricks, J., and Soloway, E., 1998, Inquiry in project-based science classrooms: Initial attempts by middle school students, *Journal of the Learning Sciences*, **7**(3&4): 313-351.

Linn, M.C, Bell, P., and Hsi, S., 1998, Lifelong Science Learning on the Internet: The Knowledge Integration Environment, *Interactive Learning Environments*, **6**(1-2): 4-38.

Mayer, R., 2001, *Multimedia Learning*, Cambridge University Press, Cambridge, UK.

Mayer, R., 2002, Cognitive theory and the design of multimedia instruction: An example of the two-way street between cognition and instruction, *New Directions for Teaching and Learning*, **89**: 55-71.

Mayer, R., 2003, The promise of multimedia learning: Using the same instructional design methods across different media. *Learning and Instruction*, **13**(2): 125-140.

Pallant, A., and Tinker, R., 2004, *Journal of Science Education and Technology*, (in press).

Pea, R., 1992, Augmenting the discourse of learning with computer-based learning environments, in: *Computer-based learning environments and problem-solving*, E. de Corte, M. Linn, and L. Verschaffel, eds., Springer-Verlag, New York, pp. 313-343.

Pea, R., 1994, Seeing what we build together: Distributed multimedia learning environments for transformative communications, *Journal of the Learning Sciences*, **3**: 285-299.

Roth, W-M., 1998, Teaching and learning as everyday activity, in: *International Handbook of Science Education*, B. Fraser and K. Tobin, eds., Kluwer, Dordrecht, pp. 169-181.

Roth, W-M., 2001, Situating cognition, *The Journal of the Learning Sciences*, **10**: 27-61.

Roth, W-M., and Bowen, G., 1999, Complexities of graphical representations during lectures: A phenomenological approach, *Learning and Instruction*, **9**: 235-255.

Russell, J., 2002, Can Software Help Chemistry Students Utilize Multiple Visualizations to "Think" Chemistry?, 17th Biennial Conference on Chemical Education, CHED 419, Bellingham, WA.

Russell, J., 2004, Enhancing chemical understanding and visualization skills with SMV:Chem – Synchronized Multiple Visualizations of Chemistry. (Submitted)

Russell, J., and Geno, J., 2000, Tools for development of chemical visualization skills and tools for assessment of visualization skills, 219th American Chemical Society National Meeting, CHED 0961, San Francisco.

Russell, J., and Kozma, R., 1994, 4M:Chem – multimedia and mental models in chemistry, *Journal of Chemical Education*, **71**(8): 669-670.

Russell, J., and Kozma, R., 1996, Multimedia Questions to Assess Students' Ability to Visualize Chemical Phenomena, 14th Biennial Conference on Chemical Education, M-EV-SY-A-2, Clemson, SC.

Russell, J., Kozma, R., Becker, D., and Susskind, T., 2000, *Synchronized Multiple Visualizations of Chemistry*, John Wiley & Sons, New York.

Russell,J., Kozma, R., Jones, T., Wykoff, J., Marx, N., and Davis, J., 1997, Use of simultaneous-synchronized macroscopic, microscopic, and symbolic representations to enhance the teaching and learning of chemical concepts, *Journal of Chemical Education*, **74**(3): 330-334.

Scardimalia, M., and Bereiter, C., 1994, Computer support for knowledge-building communities, *Journal of the Learning Sciences*, **3**: 265-283.

Schank, P., and Kozma, R., 2002, Learning chemistry through the use of a representation-based knowledge building environment, *Journal of Computers in Mathematics and Science Teaching*, **21**(3): 253-279.

Stieff, M., and Wilensky, U., 2003, Connected chemistry – incorporating interactive simulations into the chemistry classroom, *Journal of Science Education and Technology*, **12**(3): 285-302.

Wilensky, U (2002), NetLogo SimpleKinetics 2 model. http://ccl.northwestern.edu/netlogo/models/SimpleKinetics2. Center for Connected Learning and Computer-based Modeling. Northwestern University, Evanston, IL, USA.

Williamson, V., 1992, *The effects of computer animation emphasizing the particulate nature of matter on the understandings and misconceptions of college general chemistry students,* Unpublished doctoral dissertation, University of Oklahoma.

Williamson, V., and Abraham, M., 1995, The effects of computer animation on the particulate mental models of college chemistry students, *Journal of Research in Science Teaching*, **32**(5): 521-534.

Xie, Q., and Tinker, R., in press, Molecular dynamic simulations of chemical reactions for use in education, *Journal of Chemical Education.*

JOHN K GILBERT

ENDPIECE: RESEARCH AND DEVELOPMENT ON VISUALIZATION IN SCIENCE EDUCATION

Institute of Education, The University of Reading, UK

RESEARCH, DEVELOPMENT AND METHODOLOGY

Much recent work in visualization in science education has consisted of the development and implementation of systems of external visualizations, particularly those based on computers, for use by students, especially at university level. On some occasions, the implementation of these innovations has been accompanied by evaluations into their usefulness. There has been an overdue reliance on anecdotal reports of success. The situation has been complicated because the visualization tools themselves are often still undergoing development when in use. The 'pre-test /post-test' design often used is a very blunt instrument, not least because 'control' groups are virtually impossible to identify. On a few occasions, pure research has taken place, but this has been rarely followed up by development. It would seem desirably, following the practice that is normal in the world of science and technology, for the two to be as tightly coupled together as possible. The 'learning experiment', in which an innovation is introduced and its effects closely monitored during that process, has much to offer this emergent field, as have 'talk aloud' case studies of students learning with the aid of external visualizations and / or generating their own internal visualizations. The codification of the natures of useful methodologies would greatly support work in the field.

SOME POSSIBLE RESEARCH QUESTIONS

In any field, an infinite number of research questions is possible. However, what has been written in this book suggests some of particular importance that can be grouped by theme:

The nature of visualization

- What are the codes of representation associated with each of the major modes of representation?
- What are the significant sub-modes of representation associated with each major mode?

- What are the scope and limitations of each of the major and significant sub-modes of representation?
- What generic forms of 2D and of 3D representation are there?

The design of external visualizations

- What valuable structures are there for 'visual narratives'?
- What designs are especially valuable for addressing particular educational concerns e.g. the elimination of students' 'alternative conceptions'?
- What combinations of static and dynamic visualizations provide the most support for mental model building?

The integration of visualization into learning systems

- What treatment of visualization is given in science textbooks at different grade levels, up to and including university level?
- How widely and with what significance are such treatments drawn on elsewhere in these textbooks?
- How can exercises to develop the capacity to visualize best be integrated with field /laboratory work that requires the use of those visualizations/

Learning with visualizations

- What problems to students have in understanding the codes of visualization associated with the various modes of representation?
- To what extent are students able to adapt their visualizations when moving between several of the modes of representation?
- how effective are puzzles in promoting the development of the skills of rotation, inverting, reflecting, images?
- Is there a relationship between students' preferred modes of learning and the capability to engage in visualization?
- Is there a direct link between students' capabilities for visualization and their learning in science?
- How do students evaluate their internal visualizations?

Teacher development

- What activities are most effective in developing science teachers' knowledge of and skills in visualization?
- Having engaged in such activities, how widely and in what way do teachers pass on their knowledge and skills to their students?

And finally---the identification of metavisual capacity

It has been suggested that research into metacognition in general has addressed three areas (Chmiliar, 1997). First, into people's 'knowledge about strategies': knowing why, when and where particular metacognitive strategies may be used. Second, into people's 'strategies of use': their capacity to use these strategies without prompts. Third, into people's 'cognitive monitoring': the personal approaches that are used to monitor cognitive strategies and to change those in use at any given time. In the longer run, it should be possible to build on the development and associated research reported in this book to address these themes in respect of metavisualization.

REFERENCE

Chmiliar, L. (1997). Metacognition and giftedness. *AGATE, 11*(2), 28-34.

APPENDICES TO CHAPTER 13

Table A1: Parallel presentation of tasks and the help tools used in task section 'Perception' for the tests $MVT1$ and $MVT2s$

'Molecular Visualisation Test 1'	'Molecular Visualisation Tests 2'	$MVT2$ help tools
Task 1 Take a look at the model of molecule A on the computer screen (Picture 1). Write on the enclosed answer-sheet, which among the marked atoms (C1, C2, H3, Cl4, Cl5, Cl6, C7, C8) are the closest to you, which are a bit further, and which are the furthest from you.	**Task 1** Take a look at the model of molecule A on the computer screen (Picture 1). With help of the plastic molecular model (on the table) and the interactive molecular model (on the right side of the screen) write on the enclosed answer-sheet, which among the marked atoms (C1, C2, H3, H4, H5, H6, H7, H8) are the closest to you, which are a bit further, and which are the furthest from you.	Photo of $3d$ molecular model used in $MVT2$-M and $MVT2$-$3dv$
Answer: The atoms closest to me are: _____ The atoms a bit further from me are: _____ The atoms furthest from me are: _____	*Answer:* The atoms closest to me are: _____ The atoms a bit further from me are: _____ The atoms furthest from me are: _____	Pseudo-$3d$ molecular model used in $MVT2$-v and $MVT2$-$3dv$
Picture 1: Molecule A	Picture 1: Molecule A	
Task 4 Take a look at the model of molecule D on the computer screen (Picture 7). Write on the enclosed answer-sheet, which among the marked atoms (C1, C2, H3, O4, H5, H6, H7, H8, H9) are the closest to you, which are a bit further, and which are the furthest from you.	**Task 4** Take a look at the model a: the model of molecule D on the computer screen (Picture 7). With help of the plastic molecular model (on the table) and the interactive molecular model (on the right side of the screen) write on the enclosed answer-sheet, which among the marked atoms (C1, C2, H3, O4, H5, H6, H7, H8, H9) are the closest to you, which are a bit further, and which are the furthest from you.	Photo of $3d$ molecular model used in $MVT2$-$3d$ and $MVT2$-$3dv$
Answer: The atoms closest to me are: _____ The atoms a bit further from me are: _____ The atoms furthest from me are: _____	*Answer:* The atoms closest to me are: _____ The atoms a bit further from me are: _____ The atoms furthest from me are: _____	Pseudo-$3d$ molecular model used in $MVT2$-v and $MVT2$-$3dv$
Picture 7: Stereo-chemical formula of the molecule D	Picture 7: Stereo-chemical formula of the molecule D	

Green coloured text – $3d$ models in $MVT2$-$3d$ Blue colored text – v in $MVT2$-v Red underlined text – $3d$ and v in $MVT2$-$3dv$

338 — APPENDICES TO CHAPTER 13

Table A2: Parallel presentation of tasks and the help tools used in task section 'Perception and Rotation' for the tests and

'Molecular Visualisation Test 1'	'Molecular Visualisation Test 2'	MVT2 help tools
Task 7 Imagine the system of coordinates with x, y and z axes represented below and a molecule G represented below and a molecule G in this system (Picture 13). Among the molecular models a, b, c, d (Picture 14), choose the one that you would get after rotating molecule G around axis Z only. Picture 12: System of coordinates Picture 13: Molecule G Picture 14: Molecular models	**Task 7** Imagine the system of coordinates with x, y and z axes represented below and a molecule G in this system (Picture 13). Among the molecular models a, b, c, d (Picture 14), with help of the plastic molecular model (on the table) and the interactive molecular model (on the right side of the screen) choose the one that you would get after rotating molecule G around axis Z only. Picture 12: System of coordinates Picture 13: Molecule G Picture 14: Molecular models	Photo of 3d molecular model used in MVT2-3d Pseudo-3d molecular model used in MVT2-v and MVT2-3dv
Task 2 Imagine the system of coordinates with x, y and z axes represented below and a molecule B in this system (Picture 3). Among the stereochemical formulas of the molecules a, b, c, d (Picture 4), choose the one which you would get after rotating molecule B around axis Z only. Picture 2: System of coordinates Picture 3: Molecule B Picture 4: Stereo-chemical formulas of the molecules	**Task 2** Imagine the system of coordinates with x, y and z axes represented below and a molecule B in this system (Picture 3). Among the stereochemical formulas of the molecules a, b, c, d (Picture 4), with help of the plastic molecular model (on the table) and the interactive molecular model (on the right side of the screen) choose the one, which you would get after rotating molecule B around axis Z only. Picture 2: System of coordinates Picture 3: Molecule B Picture 4: Stereo-chemical formulas of the molecules	Photo of 3d molecular model used in MVT2-3d Pseudo-3d molecular model used in MVT2-v and MVT2-3dv

Green colored text – 3d models in MVT2-3d   Blue colored text – v in MVT2-v   Red underlined text – 3d and v in MVT2-3dv

APPENDICES TO CHAPTER 13   339

Table A3: Parallel presentation of tasks and the help tools used in task section 'Perception and Reflection' for the tests *MVT1* and *MVT2s*

'Molecular Visualisation Test 1'	'Molecular Visualisation Tests 2'	*MVT2* help tools
Task 3 Among the molecular models a, b, c, d (Picture 6) choose the one, which you would get after reflecting molecule C (Picture 5) with the marked mirror.	**Task 3** Among the molecular models a, b, c, d (Picture 6) with help of the plastic molecular model (on the table) and the interactive molecular model (on the right side of the screen) choose the one which you would get after reflecting molecule C (Picture 5) with the marked mirror.	Photo of *3d* molecular model used in *MVT2-3d* and *MVT2-3dv*
Picture 5: Molecule C Mirror	Picture 5: Molecule C Mirror	Pseudo-*3d* molecular model used in *MVT2-v* and *MVT2-3dv*
a b c d Picture 6: Stereo-chemical formulas of the molecules	a b c d Picture 4: Stereo-chemical formulas of the molecules	
Task 6 Among the stereochemical formulas a, b, c, d (Picture 11) choose the one that you would get after reflecting molecule C (Picture 10) with the marked mirror.	**Task 6** Among the stereochemical formulas a, b, c, d (Picture 11) with help of the plastic molecular model (on the table) and the interactive molecular model (on the right side of the screen) choose the one that you would get after reflecting molecule C (Picture 10) with the marked mirror.	Photo of *3d* molecular model used in *MVT2-3d* and *MVT2-3dv*
Picture 10: Molecule F Mirror	Picture 10: Molecule C Mirror	Pseudo-*3d* molecular model used in *MVT2-v* and *MVT2-3dv*
a b c d Picture 11: Stereo-chemical formulas of the molecules	a b c d Picture 11: Stereo-chemical formulas of the molecules	

Green colored text – *3d models* in *MVT2-3d*   Blue colored text – *v* in *MVT2-v*   Red underlined text – *3d* and *v* in *MVT2-3dv*

340 APPENDICES TO CHAPTER 13

Table A4: Parallel presentation of tasks and the help tools used in task section 'Perception, Rotation and Reflection' for the tests *MVT1* and *MVT2s*

'Molecular Visualisation Test 1'	'Molecular Visualisation Tests 2'	*MVT2* help tools
Task 5 Three of the four molecular models a, b, c, d (Picture 9) can be derived by rotation of the molecule E (Picture 8), however the fourth one can not be obtained in this manner, because it is the mirror image of the same molecule. Find out which among the molecular models a, b, c, d is the mirror image of the molecule E. Picture 8: Model of the molecule E Picture 9: Molecular models	**Task 5** Three of the four molecular models a, b, c, d (Picture 9) can be derived by rotation of the molecule E (Picture 8), however the fourth one can not be obtained in this manner, because it is the mirror image of the same molecule. With help of the plastic molecular model (on the table) and the interactive molecular model (on the right side of the screen) find out which among the molecular models a, b, c, d is the mirror image of the molecule E. Picture 8: Model of the molecule E Picture 9: Molecular models	Photo of *3d* molecular model used in *MVT2-3d* and *MVT2-3dv* Pseudo-*3d* molecular model used in *MVT2-v* and *MVT2-3dv*
Task 8 Three of the four stereo-chemical formulas a, b, c, d (Picture 16) can be derived by rotation of the molecule H (Picture 15), however the fourth one can not be obtained in this manner, because it is the mirror image of the same molecule. Find out which among the stereo-chemical formulas a, b, c, d is the mirror image of the molecule H. Picture 15: Stereo-chemical formula of the molecule E Picture 16: Stereo-chemical formulas of the molecules	**Task 8** Three of the four stereo-chemical formulas a, b, c, d (Picture 16) can be derived by rotation of the molecule H (Picture 15), however the fourth one can not be obtained in this manner, because it is the mirror image of the same molecule. With help of the plastic molecular model (on the table) and the interactive molecular model (on the right side of the screen) find out which among the stereo-chemical formulas a, b, c, d is the mirror image of the molecule H. Picture 15: Stereo-chemical formula of the molecule E Picture 16: Stereo-chemical formulas of the molecules	Photo of *3d* molecular model used in *MVT2-3d* and *MVT2-3dv* Pseudo-*3d* molecular model used in *MVT2-v* and *MVT2-3dv*

Green colored text – *3d models* in *MVT2-3d*   Blue colored text – *v* in *MVT2-v*   Red underlined text – *3d* and *v* in *MVT2-3dv*

Structured Interview

Construction of molecular models and writing of stereo-chemical formulas

This group of twelve questions is dealing with the construction of molecular models using the computer program eChem[1], manual construction of three-dimensional molecular models and writing of stereo-chemical formulas. Students' strategies for the construction of molecular models and for writing stereo-chemical formulas (Questions 1-3, 5-7, 9-11) were determined on the basis of students' comments and observation of the accompanying activities. The purpose of these questions was to find out whether students use certain strategies for constructing molecular models/writing stereo-chemical formulas, and whether they are related to the use of certain kinds of manipulatives or are more a characteristic of the student.

In the following questions students were asked to set their priorities concerning the initial type of molecular structure representation when constructing models with different tools (Question 4 - computer program eChem, Questions 8 -molecular models set) or writing a stereo-chemical formula (Question 12). The purpose of these questions was to find out whether the priorities among molecular structure representation are a characteristic of the student regardless of the manipulatives used (so it is possible to set a rule that defines her/his priorities), or whether it is more likely to be related to the use of certain kinds of manipulatives.

Perception of molecular structures

- General questions

    Students were asked to describe on the basis of which properties of a certain molecular structure representation they perceive information about the structure of the molecule (Questions 13-16). Then they were asked to compare different molecular structure representations, used in previous tasks (Questions 17-22). The purpose of these questions was to reveal the reasons for students' priorities towards perception of a molecular structure from certain representations.

- Specific questions

    During the continuation of the interview, students were instructed to solve systematically all of the *MVT* tasks. The procedure was the same in all *MVT* task sections ('Perception', Perception and Rotation', 'Perception and Reflection', 'Perception, Rotation and Reflection'). In the beginning students were left to solve tasks mentally, then they were asked whether they would like to use help tools, and why (Table A5, rows A and B). Afterwards they were asked whether they preferred certain initial molecular structure representations (static image, stereo-chemical formula) during solving of this particular task type, and why (Table A5, row D). Students were also asked whether one of the accessible help-tools (three-dimensional molecular models, computer pseudo three-dimensional molecular models) was more useful than on other, and why (Table A5, row E).

---

[1] The approval of the authors of eChem (dr. Soloway and his research group at the University of Michigan) to use the program for research purposes was gained prior to testing.

The purpose of these tasks was to find out whether preferences towards a certain initial molecular structure representation and help-tool are dependent on the kind and number of processes involved.

Table A5: Specific questions that are similar in several task sections

Groups of questions	Task Section			
	'Perception'	'Perception and Rotation'	'Perception and Reflection'	'Perception, Rotation and Reflection'
A	Q23	Q29	Q35	Q42
B	Q24	Q30	Q36	Q43
C			Q37, Q38	Q44, Q45
D	Q25	Q31	Q39	Q46
E	Q26	Q32	Q40	Q47

Rotation of molecular structures

- General questions about rotation

  Students were asked to describe the rotation process (Question 27). The purpose of this question was to find out whether students understand the rotation of molecules. To get the information about their strategy of solving this kind of task students were asked to comment how they are solving the tasks (Question 28).

- Specific questions

  See Table A5 and comments in the paragraph *Perception of molecular structures*.

Reflection of molecular structures

- General questions about reflection

  To find out if students understand the reflection of molecules they were asked to describe this process (Question 33). Students were additionally asked to comment on how they are solving the tasks to get information about their strategy (Question 34).

- Specific questions for particular task section

  See Table A5 and comments in the paragraph *Perception of molecular structures*.

Rotation and reflection of molecular structures

- General questions about reflection

  To get the information about students' strategy of solving tasks, which involve a combination of the processes perception, reflection and rotation, students were asked to comment on how they are solving the tasks (Question 41).

- Specific questions for particular task section

See Table A5 and comments in the paragraph *Perception of molecular structures*.

ABOUT THE AUTHORS

Robert Bateman received a BS from Louisiana State University and a PhD from the University of North Carolina at Chapel Hill, both in biochemistry. After a postdoctorate at the University of Texas Southwestern Medical Center, he took a faculty appointment at the University of Southern Mississippi where he is currently Professor and Chair of the Department of Chemistry and Biochemistry. His traditional research interests include biosensor development and the enzymology of peptide processing. He has been using computer graphics to teach undergraduate biochemistry for over ten years.

John Belcher is the Class of 1922 Professor of Physics in the Astrophysics Division of the Department of Physics at MIT, and a MacVicar Teaching Fellow at the Institute. He took a BS in physics and mathematics (Rice University) and a Ph.D. (Cal Tech) in physics. He has been a leader in the Technology Enabled Active Learning (TEAL) Project at MIT, a program to transform almost all freshman physics at MIT to an interactive engagement format. Professor Belcher and his colleagues have developed original methods for the visualization and animation of electromagnetic fields. He has been the Principal Investigator on the Voyager Plasma Science Experiment on the Voyager Mission to the Outer Planets and the Interstellar Medium, and has twice won the NASA Exceptional Scientific Achievement Medal in connection with this mission.

George Bodner is the Arthur E. Kelly Professor of Chemistry, Education, and Engineering and Head of the Division of Chemical Education at Purdue University. Over a 30-year career in chemical education, he has published more than 100 papers and six general chemistry textbooks. Recent work in his group has ranged from studies of the learning of quantum mechanics by advanced undergraduates and graduate students to the process of acculturation by which students of organic chemistry become practicing organic chemists. A significant fraction of this work has dealt with the process by which students visualize the structure of organic compounds.

Galit Botzer is currently a postdoctoral fellow at Haifa University, Israel. She recently graduated from a PhD program in physics education at the Department of Education in Science and Technology, Technion, Israel. Her dissertation focused on the analysis of mental models in mechanics, constructed through haptic and visual interaction in a computerized environment. She used a preliminary virtual environment in which subjects interacted with objects both through manual touch and visual information in order to learn physics. Her main interests are in the visual representation of physical phenomena and in the cognitive mechanisms of physics learning.

Mike Briggs is currently Professor of Chemistry at Indiana University of Pennsylvania, USA. His B.S. degree was in chemistry from Akron University (Ohio) and his Ph. D. and an M.S. degree, also in chemistry, from Purdue University. He also received an M.S. degree in chemistry and a secondary teaching certificate from Indiana University of Pennsylvania. His current research interests center on the structure of and processes in the mind that account for the ability to learn chemistry.

Melanie Busch received her BS in geology from the University of North Carolina at Chapel Hill and her MS degree in geological sciences from Arizona State University. Her Master's project dealt with introductory geology student misconceptions about sedimentary environments and landscape development. She has created, piloted, and assessed several large geoscience-education websites and learning modules. She is currently a geology instructor at a community college in Tennessee, where she also works with prospective K-8 teachers. She also plans to pursue a PhD in geoscience education.

John Clement is currently Professor of Science Education in the School of Education and in the Scientific Reasoning Research Institute, U. of Massachusetts, Amherst. He earned an A.B. in Physics from Harvard U. and an Ed.D. in Mathematics and Science Education from the U. of Massachusetts. He began his career as a curriculum development specialist for the School District of Philadelphia. His early research work focused on students' alternative conceptions in science and mathematics. Over the last ten years he has concentrated on processes of scientific model construction in both students and experts, funded by grants from the National Science Foundation. He is coauthor of the book Preconceptions in Mechanics.

Joshua Coyan is currently an MS student at Arizona State University, where he also received his Bachelor's degree. For the past three years he has created animations and learning modules that focus on geophysics, geology, and topography. His Master's degree includes the study of the subsurface basin geometry in Phoenix, Arizona. He has taught classes in Structural Geology and Field Geology.

Yehudit Judy Dori is Associate Professor and Graduate Studies Coordinator in the Department of Education in Technology and Science at the Technion, Israel Institute of Technology, Haifa, Israel. She received her B.Sc. in chemistry from Hebrew University, Jerusalem, in 1975, M.Sc. in Life Sciences in 1981, and Ph.D. in Science Education in 1988, both from Weizmann Institute of Science, Rehovot, Israel. Since 1999 she is also a Research Scholar at the Center for Educational Computing Initiatives at Massachusetts Institute of Technology. Since 2000, Prof. Dori has been an Assessment Leader in the Technology Enabled Active Learning Project (TEAL) - a long-term educational experiment for re-designing the freshman MIT physics courses. Her research interests are focused on models and computerized molecular modeling, education through technology, and assessment. In 2003 she received the Salomon Simon Mani Award for Excellence in Teaching at the Technion. Since 2003, Prof. Dori is the Head of the Technion MALAM – the Israeli National Center for Science Education and the Chairperson of National Committee for Chemistry Curriculum, appointed by the Ministry of Education, Israel.

John Gilbert is currently Professor of Education at The University of Reading, UK, and Editor-in-Chief of the International Journal of Science Education. He took a BSc (Leicester), a D.Phil. (Sussex), and a Postgraduate Certificate in Education

(London), all in chemistry. He then taught chemistry in secondary (high) schools before moving into science education research at both school and university level. His early work on students' 'alternative conceptions' has evolved into a focus on models and modelling in science education and hence into an interest in visualization. In 2001, he was the recipient of the award for 'Distinguished Contributions to Science Education through Research' from the National Association for Research in Science Education.

Janice Gobert is Senior Research Scientist at The Concord Consortium and North American Editor of the International Journal of Science Education. She received her Ph.D. in 1994 from the Center for Applied Cognitive Science at the University of Toronto and held a post-doctoral position at the University of Massachusetts-Amherst. Her research lies primarily in the area of model-based reasoning in sematically-rich domains such as architecture, geology, physics, and biology. Janice is currently Co-Principal Investigator and Research Director of the 'Modeling Across the Curriculum' (MAC) project, a large-scale research and implementation project funded by the (USA) Interagency Education Research Initiative. She is also the Internship Director for the National Science Foundation funded 'Technology Enhanced Learning in Science' (TELS) center.

Julia Johnson is a faculty instructor in geological sciences at Arizona State University, where she teaches large introductory geology courses for non-majors. Before that, she was a faculty member at Glendale Community College. She received her BS and MS in geological sciences from Arizona State University and is currently working on a PhD in Curriculum and Instruction at ASU. She has received teaching awards for her innovative methods of involving students in large-classroom settings, and has coauthored a geology lab manual, a number of research papers, and websites on visualization and geoscience teaching.

Robert Kozma is an Emeritus Director and Principal Scientist and Fulbright Senior Specialist at the Center for Technology in Learning at SRI International. For 20 years prior to this, he was a professor at the School of Education, the University of Michigan. His expertise includes international educational technology research, the evaluation of technology-based reform, technology in science learning, and the design of advanced interactive multimedia systems. He has directed or co-directed over 25 projects and authored or co-authored more than 60 articles, chapters, and books. He has consulted with Ministries of Education in Singapore, Thailand, Norway, Egypt, and Chile on the use of technology to improve educational systems.

Debbie Leedy is a post-doctoral research associate in science education at Arizona State University. She received Bachelor's degrees in Chemistry, Mathematics, and Secondary Education from Texas Christian University and a Master's in Chemistry and PhD in Curriculum and Instruction (Science Education) both from Arizona State University. Throughout graduate school, her research focused on the use of visualization in chemistry and geology and research methods in science education. Her dissertation research compared the effectiveness of two- and three-dimensional animations in Coordination Chemistry and involved the development and evaluation of these animations. Her current work in geology education has resulted in the development of QTVR files for an interactive, animated mineralogy module and ongoing work animating the interactions of layers and folds with topography.

James M. Monaghan is Director of the Office of Distributed Learning and an Associate Professor of Instructional Technology at California State University, San Bernardino. He is Project Director for a $2 Million Title V distance learning infrastructure grant. He is spearheading the rapid development of online degree completion programs and the integration of services at 5 campuses. He also directed the Partnering to Prepare Tomorrow's Teachers to Use Technology project and numerous collaborative projects with regional school districts. Dr. Monaghan conducts research on uses of computer simulations and on innovative uses of the Internet in distance learning.

Michael Piburn is an Emeritus Professor of Science Education at Arizona State University and is currently serving as a Program Officer in Elementary, Secondary and Informal Education (ESIE) at the National Science Foundation. He is co-editor, with Dr. Dale Baker, of the Journal of Research in Science Teaching. He received a B.S. in geology from the University of California at Davis, and a Ph.D. in the same subject at Princeton University. His doctoral dissertation was conducted as part of a research program in the Caribbean area that was conceived and executed by Harry Hammond Hess. He has taught at Rutgers University, Curtin University in Western Australia, Westminster College of Salt Lake City, and Arizona State University. He served as internal evaluator for the Arizona Collaborative for Excellence in the Preparation of Teachers, and was one of the authors of the Reformed Teaching Observation Protocol (RTOP). His most recent work has been on visualization in the geological sciences.

David N. Rapp is an Assistant Professor in the Department of Educational Psychology at the University of Minnesota. He currently serves on the consulting board of Computers in Human Behavior. In 2002, he received the Jason Albrecht Outstanding Young Scientist Award from the Society for Text and Discourse. Dr. Rapp's research, broadly defined, focuses on memory and language. Specifically, his work examines text processing, reader-guided processes in comprehension, multimedia experiences, spatial cognition, and updating prior knowledge in memory.

Miriam Reiner is head of the group of Virtual Reality, Sensory Information, and Learning in the Department of Education in Science and Technology, at the Technion, Israel. Her main research is on the transition between sensory information and mental models and associated mental mechanisms such as thought experimentation. Her questions focus on how consistent/inconsistent sensory information is translated into knowledge about behavior of objects, i.e. naïve physics, and how visual and touch cues are used for visualization and construction of knowledge. She found that touch cues are central in 'seeing' with the mind's eye. Touch patterns constitute a 'language' that conveys meaning, especially in highly somato-conceptual tasks such as surgery. She uses data from fMRI or EEG of brain activities in order to study learning through sensory information.

Stephen Reynolds is Professor of Geological Sciences at Arizona State University. He received a BS in geology from the University of Texas at El Paso and MS and PhD degrees in geosciences from the University of Arizona. He worked for 10 years at the Arizona Geological Survey, where he completed a new Geologic Map of Arizona. For the past 10 years, he has been immersed in curriculum development and research about the role of visualization in undergraduate geosciences courses. He has received a number of teaching awards and was a Distinguished Speaker for the National Association of Geoscience Teachers. He has written textbooks and books for the general public, and his work is featured in videos and displays at the American Museum of Natural History. His award-winning website about visualization is used in classes from Arizona to Papua New Guinea.

Joel Russell is a Professor of Chemistry at Oakland University, Rochester, MI, USA. He received an A.B. in chemistry from Northwestern University and a Ph.D. in physical chemistry from the University of California, Berkeley. After a postdoctoral fellowship at the University of Minnesota he joined the Oakland University faculty. He has held visiting appointments at the Australian National University, Southampton University (UK), and the University of Michigan. For the past sixteen years his research has focused on the development and assessment of chemical visualization software and uses of technology to enhance the teaching and learning of general chemistry. He is the lead author of Synchronized Multiple Visualizations of Chemistry, (SMV:Chem) published on four CDs and currently being converted to a web-based product.

Vesna Ferk Savec is a Teaching Assistant and Researcher in Chemical Education at The University of Ljubljana, Slovenia. She took a B.Sc. and M.Sc. in chemistry and a Ph.D. in chemical education, all at The University of Ljubljana. Her current research interests are in evaluation of methods for teaching and learning science, including the usefulness of help-tools and associated teaching materials.

Mike Stieff is Assistant Professor of Science Education in the School of Education at the University of California, Davis. He holds a Ph.D. in learning sciences and both a B.S and an M.S. in chemistry. He taught chemistry at the secondary and undergraduate levels before pursuing science education research. His research focuses on the use of visualization as a problem solving strategy throughout long-term instruction in chemistry and the other sciences. His work also concerns the design of simulation-based visualization tools to teach chemistry at the secondary and undergraduate levels.

Kathy Takayama is currently a Senior Lecturer in the School of Biotechnology and Biomolecular Sciences at The University of New South Wales, Sydney, Australia. She holds a BSc (Biology) from The Massachusetts Institute of Technology and a PhD (Biochemistry and Molecular Biology) from UMDNJ Robert Wood Johnson Medical School. She has taught microbiology and molecular biology at the university level, and has developed, and worked extensively with, school outreach programs. Her interest in visualizations in the biological sciences includes work in both the online and classroom environments, as well as collaborations with artists. In 2003 she was selected as a Carnegie Scholar by the Carnegie Academy for the Scholarship of Teaching and Learning. She also received The University of New South Wales Vice-Chancellor's Award for Teaching Excellence in 2003, and in 2004 she was a recipient of the Australian College of Educators New South Wales Quality Teaching Award.

Barbara Tversky is Professor of Psychology at Stanford University. Her previous appointment was at Hebrew University in Jerusalem, with visiting appointments at University of Michigan, University of Oregon, Harvard, New York University, and the Russell Sage Foundation. Her degrees in Cognitive Psychology were from the

University of Michigan. She is a Fellow of the Cognitive Science Society and the American Psychological Society, was elected to the Society of Experimental Psychologists and the Executive Committee of the International Union of Scientific Psychology. She serves on the Governing Boards of the Psychonomic Society and the Cognitive Science Society, on the National Academy of Sciences US National Committee, on the Editorial Boards of four journals, and a number of other national committees. She was awarded the EDUCOM/NCRIPTAL Distinguished Software Award and the Phi Beta Kappa Excellence in Teaching Award. Her research on spatial thinking and language, memory, and categorization has taken her to many interdisciplinary collaborations investigating visualizations, design, human-computer interaction, and eyewitness testimony.

David Uttal is Associate Professor in the Department of Psychology and in the School of Education and Social Policy at Northwestern University. He received his Ph.D. in Developmental Psychology from the University of Michigan and was post-doctoral Fellow at the University of Illinois at Urbana-Champaign. His research interests are in the development of symbolic and spatial thinking, both in young children and in college students. Much of his work focuses on the development of conceptions of various forms of representations, including maps and computer-based representations in the natural sciences. David Uttal is a Fellow of the American Psychological Association and serves on the editorial board of three journals.

Margareta Vrtacnik is currently professor of chemical education and informatics at The University of Ljubljana, Slovenia. She took a B.Sc. in chemistry, an M.Sc. in chemical education, and a Ph.D. in chemistry, all at The University of Ljubljana. She taught chemistry at a high school in Ljubljana before joining the staff of The University of Ljubljana in 1977, being successively assistant, assistant professor, and associate professor, prior to her current position, during which time she had a year working with Professor Marjorie Gardner at University of Maryland, USA. Her chemical research has been on the biodegradation and photolysis of organic pollutants in the atmosphere. Her chemical education research is into the development of information and communication technologies to support critical thinking, including the use of visualization tools to promote visual literacy in chemistry. Professor Vrtacnik has also been centrally involved in a UNESCO programme to support university-industry cooperation.

Aletta Zietsman is currently assistant professor of Physics and Science Education at Western Michigan University, Kalamazoo MI, USA. She holds a B.Sc. (Stellenbosch, SA), Ed.D. (Massachusetts, USA) and a Postgraduate Certificate in Education (Witwatersrand, SA). She taught physics and mathematics in high schools and at the undergraduate level in universities, before starting masters and doctoral work in physics education. She now teaches physics at undergraduate level and graduate courses on the learning of science. Her research has focused on students' use of extreme case reasoning in their construction of explanatory, intuitively anchored, models.

INDEX

A
Animations, 137-141

B
Ball-and-stick models, 270

C
Chemistry
  enhancing visualization in, 300
Cognitive engagement
  meaning of, 49
Conceptual development
  use of extreme cases in, 170

D
Diagrams
  conveying processes with, 38
  conveying structures with, 38
Dual Coding Theory, 189

E
eChem, 98
Electromagnetism,
  enhancing visualization in, 187
Exemplar phenomena
  definition of, 10

F
Field work
  as validation for simulations, 262

G
Genomics
  nature of, 218
  role of visualization in, 219, 222-223
Geology,
  enhancing visualization in, 253

H
Help tools, 270
  value of, 294

I
Imagery
  cognitive science research into, 148
  creativity and, 152
    educational research into, 148
    paradigms of, 153
  use by experts, 172
Inquiry
  collaborative research project, 235
Interactivity,
  meaning of, 49

L
Learning
  factors influencing, 49-51
  from multimedia, 51, 300
  spatial information, from, 105, 107

M
Maps, 29-30
Memory
  model of, 16
Mental imagery
  definition of, 9
Mental model
  characteristics of, 45-46
  components of, 63-66
  development of, 94, 170
  definition of, 12, 44-45, 61
  history of evidence for, 46-49, 148
Metavisualization
  definition of, 15
  development of, 18-21
  evaluating the existence of, 23-24

evidence for, 15-18
in genomics research, 224

Microcomputer-based laboratory
   in genomics, 217
   in physics, 190-194
Model
   based learning, 78
   everyday meaning of, 10
   ontological statuses for, 12
   levels of representation of, 14
   modes of representation of, 12
      codes of interpretation for, 13
   role in science, 10
   role in science education, 10
   scope of, 10,11
   types of, 270
modelling,
   analogies in, 169
   animations, 137
   extreme cases in, 169
   molecular, 135
   simulations, 137
   teachers' role in, 245
modelling with computer software, 79
   projects to support, 80-83
   scaffolding supports for, 84-87
Multimedia, 51
   multimedia learning, 128-129

N
NetLogo, 97

O
Observation of imagery-related categories
   personal action projections, 174
   depictive hand motions, 174
   imagery reports, 174

P
Phenomography
   definition of, 62
Problem solving
   expert/novice studies, 190

Q
Qualitative research, 62
QuickTime Virtual Reality products, 256

R
Representation,
   external, 63
      forms of, 75-76
   internal, 63
      depictive hand motions, 174
      sensory-based, 157
      pure-imagery based, 159
      formalism-based, 160
      imagery reports, 174
      integration between, 161
      incidence found in learning experiment, 162
      personal action projections, 174
      levels of representation in chemistry, 269
Representational competence, 131-132
   development of, 134-135
   levels of, 133
Research
   future directions for, 141, 295, 329, 332-335
Research techniques
   ethnography, 125-126
   interactive analysis, 242
   interview, 170, 206-213, 274
   learning experiment, 155-164

novice-expert studies, 126-127, 166
questionnaire, 274
think-aloud protocol, 62-63
Rotation
  activities involved in, 69
  study of, 68

S

Simulations, 137
  in developing visualization, 177
  in physics education, 190-192
Situative theory, 129-30, 299
Spatial imagery
  definition of, 9
Spatial metaphors,
  sources of, 30-31
Spatial relations
  elements types for depiction of, 31-36
  use in representing phenomena, 36
Spatial visualization
  test for, 258

T

Technology-Enabled Active Learning (TEAL) Project, 188,
  description of, 194-200
  impact on student learning of, 202-213
Tests
  Geospatial, 264
  of value of help tools, 269
  Molecular Visualization, 273
  for visualization skills, 317
Thought experimentation, 175
Three-dimensional structures, 93, 271
  relation to 2D structures, 233
Two-dimensional structures, 270
  Relation to 3D structures, 270

V

Virtual models, 270
Visualization
  as external representation, 73
    value in science education, 74
  as internal representation, 73
  definition of, 9
  development of, 77, 234-247
  in electromagnetism, 187
  in genomics, 217
  in geology, 253
  future research into, 55-56
  principles for designing effective, 38-39
  role in levels of representation, 14
  role in learning, 52-54
    chemistry, 14, 95, 123-125, 135
    genomics, 219, 220-221, 225-233
    geology, 254
    physics, 149-152
  range of information within, 30
  use in providing explanations, 172
Visualization tools, 93
  content-specific, definition of 96
  general learning environment, definition of 96
  examples of
    eChem, 99
    Electros, 191
    ChemDiscovery, 315
    ChemSense, 312
    4M:Chem, 301
    SMV:Chem, 300
    Geowall, 264
    Kinemage, 112-116
    Molecular Workbench, 310
    Netlogo, 97

WorldWatcher, 98
   principles for the design of, 102-105, 109-110
Visual imagery
   definition of, 9
   processes common with visual perception, 10
Visual literacy
   development of, 21-23
Visual perception
   definition of, 9
   processes common with visual imagery, 10
Visual narratives, 40
Visuo-spatial thinking, 190

W

Worldwatcher, 98

Made in the USA
Lexington, KY
19 September 2011